NATURAL HISTORY
UNIVERSAL LIBRARY

西方博物学大系

主编：江晓原

THE BIRDS OF EUROPE

欧洲鸟类志

[英] 约翰 · 古尔德 著

华东师范大学出版社

图书在版编目(CIP)数据

欧洲鸟类志 = The birds of Europe : 英文 / (英) 约翰 · 古尔德著. — 上海 : 华东师范大学出版社, 2018
(寰宇文献)
ISBN 978-7-5675-7715-2

Ⅰ. ①欧… Ⅱ. ①约… Ⅲ. ①鸟类-欧洲-英文
Ⅳ. ①Q959.7

中国版本图书馆CIP数据核字(2018)第096293号

欧洲鸟类志
The birds of Europe
(英) 约翰 · 古尔德著

特约策划 黄曙辉 徐 辰
责任编辑 庞 坚
特约编辑 许 倩
装帧设计 刘怡霖

出版发行 华东师范大学出版社
社 址 上海市中山北路3663号 邮编 200062
网 址 www.ecnupress.com.cn
电 话 021-60821666 行政传真 021-62572105
客服电话 021-62865537
门市（邮购）电话 021-62869887
地 址 上海市中山北路3663号华东师范大学校内先锋路口
网 店 http://hdsdcbs.tmall.com/

印 刷 者 虎彩印艺股份有限公司
开 本 16开
印 张 62.75
版 次 2018年6月第1版
印 次 2018年6月第1次
书 号 ISBN 978-7-5675-7715-2
定 价 1680.00元（精装全二册）

出 版 人 王 焰

总　目

《西方博物学大系》总序

江晓原

《西方博物学大系》收录博物学著作超过一百种，时间跨度为15世纪至1919年，作者分布于16个国家，写作语种有英语、法语、拉丁语、德语、弗莱芒语等，涉及对象包括植物、昆虫、软体动物、两栖动物、爬行动物、哺乳动物、鸟类和人类等，西方博物学史上的经典著作大备于此编。

中西方“博物”传统及观念之异同

今天中文里的“博物学“一词，学者们认为对应的英语词汇是Natural History，考其本义，在中国传统文化中并无现成对应词汇。在中国传统文化中原有“博物”一词，与“自然史”当然并不精确相同，甚至还有着相当大的区别，但是在“搜集自然界的物品”这种最原始的意义上，两者确实也大有相通之处，故以“博物学”对译Natural History一词，大体仍属可取，而且已被广泛接受。

已故科学史前辈刘祖慰教授尝言：古代中国人处理知识，如开中药铺，有数十上百小抽屉，将百药分门别类放入其中，即心安矣。刘教授言此，其辞若有憾焉——认为中国人不致力于寻求世界“所以然之理”，故不如西方之分析传统优越。然而古代中国人这种处理知识的风格，正与西方的博物学相通。

与此相对，西方的分析传统致力于探求各种现象和物体之间的相互关系，试图以此解释宇宙运行的原因。自古希腊开始，西方哲人即孜孜不倦建构各种几何模型，欲用以说明宇宙如何运行，其中最典型的代表，即为托勒密（Ptolemy）的宇宙体系。

比较两者，差别即在于：古代中国人主要关心外部世界“如何”运行，而以希腊为源头的西方知识传统（西方并非没有别的知识传统，只是未能光大而已）更关心世界“为何”如此运行。在线性发展无限进步的科学主义观念体系中，我们习惯于认为“为何”是在解决了“如何”之后的更高境界，故西方的分析传统比中国的传统更高明。

然而考之古代实际情形，如此简单的优劣结论未必能够成立。例如以天文学言之，古代东西方世界天文学的终极问题是共同的：给定任意地点和时刻，计算出太阳、月亮和五大行星（七政）的位置。古代中国人虽不致力于建立几何模型去解释七政“为何”如此运行，但他们用抽象的周期叠加（古代巴比伦也使用类似方法），同样能在足够高的精度上计算并预报任意给定地点和时刻的七政位置。而通过持续观察天象变化以统计、收集各种天象周期，同样可视之为富有博物学色彩的活动。

还有一点需要注意：虽然我们已经接受了用“博物学”来对译 Natural History，但中国的博物传统，确实和西方的博物学有一个重大差别——即中国的博物传统是可以容纳怪力乱神的，而西方的博物学基本上没有怪力乱神的位置。

古代中国人的博物传统不限于“多识于鸟兽草木之名”。体现此种传统的典型著作，首推晋代张华《博物志》一书。书名“博物”，其义尽显。此书从内容到分类，无不充分体现它作为中国博物传统的代表资格。

《博物志》中内容，大致可分为五类：一、山川地理知识；二、奇禽异兽描述；三、古代神话材料；四、历史人物传说；五、神仙方伎故事。这五大类，完全符合中国文化中的博物传统，深合中国古代博物传统之旨。第一类，其中涉及宇宙学说，甚至还有“地动”思想，故为科学史家所重视。第二类，其中甚至出现了中国古代长期流传的“守宫砂”传说的早期文献：相传守宫砂点在处女胳膊上，永不褪色，只有性交之后才会自动消失。第三类，古代神话传说，其中甚至包括可猜想为现代“连体人”的记载。第四类，各种著名历史人物，比如三位著名刺客的传说，此三名刺客及所刺对象，历史上皆实有其人。第五类，包括各种古代方术传说，比如中国古代房中养生学说，房中术史上的传说人物之一“青牛道士封君达”等等。前两类与西方的博物学较为接近，但每一类都会带怪力乱神色彩。

"所有的科学不是物理学就是集邮"

在许多人心目中，画画花草图案，做做昆虫标本，拍拍植物照片，这类博物学活动，和精密的数理科学，比如天文学、物理学等等，那是无法同日而语的。博物学显得那么的初级、简单，甚至幼稚。这种观念，实际上是将"数理程度"作为唯一的标尺，用来衡量一切知识。但凡能够使用数学工具来描述的，或能够进行物理实验的，那就是"硬"科学。使用的数学工具越高深越复杂，似乎就越"硬"；物理实验设备越庞大，花费的金钱越多，似乎就越"高端"、越"先进"……

这样的观念，当然带着浓厚的"物理学沙文主义"色彩，在很多情况下是不正确的。而实际上，即使我们暂且同意上述"物理学沙文主义"的观念，博物学的"科学地位"也仍然可以保住。作为一个学天体物理专业出身，因而经常徜徉在"物理学沙文主义"幻影之下的人，我很乐意指出这样一个事实：现代天文学家们的研究工作中，仍然有绘制星图，编制星表，以及为此进行的巡天观测等等活动，这些活动和博物学家"寻花问柳"，绘制植物或昆虫图谱，本质上是完全一致的。

这里我们不妨重温物理学家卢瑟福(Ernest Rutherford)的金句："所有的科学不是物理学就是集邮（All science is either physics or stamp collecting）。"卢瑟福的这个金句堪称"物理学沙文主义"的极致，连天文学也没被他放在眼里。不过，按照中国传统的"博物"理念，集邮毫无疑问应该是博物学的一部分——尽管古代并没有邮票。卢瑟福的金句也可以从另一个角度来解读：既然在卢瑟福眼里天文学和博物学都只是"集邮"，那岂不就可以将博物学和天文学相提并论了？

如果我们摆脱了科学主义的语境，则西方模式的优越性将进一步被消解。例如，按照霍金（Stephen Hawking）在《大设计》（*The Grand Design*）中的意见，他所认同的是一种"依赖模型的实在论（model-dependent realism）"，即"不存在与图像或理论无关的实在性概念（There is no picture- or theory-independent concept of reality）"。在这样的认识中，我们以前所坚信的外部世界的客观性，已经不复存在。既然几何模型只不过是对外部世界图像的人为建构，则古代中国人干脆放弃这种建构直奔应用（毕竟在实际应用中我们只需要知道七政“如何”运行），又有何不可？

传说中的“神农尝百草”故事，也可以在类似意义下得到新的解读：“尝百草”当然是富有博物学色彩的活动，神农通过这一活动，得知哪些草能够治病，哪些不能，然而在这个传说中，神农显然没有致力于解释“为何”某些草能够治病而另一些则不能，更不会去建立“模型”以说明之。

“帝国科学”的原罪

今日学者有倡言“博物学复兴”者，用意可有多种，诸如缓解压力、亲近自然、保护环境、绿色生活、可持续发展、科学主义解毒剂等等，皆属美善。编印《西方博物学大系》也是意欲为“博物学复兴”添一助力。

然而，对于这些博物学著作，有一点似乎从未见学者指出过，而鄙意以为，当我们披阅把玩欣赏这些著作时，意识到这一点是必须的。

这百余种著作的时间跨度为15世纪至1919年，注意这个时间跨度，正是西方列强“帝国科学”大行其道的时代。遥想当年，帝国的科学家们乘上帝国的军舰——达尔文在皇家海军“小猎犬号”上就是这样的场景之一，前往那些已经成为帝国的殖民地或还未成为殖民地的“未开化”的遥远地方，通常都是踌躇满志、充满优越感的。

作为一个典型的例子，英国学者法拉在（Patricia Fara）《性、植物学与帝国：林奈与班克斯》（*Sex, Botany and Empire, The Story of Carl Linnaeus and Joseph Banks*）一书中讲述了英国植物学家班克斯（Joseph Banks）的故事。1768年8月15日，班克斯告别未婚妻，登上了澳大利亚军舰“奋进号”。此次“奋进号”的远航是受英国海军部和皇家学会资助，目的是前往南太平洋的塔希提岛（Tahiti，法属海外自治领，另一个常见的译名是“大溪地”）观测一次比较罕见的金星凌日。舰长库克（James Cook）是西方殖民史上最著名的舰长之一，多次远航探险，开拓海外殖民地。他还被认为是澳大利亚和夏威夷群岛的“发现”者，如今以他命名的群岛、海峡、山峰等不胜枚举。

当“奋进号”停靠塔希提岛时，班克斯一下就被当地美丽的土著女性迷昏了，他在她们的温柔乡里纵情狂欢，连库克舰长都看不下去了，“道德愤怒情绪偷偷溜进了他的日志当中，他发现自己根本不可能不去批评所见到的滥交行为”，而班克斯纵欲到了“连嫖妓都毫无激情”的地步——这是别人讽刺班克斯的说法，因为对于那时常年航行于茫茫大海上的男性来说，上岸嫖妓通常是一项能够唤起“激情”的活动。

而在“帝国科学”的宏大叙事中，科学家的私德是无关紧要的，人们关注的是科学家做出的科学发现。所以，尽管一面是班克斯在塔希提岛纵欲滥交，一面是他留在故乡的未婚妻正泪眼婆娑地“为远去的心上人绣织背心”，这样典型的“渣男”行径要是放在今天，非被互联网上的口水淹死不可，但是“班克斯很快从他们的分离之苦中走了出来，在外近三年，他活得倒十分滋润”。

法拉不无讽刺地指出了“帝国科学”的实质：“班克斯接管了当地的女性和植物，而库克则保护了大英帝国在太平洋上的殖民地。”甚至对班克斯的植物学本身也调侃了一番：“即使是植物学方面的科学术语也充满了性指涉。……这个体系主要依靠花朵之中雌雄生殖器官的数量来进行分类。”据说“要保护年轻妇女不受植物学教育的浸染，他们严令禁止各种各样的植物采集探险活动。”这简直就是将植物学看成一种“涉黄”的淫秽色情活动了。

在意识形态强烈影响着我们学术话语的时代，上面的故事通常是这样被描述的：库克舰长的“奋进号”军舰对殖民地和尚未成为殖民地的那些地方的所谓“访问”，其实是殖民者耀武扬威的侵略，搭载着达尔文的“小猎犬号”军舰也是同样行径；班克斯和当地女性的纵欲狂欢，当然是殖民者对土著妇女令人发指的蹂躏；即使是他采集当地植物标本的“科学考察”，也可以视为殖民者“窃取当地经济情报”的罪恶行为。

后来改革开放，上面那种意识形态话语被抛弃了，但似乎又走向了另一个极端，完全忘记或有意回避殖民者和帝国主义这个层面，只歌颂这些军舰上的科学家的伟大发现和成就，例如达尔文随着“小猎犬号”的航行，早已成为一曲祥和优美的科学颂歌。

其实达尔文也未能免俗，他在远航中也乐意与土著女性打打交道，当然他没有像班克斯那样滥情纵欲。在达尔文为“小猎犬号”远航写的《环球游记》中，我们读到：“回程途中我们遇到一群黑人姑娘在聚会，……我们笑着看了很久，还给了她们一些钱，这着实令她们欣喜一番，拿着钱尖声大笑起来，很远还能听到那愉悦的笑声。”

有趣的是，在班克斯在塔希提岛纵欲六十多年后，达尔文随着“小猎犬号”也来到了塔希提岛，岛上的土著女性同样引起了达尔文的注意，在《环球游记》中他写道：“我对这里妇女的外貌感到有些失望，然而她们却很爱美，把一朵白花或者红花戴在脑后的髮髻上……”接着他以居高临下的笔调描述了当地女性的几种发饰。

用今天的眼光来看，这些在别的民族土地上采集植物动物标本、测量地质水文数据等等的“科学考察”行为，有没有合法性问题？有没有侵犯主权的问题？这些行为得到当地人的同意了吗？当地人知道这些行为的性质和意义吗？他们有知情权吗？……这些问题，在今天的国际交往中，确实都是存在的。

也许有人会为这些帝国科学家辩解说：那时当地土著尚在未开化或半开化状态中，他们哪有“国家主权”的意识啊？他们也没有制止帝国科学家的考察活动啊？但是，这样的辩解是无法成立的。

姑不论当地土著当时究竟有没有试图制止帝国科学家的“科学考察”行为，现在早已不得而知，只要殖民者没有记录下来，我们通常就无法知道。况且殖民者有军舰有枪炮，土著就是想制止也无能为力。正如法拉所描述的：“在几个塔希提人被杀之后，一套行之有效的易货贸易体制建立了起来。”

即使土著因为无知而没有制止帝国科学家的“科学考察”行为，这事也很像一个成年人闯进别人的家，难道因为那家只有不懂事的小孩子，闯入者就可以随便打探那家的隐私、拿走那家的东西、甚至将那家的房屋土地据为已有吗？事实上，很多情况下殖民者就是这样干的。所以，所谓的“帝国科学”，其实是有着原罪的。

如果沿用上述比喻，现在的局面是，家家户户都不会只有不懂事的孩子了，所以任何外来者要想进行“科学探索”，他也得和这家主人达成共识，得到这家主人的允许才能够进行。即使这种共识的达成依赖于利益的交换，至少也不能单方面强加于人。

博物学在今日中国

博物学在今日中国之复兴，北京大学刘华杰教授提倡之功殊不可没。自刘教授大力提倡之后，各界人士纷纷跟进，仿佛昔日蔡锷在云南起兵反袁之“滇黔首义，薄海同钦，一檄遥传，景从恐后”光景，这当然是和博物学本身特点密切相关的。

无论在西方还是在中国，无论在过去还是在当下，为何博物学在它繁荣时尚的阶段，就会应者云集？深究起来，恐怕和博物学本身的特点有关。博物学没有复杂的理论结构，它的专业训练也相对容易，至少没有天文学、物理学那样的数理“门槛”，所以和一些数理学科相比，博物学可以有更多的自学成才者。这次编印的《西方博物学大系》，卷帙浩繁，蔚为大观，同样说明了这一点。

最后，还有一点明显的差别必须在此处强调指出：用刘华杰教授喜欢的术语来说，《西方博物学大系》所收入的百余种著作，绝大部分属于“一阶”性质的工作，即直接对博物学作出了贡献的著作。事实上，这也是它们被收入《西方博物学大系》的主要理由之一。而在中国国内目前已经相当热的博物学时尚潮流中，绝大部分已经出版的书籍，不是属于“二阶”性质（比如介绍西方的博物学成就），就是文学性的吟风咏月野草闲花。

要寻找中国当代学者在博物学方面的“一阶”著作，如果有之，以笔者之孤陋寡闻，唯有刘华杰教授的《檀岛花事——夏威夷植物日记》三卷，可以当之。这是刘教授在夏威夷群岛实地考察当地植物的成果，不仅属于直接对博物学作出贡献之作，而且至少在形式上将昔日“帝国科学”的逻辑反其道而用之，岂不快哉！

2018 年 6 月 5 日

于上海交通大学

科学史与科学文化研究院

约翰·古尔德
（1804-1881）

英国著名鸟类学家兼鸟类画家约翰·古尔德（John Gould）生于多塞特郡，其父是温莎城堡御花园的园丁主管。由于家庭关系，古尔德幼年常往来于御花园，十四岁时正式成为这里的园丁学徒。这份工作培养了他观察自然界生物的本领和剥制标本的手艺。1827年，古尔德即被位于伦敦的动物学会博物馆聘为首席标本剥制师。

新的职务给了古尔德接触国内顶尖博物学家的大好机会，送到动物学会的鸟类标本，也大都由他第一时间查验过目。1830年，他根据一批送到学会的喜马拉雅山鸟类标本，出版了著作《喜马拉雅百年鸟类志》。之后，查尔斯·达尔文完成第二次小猎犬号科考航行之后，也将收集到的鸟类样本交给古尔德研究，其成果刊发于达尔文主编的《小猎犬号科考动物志》（1839）。

自1838年起，古尔德对澳大利亚及南太平洋诸岛的鸟类进行了实地研究，也是第一个系统、完整记录该地区鸟类生态活动的科学家。为进行这项研究，古尔德和收藏家约翰·吉尔伯特一道先前往塔斯马尼亚岛，征求该地总督约翰·富兰克林爵士的许可。获准之后，二人在岛上进行了四个多月的调查、收集和研究。翌年2月，古尔德搭船前往悉尼进行了两个月的研究。同年5月，又到阿德莱德和探险家查尔斯·斯图尔特一起远行，1840年5月才回到英国。之后，他刊行了《澳大利亚鸟类志》，并投注大量精力绘制自己钟爱的蜂鸟及英国鸟类。

古尔德享寿77岁，一生共留下近三千幅插画，其中最被人称道的是鸟类画。这部《欧洲鸟类志》是约翰·古尔德继《喜马拉雅百年鸟类志》成功后，于1832年开始推出的著作，同样很受欢迎，至1837年出齐全部五卷。

当时，新科画家爱德华·利尔也为本书提供了部分鹦鹉插画。后来古尔德还斥资买下利尔的42幅鹦鹉画作，以解后者经济之忧。本次影印，在《欧洲鸟类志》后附有这套利尔绘制的鹦鹉画作，即《鹦鹉科》（*Illustrations of the Family of Psittacidae, or Parrots*）。

THE

BIRDS OF EUROPE.

BY

JOHN GOULD, F.L.S., &c.

IN FIVE VOLUMES.

VOL. I.

RAPTORES.

LONDON:

PRINTED BY RICHARD AND JOHN E. TAYLOR, RED LION COURT, FLEET STREET.

PUBLISHED BY THE AUTHOR, 20 BROAD STREET, GOLDEN SQUARE.

1837.

TO

THE RIGHT HONOURABLE

THE EARL OF DERBY,

PRESIDENT,

AND

THE NOBLEMEN AND GENTLEMEN

FORMING

THE COUNCIL OF

THE ZOOLOGICAL SOCIETY OF LONDON.

My Lords and Gentlemen,

FAVOURED by your kind permission, it is with feelings of mingled gratitude and pleasure, natural to one so long connected with the Society over which you preside, that I dedicate to you this Work on "The Birds of Europe," *and I have the honour to subscribe myself,*

My Lords and Gentlemen,

Your most obedient and obliged Servant,

JOHN GOULD.

PREFACE.

IT has been frequently remarked that the productions of distant countries have received a much larger share of attention than those objects by which we are more immediately surrounded; and it is certainly true, that while numerous and costly illustrations have made us acquainted with the Ornithology of most other parts of the world, the Birds of Europe, in which we are, or ought to be, most interested, have not received that degree of attention which they naturally demand. The present work has been undertaken to supply this deficiency, and I cannot but feel highly gratified that a number of concurring circumstances have enabled me to carry my intentions into effect with comparative facility, and I hope to the entire satisfaction of the whole of the Subscribers; this I am induced to believe is the case from the patronage with which the work has been honoured since its commencement.

My reasons for undertaking a work comprising the ornithology of the whole of Europe are sufficiently obvious: in the first place no publication of a similar kind had been completed, although several had been commenced; and secondly, a work exclusively confined to the British Fauna would never be perfect for any length of time, owing to the frequent accession of other species from various European localities; besides which I was desirous of rendering my work interesting to the continental ornithologists as well as to those of our own country. I have used my best exertions to render it as perfect as possible up to the present time, for which purpose I have visited nearly all the continental collections with the view of examining personally every bird before figuring it; but should additional species reward the zeal and ardour with which natural history is now cultivated (which I doubt not will be the case), I propose to publish them from time to time in the form of a Supplement, and by this means keep the work as nearly complete as possible. I have omitted a few of the species enumerated in the lists of

b

continental authors, in consequence either of my never having met with specimens of the birds; or because I am doubtful as to the propriety of their being separated from other known species*.

In my arrangement of the species I have followed with some very slight modifications the views of Mr. Vigors, which are now so generally adopted in this country. In the subdivision of the genera I have perhaps gone further than most other ornithologists, but at the same time I feel convinced that these subdivisions are naturally indicated; that they tend to facilitate the studies of the naturalist; and that some of the groups might have been still further divided with propriety. In a few instances the characters of some of the minor sub-genera have been inadvertently omitted; but this omission will not be found of any material consequence. The generic characters accompanying *Totanus hypoleucus* should be cancelled, as they have been previously given with *Totanus fuscus*. I am also aware that some other trifling errors have occurred, but I trust they will be looked upon with leniency when it is taken into consideration that the work has been commenced and completed in the short space of five years amidst numerous other avocations.

It would argue a want of gratitude in me were I to omit acknowledging the facilities which the Council of the Zoological Society have afforded me in this undertaking, and their kindness in permitting me to dedicate the work to them. I also conceive that I am considerably indebted for the success of my publication to the more general diffusion of a taste for natural history, towards which that admirable Society has so largely and successfully contributed, and to which in a great measure must be attributed the favourable reception which works of Natural History now so universally meet with; and I am also deeply indebted to the Council for the liberality with which I have at all times been allowed to avail myself of the treasures contained in the Society's Museum. To the national establishment, the British Museum, I am similarly

* *Viz.*—Corvus leucophæus, a variety of Corvus corax.
——— spermologus, a variety of Corvus monedula?
Alauda Dupontii (Certhilauda bifasciata?).
——— Kollyii.
Emberiza provincialis.
Fringilla incerta.
——— borealis.
Hemipodius lunatus, synonymous with Hem. tachydromus; and
Larus capistratus.

indebted, and would here beg to express my thanks to its officers for the many attentions I have received from them. From most of the public Museums of the Continent I have received much assistance. To M. Temminck, Director of the Royal Museum at Leyden, so well known for his valuable works on Natural History, I beg to offer my most grateful acknowledgments for the assistance he has afforded me, without which my work would necessarily have been long delayed, and in fact could never have been brought to that complete state in which I am happy to say it now stands: so great indeed has been the liberality of this eminent naturalist, that he has even confided new species to my care, and allowed me to figure and describe them in my work before including them in his own; and his liberality will, I doubt not, be duly appreciated by the scientific public. In Professor Lichtenstein of Berlin I have met with another kind and liberal friend, to whom I am indebted for the use of several of the rarer European Birds, among which were some of the original specimens collected by the celebrated Pallas. To M. Schreibers and MM. Natterer of Vienna I am also under great obligations, as well as to the gentlemen connected with the collections of Paris and Frankfort, who readily afforded me every assistance I required. To the collection of the Baron Feldegg of Frankfort I have had free access, and to this gentleman I am indebted for the use of many rarities, and of some species entirely new to science. To William Yarrell, Esq., I shall ever feel deeply grateful, for the judicious and kind assistance which he has at all times rendered me, and for the use of many valuable specimens from his excellent collection. My ever lamented friend, E. T. Bennett, Esq., was at all times much interested in my publications; I am therefore proud to add my grateful testimony to his varied talents and kind and amiable conduct upon every occasion; and of the numerous individuals honoured by his friendship there is none who more sincerely deplores his untimely decease than myself. To the Earl of Derby, the Honourable W. T. T. Fiennes, Sir William Jardine, Bart., N. A. Vigors, Esq., W. H. Rudston Read, Esq., T. B. L. Baker, Esq., J. J. Audubon, Esq., Captain S. E. Cook, A. Waterhouse, Esq., of Liverpool, Dr. De Jersey, E. H. Reynard, Esq., and to several other gentlemen my thanks are likewise due, for the warm interest which they have at all times taken in the present work. Neither must the valuable assistance afforded me by Mr. Martin of the Zoological Society be forgotten. In conclusion I would beg leave to return my grateful thanks to the whole of the Subscribers for the support with which they have been pleased to favour me.

Perhaps I may be allowed to add, that not only by far the greater number of the Plates of this work, but all those of my "Century of Birds," of the "Monograph of the Trogons," and at least three fourths of the "Monograph of the Toucans" have been drawn and lithographed by Mrs. Gould, from sketches and designs by myself always taken from nature. The remainder of the drawings have been made by Mr. Lear, whose abilities as an artist are so generally acknowledged that any comments of my own are unnecessary. With the opportunities still in my power I should consider myself to blame were I not to continue in the course I have hitherto pursued in the study and illustration of subjects in ornithology; and it is my intention, so long as permitted, steadily to use my humble efforts to advance this delightful branch of natural science. My thanks are due to Mr. Bayfield, under whose direction the whole of the Plates have been carefully and accurately coloured, and in fact too much praise cannot be accorded to the unceasing attention with which he has at all times afforded his assistance. The plates have been printed by Mr. Hullmandel, and the letter-press by Mr. R. Taylor; and these portions of the work have I trust been satisfactorily executed.

August 1, 1837.

JOHN GOULD.

INTRODUCTION.

IF we examine the geographical situation of the British Islands in relation to continental Europe, we cannot but perceive the advantages offered, as a point of observation to the naturalist, wherein to study, among other interesting facts connected with the habits of the feathered race indigenous in our portion of the globe, the periodical migrations undertaken by so many species, the time of their arrival and retreat from our shores, together with the ends to be answered both by their visit and departure. As regards temperature, no less than relative situation, are these islands favourable for a series of such observations: we need scarcely say that, placed to the westward of Europe, they occupy a medium station between the extremes of heat and cold: no portion indeed of the European continent advances within the line of the intertropics; still, however, the southern shores of Spain, Italy, and Turkey in Europe, together with the minor islands of the Grecian Archipelago, participate so nearly in the temperature of the hotter portions of the globe as to present us with many natural productions whose congenial habitat is exclusively beneath a sultry sky. But the summer heat of England never rises above a moderate degree of temperature, and the severities of winter are mild in proportion; on the other hand, if we visit the extreme north of the European continent, we there find a climate, the severities of which in winter are extreme, while the summer, though hot while it lasts, endures but for a short period.

We will not attempt to discuss the subject of the universal law of migration further than to observe that its immediate intention is the well-being of such species as would be deprived of their natural food were they to remain stationary in any given locality; in addition to which it is essential in another point of view, inasmuch as by its operation there is secured both a temperature congenial to the young, and an abundant supply of food suitable for their nourishment; we here allude more particularly to our summer visitants which have left the climate of Africa, too hot to be borne in summer, but well adapted for their winter retreat. A reverse of these circumstances takes place among our winter visitors; the high polar latitudes are their summer residence; but on the setting in of the cold in those regions the supply of food necessarily fails them, and this, with the extreme rigour of the climate, forces them to sojourn for a while in more temperate latitudes; hence while the Swallow, the Cuckoo, the Nightingale, many species of soft-billed Warblers, and numerous others visit us in spring from the south, for the purpose of nidification, and leave us on the approach of winter; the Fieldfare, the Redwing, the Woodcock, and various aquatic birds find a winter asylum with us, and depart again in spring to make room for a new succession of visitors. Independently, however, of the numerous migratory birds which are only temporary residents, a large number of species permanently remain in our latitudes; yet strange to say, of many of these the number is greatly augmented, especially during winter, by accessions from the north, among which latter are some of the smallest and most delicate of their race; we may mention the Golden-crested Wren as an example in point. Instances are not wanting of the arrival of multitudes of this species on our shores, but in such a state of exhaustion as to be almost powerless. In the case of the Lark and the Thrush, which also visit us in great numbers, the performance of a flight across the German Ocean does not much surprise us; but when we examine this little bird, which is by no means adapted for long-sustained aërial progression, we are at a loss to conceive how such a migration could have been performed. It is, however, only one amongst the many wonders of nature which are continually forced upon the attention of the naturalist.

So much has already been written on the structural adaptation of birds to their respective habits, and on their periodical changes of plumage, that we may be readily excused if we omit any detail connected with these points, more especially as they rather belong to the physiology of the feathered race, than to the natural history of the species of one quarter of the globe.

While the strictly tropical climates of the world abound with species infinitely diversified in form, and often adorned with the richest hues, the Birds of Europe are not only far less specifically numerous, but with the exception of a very few, arrayed also in the most sombre livery, an inferiority however amply compensated by their superiority of song.

At the present time the Fauna of Europe may be fairly stated to contain four hundred and sixty-two species, of which three hundred and ten may be regarded as British; of the latter number about one hundred and seventy are permanent residents in our islands; eighty-five are summer birds of passage, visiting us from the south; and forty-five from the north make our shores their winter residence.

In our arrangement we have classed and subdivided the groups (as nearly as may be) after the plan proposed by Mr. Vigors. They form five volumes, the first of which comprises the whole of the birds of the Raptorial Order, an order containing, as implied by the name, the sanguinary and ferocious of the feathered race, among which are included not only the large tyrants of the air, the Eagles, the Falcons, the Owls, &c., which make living animals their prey; but also the Vultures, which gorge upon any loathsome carrion that chance throws in their way.

Our second and third volumes comprise the species contained in the second order, termed Insessores, or perching birds; an extensive order, in which are included birds varying in their powers of flight and in their habits no less than in their food; some, like the Swallow, taking their insect prey on the wing, others pursuing it among the branches of trees and thickets; others feed indifferently upon insects, their larvæ, and upon grain; and others, eminently arboreal, (such as the Woodpeckers) search for their food among the crevices of the bark of trees, for which purpose they are expressly and beautifully organized.

To these succeed the Rasorial and Grallatorial Orders, both of which are included in our fourth volume. The Pigeons, at the head of the Rasores, as their perching habits indicate, naturally lead from the last order to the more typical of the Gallinaceæ, which are well represented by the Pheasant, Partridge, and Grouse, whose food and habits it is unnecessary to describe; and from thence to the Grallatores, represented by the Cranes, Herons, Storks, Sandpipers and Gallinules. The first of these, the Cranes, which are more granivorous in their habits, may be regarded as exhibiting some degree of affinity to the previous order, while at the same time they distinctly lead through a succession of forms, such as the Spoonbills, Storks, Herons, and Gallinules, to the true water birds, to which in their aquatic habits the Gallinules very beautifully approximate.

Our fifth volume contains the birds of the Natatorial Order, or swimming birds, which are represented by the Ducks, Grebes, Divers, Auks, Pelicans and Gulls; the aerial representative of one of these latter groups, viz. the *Tachypetes*, or Frigate Bird, among the Pelicans, exhibits several traits of resemblance to the Raptores, or birds of prey.

SUBSCRIBERS.

HIS MOST GRACIOUS MAJESTY KING WILLIAM IV.

HIS IMPERIAL MAJESTY THE EMPEROR OF AUSTRIA.

HIS MAJESTY THE KING OF PRUSSIA.

HIS IMPERIAL AND ROYAL HIGHNESS THE GRAND DUKE OF TUSCANY.

AMSINCK, Monsr. John. *Hamburgh.*
Artaria and Fontaine, Messrs. *Mannheim.*
Audubon, J. J., Esq., F.R.S.L. & E., F.L.S., F.Z.S. *Wimpole-street, Cavendish-square.*
Aylesford, the Right Hon. the Earl of. *Lower Grosvenor-street; Packington Hall, near Coventry, Warwickshire; and Aylesford, Kent.*
Baker, T. B. L., Esq., F.G.S., F.Z.S. *Hardwicke Court, Gloucestershire.*
Baker, W., Esq., F.Z.S. *Bayfordbury, Hertfordshire.*
Banks, D., Esq. *Adelphi-terrace; and Banks-town, Sheerness.*
Bath, the Most Noble the Marquess of. *Longleat, Wiltshire.*
Beach, Sir Michael H. H., Bart., F.Z.S. *Williamstrip-park, Fairford, Gloucestershire.*
Bedford, His Grace the Duke of, K.G., F.A.S., F.L.S., F.H.S., F.G.S., F.Z.S., M.R.I. *Belgrave-square; Woburn Abbey, Bedfordshire; and Tavistock House, Devonshire.*
Bell, J., Esq., F.L.S., F.Z.S., M.R.I. *Oxford-street.*
Bell, J., Esq. *Thirsk, Yorkshire.*
Bell, T., Esq., F.R.S., F.L.S., F.G.S., F.Z.S. *New Broad-street.*
Beresford, the Right Hon. Lord George, F.H.S., F.Z.S. *Charles-street, St. James's.*
Berkley, T., Esq. *Tottenham, Middlesex.*
Bevan, D. B., Esq. *Upper Harley-street; and Belmont, East Barnet, Hertfordshire.*
Bickersteth, R., Esq. *Liverpool.*
Blyth, the Rev. T., M.A., F.S.A. *The Rectory, Knowle, Warwickshire.*
Bohn, J., Esq. *Henrietta-street, Covent-garden.*
Boone, T. & W., Messrs. *New Bond-street;* 5 copies.
Bosville, A., Esq. *Thorpe Hall, Burlington, Yorkshire.*
Braddyll, E. R. G., Esq. *Conishead Priory, Ulverstone, Lancashire.*
Brewin, R., Esq. *Leicester.*
Bromley, Admiral Sir Robert Howe, Bart. *Stoke Hall, Newark, Nottinghamshire.*
Brown, the Rev. J. R. *Presteigne, Radnorshire, North Wales.*
Buccleuch, His Grace the Duke of, K.T., D.C.L., F.R.S.L. & E., F.L.S., F.Z.S. *Whitehall Gardens; Boughton House, Kettering, Northamptonshire; Richmond, Surrey; Dalkeith, Edinburgh; Drumlanrig Castle, and Langholm Lodge, Dumfriesshire; and Bowhill, Selkirk, North Britain.*
Burlington, the Right Hon. the Earl of, F.R.S., F.Z.S., M.R.I. *Belgrave-square; Latimers, Cheshunt, Buckinghamshire; Compton-place, Eastbourn, Sussex; and Holkar Hall, Milnthorpe, Westmoreland.*
Cabbell, B. B., Esq., F.A.S., F.H.S., F.Z.S., V.P.R.I. *Brick Court, Temple.*
Cambridge University, The.
Campbell, W. F., Esq., M.P., F.H.S., F.Z.S. *Woodhall, Lanarkshire; and Islay House, Isle of Islay, North Britain.*
Carbery, the Right Hon. Lord, F.H.S., F.Z.S. *Belgrave-square; Laxton Hall, Stamford, Lincolnshire; and Castle Freke, Cork, Ireland.*
Carlisle, the Rev. Robert Hodgson, Dean of, D.D., F.R.S., F.G.S., F.Z.S., &c. *Lower Grosvenor-street; and Hillingdon, Middlesex.*
Cawdor, the Right Hon. the Earl of, B.A., F.R.S., F.H.S., F.G.S., F.Z.S., Trust. Brit. Mus. *South Audley-street, Grosvenor-square; and Stackpole Court, Pembrokeshire.*
Charleville, the Right Hon. the Earl of., F.Z.S. *St. George's-place, Hyde-park Corner; and Charleville Forest, Tullamore, Ireland.*
Chearnley, R., Esq. *Lismore, Ireland.*
Cheetham Library, The. *Manchester.*
Christmas, W., Esq. *Isamore, Waterford, Ireland.*
Clark, W. B., Esq. *Budle House, Belford, Northumberland.*
Clitherow, Colonel, F.Z.S. *Boston House, Brentford, Middlesex.*
Clive, the Lady Harriet. *Lower Grosvenor-street; and Oakley Court, Ludlow, Shropshire.*
Craven, the Right Hon. the Earl of, F.Z.S. *Charles-street, Berkeley-square; Combe Abbey, near Coventry, Warwickshire; and Hampstead Park, Berkshire.*
Contes, J., Esq. *Fox-hill, Banks, Lancashire.*
Collingwood, H. J. W., Esq. *Lilburn Tower, Northumberland.*
Cooke, G. Esq. *Liverpool.*
Cooper, the Dowager Lady. *Portland-place; and Isleworth House, Middlesex.*
Cox, H. R., Esq. *Grosvenor-place; and Hillingdon, Uxbridge, Middlesex.*
Coxen, S., Esq. *Down Farm, Dartbrook, Hunter's River, New South Wales.*
Currer, Miss. *Gargrave, Skipton, Yorkshire.*
Currie and Bowman, Messrs. *Collingwood-street, Newcastle-upon-Tyne, Northumberland.*
De Jersey, C. B., Esq., F.Z.S. *Guernsey.*
De Jersey, P. F., M.D., F.L.S., F.Z.S. *Brook Cottage, Romford, Essex.*
Derby, the Right Hon. the Earl of, LL.D., Pres. Z.S., F.H.S., Trust. Brit. Mus. *Grosvenor-square; and Knowsley-park, Prescot, Lancashire.*
De Tabley, the Right Hon. Lord Warren, F.Z.S., &c. *Tabley House, Knutsford, Cheshire.*
Dixon, D., Esq. *Long Benton, Northumberland.*
Drummond, C., Esq., F.Z.S. *Grosvenor-place; and Roehampton, Surrey.*
Dynevor, the Right Hon. Lord, F.H.S., F.Z.S. *Dover-street; Barrington-park near Burford, Oxfordshire; and Dynevor Castle, Llandilo, Caermarthenshire.*
Dyson, the Rev. F. *Tidworth, Andover, Hampshire.*
Eden, Sir Robert Johnson, Bart. *Windleston, Rushyford, Durham.*
Edye, W. O., Esq., F.Z.S. *King's Bench Walk, Temple.*
Egerton, Sir Philip de Malpas Grey, Bart. M.P., F.R.S., F.G.S., F.Z.S. *Oulton-park, Tarporley, Cheshire.*
Egremont, the Right Hon. the Earl of, F.R.S., F.A.S., F.H.S., F.Z.S., M.R.I. *Grosvenor-place; and Petworth, Sussex.*
Empson, W., Esq., F.Z.S. *King's Bench Walk, Temple.*
Errington, G. H., Jun., Esq. *Colchester, Essex.*
Eyton, T. C., Esq., F.Z.S., &c. *Eyton, Wellington, Shropshire.*
Fane, J., Esq., F.Z.S. *Wormsley, Stokenchurch, Oxfordshire.*
Fanshawe, the Rev. T. L., F.Z.S. *Parsloes, Dagenham, Essex.*
Ferrand, Mrs. *Harden Grange, Bingley, Yorkshire.*
Fielden, J., Esq. *Whitton Hall, Lancashire.*
Fiennes, the Hon. W. T. T., F.L.S., F.H.S., F.Z.S. *Albany; Belvidere, Kent; and Broughton Castle, Oxfordshire.*
Finch, Lieut.-Col. the Hon. J., F.Z.S. *Audley-square.*
Folliott, G., Esq., F.Z.S. *Chester.*
Foot, J., Esq., F.Z.S. *Dorset-square.*
Forde, Col. M. *Seaforde, Clough, Ireland.*
Fowlis, Mrs. *York.*
Fox, B., Jun., Esq., F.Z.S. *Beaminster, Dorsetshire.*
France, The Royal Institute of.
Fuller, J. G., Esq., F.H.S., F.Z.S. *St. James's-street; and Streatham, Surrey.*
Gage, the Right Hon. Viscount, M.A., M.R.I., F.Z.S. *Whitehall Yard; Firle Place, Lewes, Sussex; and Westbury House, Hampshire.*
Garden, the Rev. J. L. *Kingsdown, Kent.*
Gibson, J., Esq. *Saffron Walden, Essex.*
Giffard, T. W., Esq. *Chillington-park, Somerford, Staffordshire.*
Glynne, Sir Stephen, Bart., M.P., F.A.S., F.L.S., F.Z.S. *Berkeley-square; and Hawarden Castle.*
Goodall, Rev. J., D.D., F.S.A., F.L.S., F.H.S., F.Z.S. Provost of Eton College. *Eton, Buckinghamshire.*
Grant, W. Jun., Esq., F.Z.S. *Brick Court, Temple.*
Greenaway, E., Esq., F.Z.S. *River-terrace, Islington.*
Gurney, J. H., Esq. *Earlham Hall, Norwich.*
Gurney, S., Esq. *Lombard-street; and Upton, Essex.*
Hale, R. B., Esq., M.P., F.Z.S. *Bolton-street, Piccadilly; and Cottles, Melksham, Wiltshire.*
Haslam, S. H., Esq., F.L.S. *Chesham, Bury, Lancashire.*
Hatchett, C., Esq., F.R.S.L. & E., F.S.A., F.L.S., &c. *Belle Vue House, Chelsea; and Ballington, Lincolnshire.*
Henson, the Rev. F. *Kilvington, Thirsk, Yorkshire.*
Hering, Mr. H. *Newman-street, Oxford-street.*
Herne, J., Esq., F.Z.S. *Ratcliff.*
Hewitson, W. C., Esq. *Derby.*
Heysham, T. C., Esq. *Carlisle.*
Hill, Sir Rowland, Bart., M.P., F.Z.S. *Hawkstone, Shrewsbury, Shropshire.*

Hodgson, B. H., Esq. *Nepál.*

Hoffmann, J., Esq., F.Z.S. *Hanover-terrace, Regent's-park.*

Holford, R., Esq. *Lincoln's-inn-fields; Knighton, Newport, Isle of Wight; and Kingsgate, near Margate, Kent.*

Holmesdale, the Right Hon. Viscount, M.P., F.Z.S. *Montreal, Kent.*

Howard, Colonel, the Hon. F.G., F.R.S., F.A.S., F.H.S., F.Z.S. *Grosvenor-square; Ashstead-park, Epsom, Surrey; Castle Rising, Norfolk; Elford, Lichfield, Staffordshire; and Levens Hall, Milnthorpe, Westmoreland.*

Hullmandel, C., Esq. *Great Marlborough-street.*

Hunter, W. P., Esq., F.G.S. *Albany.*

Hustler, J., Jun., Esq. *Bolton House, Bradford, Yorkshire.*

Jardine, Sir William, Bart., F.R.S.E., F.L.S., M.W.S., &c. *Jardine Hall, Lockerby, Dumfriesshire.*

Jenyns, the Rev. L., M.A., F.L.S., F.Z.S. *Swaffham-Bulbeck, Cambridgeshire.*

Kenmare, the Right Hon. the Earl of, F.Z.S. *Killarney, Ireland.*

Kennedy, the Rev. Dr., Master of the Grammar School. *Shrewsbury, Shropshire.*

Kennell, J. P., Esq. *University College, Gower-street, Bedford-square.*

Kensington, E., Jun., Esq. *Bridge-street, Blackfriars.*

Kirkpatrick, G., Esq. *Holly-dale House, Lock's Bottom, Kent.*

Lane, J., Esq., F.H.S., F.Z.S. *Goldsmiths'-hall, Foster-lane, Cheapside.*

Lawley, Sir Francis, Bart., M.P., F.H.S., F.Z.S. *Grosvenor-square; and Middleton Hall, Warwickshire.*

Leeds, His Grace the Duke of, K.G., F.H.S., F.Z.S. *St. James's-square; and Hornby Castle, Catterick, Yorkshire.*

Leigh, C., Esq., F.H.S., F.Z.S., M.R.I. *Stoneleigh Abbey, Kenilworth, Warwickshire.*

Legh, G. C., Esq., F.Z.S. *High Legh, Knutsford, Cheshire.*

Leyden University, Messrs. S. and J. Luchtmans for the.

Linnean Society of London, The.

Lisburne, the Right Hon. the Earl of, F.Z.S. *Lisburne House, Devonshire; and Crosswood, Cardiganshire.*

Liverpool Library, The.

Lomax, R. G., Esq. *Clayton Hall, Accrington, Lancashire.*

London Institution, The. *Finsbury Circus.*

Lyttelton, the Right Hon. Lord. *Hagley-park, Stourbridge, Worcestershire.*

Mackworth, Sir Digby, Bart., F.Z.S. *Cavendish Hall, Sudbury, Suffolk.*

Malcolm, N., Esq., F.Z.S. *Duntroon Castle; Poltalloch House, Argyllshire; and Lamb Abbey, Kent.*

Mann, J., Esq. *Birmingham.*

Martin, W., Esq., F.L.S., &c. *Hammersmith.*

Massena, The Prince. *Paris.*

Maxwell, P. C., Esq. *Hotham House, Market Weighton, Yorkshire.*

Meteyard, H. W., Esq. *Pump Court, Temple.*

Mexborough, the Right Hon. the Earl of, F.Z.S. *Dover-street; and Methley Park, Ferrybridge, Yorkshire.*

Miland, Mr. *Chapel-street, Belgrave-square.*

Mills, J., Esq., F.Z.S. *Clayberry Hall, Woodford Bridge, Essex.*

Milton, the Right Hon. the Viscountess. *Milton, Northamptonshire.*

Moore, M. S., M.D., F.H.S., F.Z.S., M.R.I. *Maddox-street, Hanover-square.*

Moore, W., Esq., F.Z.S. *Grimes-hill, Kirkby Lonsdale, Westmoreland.*

Morgan, O., Esq., F.H.S., M.R.I., F.Z.S. *Pall-Mall.*

Morland, W. A., Esq., F.H.S., F.Z.S. *Court Lodge, Lamberhurst, Sussex.*

Mosley, O., Esq., F.Z.S. *Cumberland-terrace, Regent's-park; and Rolleston Hall, Burton-on-Trent, Staffordshire.*

Noel, Colonel. *Wellingore, Sleaford, Lincolnshire.*

Ogle, the Rev. J. S., F.H.S., F.Z.S. *Kirkley, Northumberland.*

Ord, J. P., Esq. *Edge-hill, Derbyshire.*

Orford, the Right Hon. the Earl of, M.A., &c. *Wolterton-park, Aylsham, Norfolk.*

Ouseley, the Right Hon. Sir Gore, Bart., G.C.H., F.R.S., F.A.S., F.H.S., F.Z.S. *Upper Grosvenor-street; and Woolmers, Hatfield, Hertfordshire.*

Oxley, C. C., Esq. *Ripon, Yorkshire.*

Palmer, S., M.D. *Birmingham.*

Parker, R. T., Esq., F.H.S., F.Z.S. *Cuerdon Hall, Preston, Lancashire.*

Perkins, F. O., Esq., F.Z.S. *Addison-road, Kensington, Middlesex; and Chipstead-place, Seven Oaks, Kent.*

Perkins, H., Esq., F.L.S., F.H.S., F.G.S., F.Z.S. *Springfield, Surrey.*

Phelps, the Rev. H. D. *Snodland Rectory, West Malling, Kent.*

Phillipps, C. M., Esq., M.P., F.H.S., F.Z.S. *Piccadilly; and Garendon-park, Leicestershire.*

Pigott, J. H. S., Esq., F.A.S., F.Z.S. *Brockley Hall, Somersetshire.*

Plunkett, J., Esq., F.Z.S. *St. Maurice, Beauvais, France.*

Poulett, the Right Hon. the Dowager Countess. *Piccadilly; and Poulett Lodge, Twickenham, Middlesex.*

Preston, Mrs. *Moreby Hall, York.*

Prickett, R., Esq., M.R.I., F.Z.S. *Harley-street, Cavendish-square; Octon Lodge, Sledmere, Yorkshire; and Upton Cottage, Broadstairs, Kent.*

Pye, H. A., Esq. *Louth, Lincolnshire.*

Radcliffe Library, The. *Oxford.*

Read, the Rev. T. F. Rudston, M.A. *The Rectory, Full Sutton, near York.*

Read, W. H. Rudston, Esq., M.A., F.L.S., F.H.S., F.Z.S. *Union Club; and Hayton, near Pocklington, Yorkshire.*

Reeves, J., Esq., F.R.S., F.L.S., F.H.S., F.Z.S. *Clapham, Surrey.*

Reynard, E. H., Esq., F.Z.S. *Sunderland-wick, Driffield, Yorkshire.*

Robarts, A. W., Esq., M.P. *Hill-street, Berkeley-square; and Fitzroy Farm, Highgate.*

Rolle, the Right Hon. Lady. *Upper Grosvenor-street; Stevenstone, Torrington; Bicton, Exeter; and Bovey, Axminster, Devonshire.*

Rous, the Hon. T. *South-street, Park-lane.*

Salvin, B. J., Esq. *Burn Hall, Durham.*

Sandbach, H. R., Esq. *Woodlands, Aigburth, Liverpool.*

Scarborough, the Right Hon. the Dowager Countess of. *Portman-square.*

Sibthorp, the Rev. H. W., F.Z.S. *Washingborough, Lincolnshire.*

Skaife, J., Esq. *Blackburn, Lancashire.*

South, J. F., Esq., F.L.S., F.G.S., F.Z.S. *Adelaide-terrace, London-bridge.*

Southampton, the Right Hon. Lord, F.Z.S. *Carlton-gardens; and Whittlebury, Northamptonshire.*

Spencer, the Right Hon. Earl, F.A.S., M.R.I., F.Z.S. *Spencer House, St. James's-place; Wiseton Hall, Bawtry, Yorkshire; and Althorpe, Northamptonshire.*

Staniforth, the Rev. T. *The Rectory, Bolton by Bolland, Clitheroe, Lancashire.*

Strickland, A., Esq. *Bridlington Quay, Yorkshire.*

Sturt, Captain C., 39th Regt., F.L.S., F.R.G.S.

Suffield, the Right Hon. the Dowager Lady. *Berkeley-square; and Blickling-park, Aylsham, Norfolk.*

Sulivan, L., Esq., F.H.S., F.Z.S., Dep. Sec. at War. *War Office; and Broom House, Fulham, Middlesex.*

Sutherland, His Grace the Duke of, M.A., F.Z.S. *Stafford House, St. James's; Dunrobin Castle, Sutherlandshire; and Trentham Hall, Yorkshire.*

Taylor, Lieut.-General Sir Herbert, G.C.B., F.H.S., F.Z.S. *St. Katherine's Lodge, Regent's-park.*

Temminck, Monsieur, C. J. Chevalier de l'ordre du Lion Neerlandais; Directeur du Musée Royal des Pays-bas; Membre de l'Institut, des Académies de Stockholm et de Bonn; des Sociétés Royales de Medecine et de Chirurgie, Linnéenne et Zoologique de Londres; des Sociétés Impériales de Moscou et de celle des Naturalistes d' Utrecht, de Groningue, de Leiden, de Paris, Lausanne, Lille, Frankfort, Mayence, Halle, Marbourg, Wurzbourg, Heidelberg, Stockholm, Hanau, Batavia, Philadelphia, et Cap de Bonne Espérance. *Leyden.*

Teylerian Library, The. *Haarlam.*

Thompson, P. B., Esq., M.P., F.Z.S. *Berkeley-square; and Escrick Hall, Barlby, Yorkshire.*

Thurlow, the Rev. T., F.Z.S. *Saville-row; Baynard's-park, Guildford, Surrey; and Boxford Rectory, Suffolk.*

Tinné, J. A., Esq. *Liverpool.*

Trevelyan, Sir John, Bart., F.H.S., F.Z.S. *Wallington, Newcastle-upon-Tyne, Northumberland; and Nettlecombe, Somersetshire.*

Upton, the Hon. H.M., F.Z.S. *Whitehall-place.*

Upton, T., Esq. *Ashton Court, Bristol.*

Vienna, Messrs. Rohrmann and Schwiegerd for the Imperial Library of.

Walker, G., Esq. *Nether Green, Eastwood, Nottinghamshire.*

Walker, I., Esq. *Southgate, Middlesex.*

Walker, T., Esq. *Ravenfield Hall, Rotherham, Yorkshire.*

Wall, C. B., Esq., M.A., F.R.S., F.A.S., F.H.S., M.R.I., F.Z.S. *Berkeley-square; and Norman Court, Stockbridge, Hampshire.*

Warde, C. T., Esq. *Welcombe House, Stratford-on-Avon, Warwickshire; and Westover, Isle of Wight.*

Waterhouse, A., Esq. *Old Hall-street, Liverpool.*

Webb, Colonel, F.H.S., F.Z.S. *Adwell, Tetsworth, Oxon.*

Wells, W., Esq., F.H.S., F.Z.S. *Redleaf, Tonbridge, Kent.*

Wernerian Natural History Society of Edinburgh, The.

Wheeler, J. R., Esq., F.Z.S. *Oakingham, Berkshire.*

Wheble, J., Esq., F.H.S., F.Z.S. *Woodly Lodge, Reading, Berksh.*

Wilson, E., Esq. *Abbot's Hall, Kendal, Westmoreland.*

Winn, C., Esq., F.H.S., F.Z.S. *Nostell Priory, Wragby, Yorkshire.*

Wollaston, the Rev. F.H., F.Z.S. *Rowling, Wingham, Kent.*

Wollaston, Miss. *Clapham Common, Surrey.*

Wyndham, Captain, R.N. *Bramley House, Guildford, Surrey.*

Yarborough, the Right Hon. the Earl of, F.H.S., M.R.I., F.Z.S. *Arlington-street, Piccadilly; Brocklesby, Brigg, Lincolnshire; and Appuldercombe-park, Isle of Wight.*

Yarrell, W., Esq., F.L.S., Sec. Z.S. *Ryder-street, St. James's.*

Zoological Society of London, The.

GENERAL LIST OF PLATES.

NOTE.—As the arrangement of the Plates during the course of publication was found to be impracticable, the Numbers here given will refer to the Plates when arranged, and the work may be quoted by them.

VOLUME I.

RAPTORES.

VOLUME II.

INSESSORES.

VOLUME III.

INSESSORES.

VOLUME IV.

RASORES.

GRALLATORES.

VOLUME V.

NATATORES.

LIST OF PLATES.

VOLUME I.

NOTE.—As the arrangement of the Plates during the course of publication was found to be impracticable, the Numbers here given will refer to the Plates when arranged, and the work may be quoted by them.

RAPTORES.

GRIFFON VULTURE.

Vulture fulvus. *(Linn)*

Drawn from Life & on Stone by J. & E. Gould. *Printed by C. Hullmandel.*

Genus VULTUR, *Linn.*

GEN. CHAR. *Beak* strong, thick and deep, base covered with a cere; *upper mandible* straight until it reaches the point, where it is hooked abruptly; *under mandible* straight, rounded and becoming narrower towards the point. *Head* naked or covered with short down; *nostrils* naked and pierced diagonally at the cere. *Feet* very strong, furnished with nails slightly hooked; the middle *toe* very long and united at the base to the external toe. *Wings* long; first *quill-feather* short, the fourth the longest.

GRIFFON VULTURE.

Vultur fulvus, *Linn.*

Le Vautour Griffon.

PRE-EMINENT for size and strength, the Vultures exceed all other birds whose powers of wing are adequate to sustain continued flight. They are a race peculiar to hot climates, and their food consists of putrid animal substances, for the removal of which (where indeed a quick removal is called for,) they seem expressly appointed. Their flight is wonderfully rapid and graceful, and they are led by some faculty, not yet fully understood, (but most probably by the sense of smell,) from astonishing distances, and an elevation in the atmosphere beyond the reach of human sight, to their fœtid repast. In a tribe of birds thus characterized the Griffon Vulture is one of the most conspicuous, particularly among those individuals who inhabit the older continent.

The present species takes a wide range, inhabiting, in considerable abundance, Spain, Turkey, and the whole of the southern portion of Europe, as well as the northern portions of Africa; they also occur in the mountainous parts of the northern and central countries, but we are not aware of its having ever been seen in a wild state in the British Islands.

Like the rest of its family, except when pressed by the utmost necessity it never preys on living animals, but prefers carrion and putrid substances, and when fed to repletion is easily made captive. There is nothing, however, of ferocity or wildness in the disposition of this bird, as in that of the Eagle; hence in captivity it becomes gentle and domestic: its principal enjoyment consists in the gratification of its appetite, and that accomplished it seems perfectly contented.

It breeds among the most inaccessible precipices; its eggs are of a dull greyish-white slightly marked with spots of a pale reddish colour.

In the adult bird the head and neck are covered with short white downy feathers; the lower part of the neck is surrounded with a ruff of long slender feathers of the same colour or slightly tinged with red; on the breast there is also a space covered with white down; the whole of the upper and under surfaces, except the quill- and tail-feathers, which are blackish-brown, are of a fulvous grey, the belly having a slight tinge of rufous; beak bluish yellow; cere darker; irides hazel; feet light brown: total length about four feet. The male is, as usual with other rapacious birds, smaller than the female.

Young birds differ considerably from the adults; the downy feathers of the head and neck being dirty white varied with brown, and the rest of the plumage of a very light yellowish colour, interspersed with large markings of white or grey.

Our Plate represents an adult in perfect plumage.

CINEREUS VULTURE.

Vultur cinereus; *(Linn.)*

Drawn from Nature & on Stone by J. & E. Gould.

Printed by C. Hullmandel.

CINEREOUS VULTURE.

Vultur cinereus, *Linn.*

Le Vautour noir.

This, the largest of the European Vultures, offers to our notice, by the partially bare neck, open ears, curved claws and powerful beak, a deviation from the true or more typical Vultures as restricted by modern authors, the true Vultures having claws less curved, and a beak more lengthened and feeble, characters which render them unable to seize and carry off living prey. This striking feature was not passed over by the discriminating eye of Mr. Bennett while engaged in describing the *Vultur auricularis* of Daudin, a species inhabiting Southern Africa, which in general form and structure strictly resembles the one under consideration. In "The Gardens and Menagerie of the Zoological Society delineated," that gentleman intimates that in his opinion the bird he has described, from a fine living example in the Society's Gardens, would be found to possess characters sufficiently prominent and different from the rest of the Vultures to form the type of a new genus. Although the Cinereous Vulture has not that longitudinal fold of the skin which is so prominent a feature in the *Vultur auricularis*, still we should regard that more as a specific character than as having any influence over its natural economy; and we fully concur in Mr. Bennett's views in considering a further subdivision of the family to be necessary. The two birds in question, with the *Vultur pondicerianus* as a type, would constitute a very natural division. We refrain ourselves from assigning a generic name, or from entering more fully into the subject, as we are aware that M. Temminck is at this moment paying strict attention to this highly interesting family; and we have no doubt that with his discerning views and profound knowledge of Ornithology, he has long ere this observed the characters alluded to.

The European habitat of the Cinereous Vulture is the vast forests of Hungary, the mountainous districts of the Tyrol, the Swiss Alps, the Pyrenees, and the middle of Spain and Italy; it is seen also occasionally in other places.

M. Temminck states that its food consists of dead and putrid animals, never living ones, of which it is much afraid, even the smallest appearing to excite fear; but Bechstein informs us that "in winter it is chiefly seen in the plains, where it attacks sheep, hares, goats, and even deer. The farmers suffer severely from this bird, as it will frequently pick out the eyes of sheep; but as it is not a very shy species, it gives the huntsman some advantage, added to his being well paid for shooting so destructive an enemy."—(Latham's General History of Birds, vol. i. p. 23.)

Of its nidification and eggs nothing is known.

The whole of the plumage is of a dark chocolate brown, each feather being a little lighter on the edges; the head and upper part of the neck are covered with down, which, with a kind of beard under the throat, is of the same colour as the plumage; the basal half of the mandibles, the bare space on the front and sides of the neck, the tarsi, and the toes, are of a blueish flesh colour; the points of the mandibles and the claws black; irides dark brown.

We have figured an adult male, about one third of the natural size.

E. Lear del. et lithog.

EGYPTIAN NEOPHRON.

Neophron Percnopterus, (*Savig.*)

Printed by C. Hullmandel.

Genus NEOPHRON.

GEN. CHAR. *Beak* elongated, slender, straight, the upper mandible covered with a cere for half its length, and with a distinct hooked dertrum or tip, the lower mandible curving downwards at the point. *Nostrils* longitudinal, lateral, directed forwards, and placed near the culmen of the bill. Anterior part of the head, and the face naked. The neck covered with acuminated feathers. *Legs* of mean strength and length. *Tarsi* reticulated. *Feet* with four toes, three before, and one behind; the front toes united at the base.

EGYPTIAN NEOPHRON.

Neophron percnopterus, *Sav.*

Le Catharte alimoche.

OF the family of *Vulturidæ*, which is so extensively spread over the hotter portions of nearly every part of the globe, the present is the only species which has ever been taken in England; and of this fact, only a solitary instance is on record. It appears that the example alluded to was killed near Kilve in Somersetshire, in the month of October 1825, and is now in the possession of the Rev. A. Matthew of the same place. When first discovered, it was feeding upon the body of a dead sheep, with the flesh of which it was so gorged, as to be either incapable of flight, or, at all events, unwilling to exert itself sufficiently to effect its escape; it was therefore shot with little difficulty. Another bird, apparently of the same species, was at the same time observed in the neighbourhood, but escaped its pursuers. The circumstance of this example coming so far north, must be attributed entirely to accident, its native habitat being exclusively the southern provinces of Europe and the adjoining districts of Asia and Africa.

The traveller who visits Gibraltar, the adjacent parts of Spain, the islands of the Mediterranean, Turkey, and the northern coasts of Africa, cannot fail to have his attention attracted by this remarkable bird, one of the smallest of the *Vulturidæ*, which is there often found associating in flocks. Like the rest of its family, it is one of Nature's scavengers, being ever on the search for carrion and putrid offal, upon which it greedily feeds, seldom if ever attacking living prey.

The sexes, when adult, offer no difference in their characters or the colouring of their plumage; the young birds, on the contrary, in which state was the individual noticed as being taken in this country, offer striking contrasts. These decided opposites of colouring we have illustrated in our figures. We need scarcely remark that the young acquire their mature plumage by gradual changes, the completion of which takes two or three years; hence it arises that birds in all grades, from the dark plumage of youth to the snowy white of maturity, are continually to be met with.

It is said to build its nest in the most inaccessible parts of rocks: of its eggs nothing is correctly ascertained.

The adults have the face and cere naked, and of a fine yellow; the whole of the plumage is pure white, with the exception of the greater quill-feathers, which are black; the plumes of the occiput are long and narrow; the beak yellow, with a black horny tip; the tarsi and toes yellow; nails black; irides hazel.

The young bird of the first year has the whole of the plumage of a dark chocolate brown; the elongated feathers of the neck, as well as those situated on the shoulders and upper part of the back, are tipped with yellowish white; the cere and naked part of the face dull yellow; tarsi and feet of a dull livid yellow.

The Plate represents an adult, and a young bird of the first year, nearly half their natural size.

BEARDED VULTURE OR LŒMMER GEYER.

Gypaëtus barbatus; *(Storr)*

Drawn from Life & on Stone by J & E Gould

Printed by C Hullmandel

Genus GYPAËTUS, *Storr.*

Gen. Char. *Beak* straight, its base covered with setaceous feathers tending forward, rounded above; the *under mandible* furnished at the base with a fasciculus of stiff and elongated feathers. *Cere* clothed with feathers. *Tarsi* short, feathered.

BEARDED VULTURE OR LŒMMER-GEYER.

Gypaëtus barbatus, *Storr.*

Le Gypaëte barbu.

Ornithologists have had no little difficulty to contend with in clearing up the confusion which the numerous synonyms of this bird have occasioned; in fact, as far as our observations have extended, there is but one species comprehended in the present genus,—a genus distinguished by characters which place it intermediate between the Vultures and Eagles. The descriptions of Bruce the African traveller, and of the writers on Indian Ornithology, are all referrible to this species, whose habitat appears to extend to a certain range of elevation over the vast continents of Europe, Asia, and Africa. We have ourselves received it from the Himalaya, where it was discovered in very considerable abundance. In Europe it is confined to the highest ranges of mountains, such as the Alps and Pyrenees, but more especially those of the Tyrol and Hungary.

The habits and manners of the Lœmmer-geyer, also point out its true situation in nature to be intermediate between the Vultures and Eagles; and Authors have, according to their respective views, referred it to each of these groups. The first who pointed out its true situation was the eminent naturalist Storr, who advanced it to a genus with the expressive name of *Gypaëtus*, i. e. Vulture Eagle. The genus thus established has become now, with justice, universally adopted. Unlike the typical Vultures, which are distinguished by their bare necks, indicative of their propensity for feeding on carrion, the Lœmmer-geyer has the neck thickly covered with feathers, resembling those of the true Eagles, with which it also accords in its bold and predatory habits; pouncing with violent impetuosity on animals exceeding itself in size: hence the young Chamois, the Wild Goat, the Mountain Hare, and various species of birds find in it a formidable and ferocious enemy. Having seized its prey, the Lœmmer-geyer devours it upon the spot, the straight form of their talons disabling them from carrying it to a distance. It refuses flesh in a state of putrefaction unless sharply pressed by hunger; hence Nature has limited this species as to numbers, while on the other hand to the Vultures, who are destined to clear the earth of animal matter in a state of decomposition, and thus render the utmost service to man in the countries where they abound, she has given an almost illimitable increase.

M. Temminck informs us that it incubates on the summits of precipitous and inaccessible rocks, making no nest, but laying two eggs, on the naked surface, of a white colour marked with blotches of brown.

The adult birds offer no sexual differences of plumage, and less of size than is usual among rapacious birds.

The head and upper part of the neck are of a dull white; a black line extends from the base of the beak and passes above the eyes; another beginning behind the eyes occupies the ear-coverts; the beard is black; the lower part of the neck, the breast and under parts are of an orange-red; the upper surface of a dark greyish brown, the centre of each feather having a white longitudinal line; the quill- and tail-feathers grey with white shafts; tail long and graduated; tarsi, beak, and nails black; irides orange.

The young of the year have the head and neck dull brown; the under parts dark grey with spots of white; the upper parts blackish, with lighter spots; the irides brown, and feet olive. In this state it has been called *Vultur niger*.

We have figured an adult male one third its natural size.

IMPERIAL EAGLE.
Aquila Imperialis.

Drawn from Nature & on Stone by J. & E. Gould

Printed by C. Hullmandel

Genus AQUILA, *Briss.*

Gen. Char. *Bill* straight at the base, strong, much hooked at the point, compressed, the sides inclining upwards and forming a narrow culmen; the tomia of the upper mandibles having a faint obtuse lobe situated behind the commencement of the hook. *Nostrils* oval, lateral, placed transversely in the cereous part of the bill; space between the nostrils and eye thinly covered with radiating hairs. *Wings* ample; the fourth and fifth quill-feathers the longest. *Tarsi* thickly clothed with feathers to the toes, which are rather short, and united by a membrane at the base. *Claws* very strong, hooked, and very sharp, grooved beneath; those on the outer and hind toes the largest.

IMPERIAL EAGLE.

Aquila Imperialis, *Briss.*

L'Aigle Imperial.

The range of habitat occupied by this noble species in Europe is far more limited than that of its congener the Golden Eagle, which it closely resembles in its form, habits, and manners, being in fact exclusively confined to the eastern portions of the Continent, where it is abundant, particularly in Hungary, Dalmatia, and Turkey. In its adult state it may be readily distinguished from the Golden Eagle by the large white marks which are situated on the scapularies. It is said to give a preference to the extensive forests of mountain districts, rarely frequenting those of the plains. It always builds, says M. Temminck, either in the mountain forests or on high rocks, the female laying two or three eggs, of a dull white. The young in the plumage of the first and second year differ from the adult in having the upper part of a rufous brown, varied with large blotches of light red, and in having the scapularies merely terminated with white instead of being wholly of that colour; tail ash-coloured, spotted towards the extremity with brown, and terminating in rufous; back of the neck and all the under surface light buff, the feathers of the breast and belly bordered with bright red; beak dark ash; irides brown; tarsi olive.

In the adult the feathers on the crown of the head and back of the neck are of a lanceolate form, and of a rufous tinge bordered with a brighter tint; all the under surface of a deep blackish brown, with the exception of the belly, which is yellowish red; the upper surface is of dark glossy brown; several of the scapularies of a pure white; tail deep ash-colour irregularly banded with black, each feather having a large black bar near its extremity, which is yellowish white; irides light yellow; cere and tarsi yellow.

The Plate represents an adult and a young bird one third of the natural size.

GOLDEN EAGLE.

Aquila chrysaëta, *(Briss.)*

GOLDEN EAGLE.

Aquila chrysaëta, *Briss.*

L'Aigle Royal.

Of the two large Eagles which make the British Isles a permanent residence, the present noble species, although rather inferior in point of size, is more rapacious and sanguinary in its habits, feeding more exclusively on prey acquired by its own exertions, fawns, lambs, hares, rabbits and large birds being its usual victims : the Sea Eagle, on the contrary, feeds chiefly upon fish, large sea birds, and, not unfrequently, putrid carcases ; its habitat is consequently the mountains and craggy rocks along the sea shore, while the Golden Eagle frequents in preference the inland parts of the country, resorting to large forests and secluded situations.

The Golden Eagle appears formerly to have been by no means an uncommon bird in the British Isles ; but the increase of population and the cultivation of the land have driven it to the remoter portions of the kingdom, and it is now only to be found, and that but sparingly, among the highlands of the North, the wilder parts of Ireland, and occasionally in Wales : and although the romantic lakes and hills of Westmoreland and Cumberland, the rocky parts of Derbyshire, and the barren districts of Cornwall, were not long since among the number of its breeding-places, it is now seldom, if ever, to be found there, a bird of its size and habits not only exciting the attention, but the hostility also of the inhabitants of the surrounding districts. On the Continent it is more abundant, particularly in the northern and hilly countries, as Norway, Sweden and some parts of Russia : it is also found, but in less abundance, in Germany and France, and still less frequently in Italy or further southwards.

In the cleft of some inaccessible rock, or, as M. Temminck states, on the tops of the tallest trees of the forest, the Golden Eagle constructs its eyrie, and brings up its young, feeding them with the yet quivering flesh of the prey, whose remains are found scattered in abundance around. The eggs are two in number, sometimes three, of a dull white stained with dull red.

The young and adult of this noble bird exhibit marked differences of plumage, a circumstance which led the older writers on Ornithology to make in this instance, as in some others, two species out of one, an error which has been but lately corrected ; and we have yet much to learn respecting the laws which regulate these changes, so remarkable in this ferocious tribe.

The Ring-tailed Eagle, then, is but the immature stage of the Golden Eagle, nor is the full plumage attained but by slow degrees, two or three years being required for bringing the markings to their stationary character. When in full plumage, the feathers on the head and occiput are lancet-shaped, and of a rich gilded brown ; the rest of the body is of a dull brown approaching to chocolate brown, the feathers of the inner side of the thighs and tarsi being lighter ; tail greyish brown with transverse bands of blackish brown, with which colour it is tipped ; beak horn colour ; irides brown ; and tarsi yellow. Length three feet, the female being from four to six inches longer.

The immature birds, till the commencement of the third year, have the whole of the plumage of a reddish brown, with the under tail-coverts, inner side of the thighs and tarsi nearly white ; the tail white for three parts of its length, (whence the synonym of Ring-tailed Eagle,) the tip being brown. In proportion as the young bird advances, the colours become richer and deeper, the white of the tail contracts, and bars begin to appear. The third year is that of the assumption of perfect plumage.

The figures are a young and an adult, about one third of their natural size.

BONELLI'S EAGLE.

Aquila Bonelli.

Falco............... (Temm.)

Drawn from Nature & on stone by J & E Gould. Printed by C Hullmandel.

BONELLI'S EAGLE.

Aquila Bonelli.

L'Aigle Bonelli.

M. Temminck was, we believe, the first to make known this elegant species of Eagle as an occasional visitant in Europe. The number of examples which have come under our notice within the last few years induce us to believe it to be much more common than is generally suspected; it is, however, more particularly an African species, though its range appears to be extensive, as it may be frequently met with in collections from India. Several living specimens have at various times been received by the Zoological Society, and appeared to bear confinement equally well with others of the same genus. All these, we may observe, were received from Africa, and it is from one of them that our figure is taken; and from the circumstance of its having been an inhabitant of the menagerie for at least two or three years, we may reasonably conclude that it has nearly attained its adult colouring. There appears to be no other species of the group that exhibits so many and varied changes as the present bird, and in fact specimens are to be found of all shades, from a uniform tint of rich fulvous over the under surface to white or nearly so, with merely the centre of the feather striped with dark brown, and even much lighter than the birds represented in our Plate. M. Temminck having described these changes with considerable minuteness, we take the liberty of subjoining his description in full.

Adult male. "Upper surface brown, more or less deep, without any well-defined markings; under surface rust red, more or less bright, the shaft of all the feathers dark brown; tail uniform ash colour or slightly reddish with a terminal band of brown, or marked with distant brown bands; cere and toes yellow; irides nut brown; total length two feet."

Adult female. "Upper surface blackish brown; cheeks, throat, front and sides of the neck rust red, marked with small brown stripes along the stem of each feather; the remainder of the under surface of a duller rust red, each feather having a large longitudinal brown stripe, with the shaft black; similar markings appear on the feathers of the tarsus; inferior wing-coverts black; base of the primaries and secondaries deep grey marbled with black; tips of the primaries entirely black; tail pure ash, faintly banded or nearly uniform; all the quills with a large blackish band near the tip, which is more or less whitish; beak black at the tip, and greenish towards the base; toes yellowish; total length from two feet to two feet six inches.

"A specimen probably younger presented the following appearances: head, neck, back, scapularies and wings ash brown, marked along the shafts with blackish brown; all the great coverts, scapularies, and quills marked at intervals with very large black bands disposed in zigzags; the primaries and secondaries white, on the interior webs rayed with blackish bands; all the tail-feathers are ash brown above with nine or ten transverse bands, the intervening spaces being twice as wide as the bands; all these feathers are terminated with golden red more or less bright; beneath, the tail is whitish with a tint of red, and faint indications of transverse bands; front of the neck and chest clear red, with the stems of the feathers brown; thighs, feathers on the tarsus, abdomen, and under tail-coverts dirty white clouded with red and without spots.

We have figured an adult male about two thirds of the natural size.

SPOTTED EAGLE.

Aquila nævia: *(Meyer)*.

E. Lear del.

Printed by C. Hullmandel.

SPOTTED EAGLE.

Aquila nævius, *Meyer*.

L'Aigle criard.

This small but true Eagle receives its specific name from the spotted markings which characterize the species in its youthful dress. During the first year this feature is much more conspicuous than in the specimen from which our figure was taken, and which was in its second or third year. When in its permanent state of plumage, which is not attained till the fourth or fifth year, these markings become nearly effaced, the whole of the plumage being then of a uniform rich shining brown. In many of its habits and manners it closely resembles the Golden Eagle, and others of its genus, though in size it is far inferior to that noble bird. It is sparingly dispersed throughout Germany, the Pyrenees, and Russia; and from the circumstance of individuals having been received from India, we may naturally conclude that those found in Europe are only a scattered few, dwelling in the extreme limits of their true habitat. According to M. Temminck it is common in Africa, and especially in Egypt; hence we may infer that its range is throughout the south-eastern portions of the Old World.

It builds in high trees, and the eggs are said to be two in number, of a light colour thinly blotched with reddish brown.

Its food consists of small quadrupeds and birds; it is also well known that it feeds, particularly during the summer, upon the larger kinds of insects which abound in its native regions. We are not aware that any of the other true Eagles live upon this kind of prey, though we know it to be the case with many of the smaller genuine Falcons.

The female, although not differing in colour, has the same relative superiority in size over the male as in the *Falconidæ* generally.

In the adult, the whole of the plumage is of a fine rich glossy brown; the primaries black; the cere and toes yellow; bill black; irides brown.

The Plate represents a bird in the plumage of the second year, three fourths of the natural size.

BOOTED EAGLE.

Aquila pennata; (Steph.)

Drawn from Nature & on Stone by J. & E. Gould. *Printed by C. Hullmandel.*

BOOTED EAGLE.

Aquila pennata, *Steph.*

L'Aigle botte.

THE *Aquila pennata* may be regarded as the smallest of the true Eagles, and one of the most beautiful of its tribe; a casual glance would, however, almost lead to confound it with the Buzzards, and especially with that group which is feathered to the toes: it is smaller in size than any European species of Buzzard, nevertheless when we examine its beak, strong tarsus, and powerful claws, together with the long lanceolate feathers on the top of the head and neck, the great breadth and power of the shoulders, and the shortness of the tail, we at once recognise the characteristic features of the genuine Eagles. The eastern portions of Europe and the adjacent districts of Asia constitute its native habitat, whence it migrates annually as far as Austria, Moravia, and the eastern parts of Germany. A fine specimen of the male of this species, which was killed in the Austrian territories, was placed at our disposal by Baron de Feldegg, and of this bird our plate is a careful representation. In an interesting collection lately received by the Zoological Society from Trebizond, we observed a fine example of this species in a younger and consequently in a somewhat different state of plumage from that which we have figured, which is fully adult.

Its food consists of small quadrupeds, birds, and insects. M. Temminck informs us that it builds its nest in Hungary, near the Carpathian mountains. Its eggs are not known.

The adult has the top of the head light yellowish brown, each feather being lanceolate in form and having a dash of dark brown; the middle of the back and upper surface dark greyish brown; at the insertion of the wings is a patch, consisting of eight or ten feathers, of a pure white; a broad stripe of light yellowish brown extends from the shoulders across the wing to the secondaries, which with the quills are deep blackish brown; tail deep greyish brown, each feather having a lighter tip; under surface white, with the exception of the chest, which has the stem of each feather slightly dashed with brown; cere and claws yellow; irides hazel.

The young differs from the adult in having narrow transverse bars of sandy yellow across the breast and thighs.

The figure is of the natural size.

SEA EAGLE.

Haliæetus albicilla; *(Selby)*.

E. Lear del et lithog. Printed by C. Hullmandel.

Genus HALIÆËTUS.

Gen. Char. *Beak* elongated, strong, straight at the base, curved in a regular arc in advance of the cere to the tip, and forming a deep hook; *culmen* broad, and rather flattened; *tomia* of the upper mandible slightly prominent behind the commencement of the hook. *Nostrils* large, placed transversely in the cere, and of a lunated shape. *Wings* ample, the fourth quill-feather the longest. *Legs* having the tarsi half feathered; the front of the naked part scutellated, and the sides and back reticulated. *Toes* divided to their origin; the outer one versatile. *Claws* strong, hooked, and grooved beneath; the claw of the hind toe larger than that of the inner toe, which, again, exceeds that of the middle and outer toes.

SEA EAGLE.

Haliæëtus albicilla, *Savigny*.

L'Aigle pygargue.

Science is indebted to the observation of Mr. Selby for a knowledge of the fact that the Cinereous and Sea Eagles of the older writers are identical species, differing only in the respective stages of plumage, which depend solely upon age. The fact thus ascertained by experiment had been in some measure anticipated both by Cuvier and Temminck, but wanted that direct proof which rearing the birds from youth and preserving them to maturity could alone furnish.

It is the most common of the European Eagles, and perhaps the most widely dispersed. In the British Islands it frequently occurs along the rocky shores of England, Wales, Ireland, and Scotland and the adjacent islands, and many pairs are known annually to breed in different parts of the three last-named countries. The appetite for fish which this noble bird possesses leads it to give the preference to the margin of the sea, the shores of rivers and large lakes. Aquatic birds, small mammalia, such as hares, lambs, fawns, &c., and, when pressed by hunger, even carrion also, may be reckoned among the articles forming its diet; but like all the rapacious birds, especially the Eagles, it is capable of sustaining life for a considerable period when food cannot be obtained. Although not so alert and sprightly as the Golden Eagle, it is nevertheless vigorous and resolute, its powers of flight enabling it to soar with great majesty and ease through the upper regions of the air, whence it often precipitates itself upon its prey, or any intruder near its nest, with great force and velocity. Its range over Europe, although extensive, is limited to the more northern portions, particularly the rocky coast of Norway and Sweden, as well as that of Russia, Germany, Holland, and France. In the absence of bold precipitous rocks, which form its favourite place of nidification, it accommodates itself to the circumstances of the locality, constructing its nest on the top of the largest tree of the forest, bordering inland seas and lakes. The eggs are white, and two in number.

Three or four years at least are required to complete the state of plumage represented by the bird in the foreground of our Plate, which is that of maturity, a period characterized by the white tail, and the bright straw yellow-coloured bill.

The sexes offer little or no difference in their plumage at the corresponding periods of their age.

The adults have the bill and cere bright straw yellow; irides reddish brown; the whole head and neck are of a pale ashy brown, the feathers being long and pointed; the rest of the plumage is of a dark greyish brown, more intense on the upper surface; the tail pure white; tarsi and toes yellow.

The young have the beak and tail blackish brown, and the general plumage of a deep brown, the feathers of the head and neck being somewhat lighter than the rest.

The Plate represents an adult and a young bird of the first year, about one third of the natural size.

WHITE HEADED EAGLE.

Haliæetus leucocephalus; *(Savigny)*

E. Lear del et lithog.

Printed by C. Hullmandel.

WHITE-HEADED EAGLE.

Haliæëtus leucocephalus, *Savigny*.

L'Aigle à tête blanche.

It is not until very recently that the confusion which had existed in the instance of the Golden Eagle (gen. *Aquila*), as well as in that of the Sea Eagle, and of the present species (gen. *Haliæëtus*), has by patient observation been satisfactorily cleared away. This confusion arose from the striking difference in the plumage of the immature bird from that which characterizes it in an adult condition. Hence it was that the White-headed or Bald Eagle (as it is called by Wilson), has been universally confounded with the Sea Eagle (*H. albicilla*), a species which appears to be exclusively European. It must, indeed, be confessed that the immature birds of both species very closely resemble each other; but we believe that distinguishing characters are not wanting even at this period, though perhaps not very apparent upon a superficial examination; the tail, for example, is longer in the White-headed Eagle, and the plumage is less regularly varied with brown.

Sir W. Jardine, in his notes on Wilson's description of this species, observes, that having had both the White-headed and the Sea Eagle in his possession for several years, he has observed their respective manners to be also different, the White-headed being "more active and restless in disposition," "constantly in motion," and incessantly uttering "its shrill barking cry." It is also more fierce and untameable.

The adult of the present species cannot be mistaken, but the white of the head and tail is not acquired in its full purity till the third year. The first moult gives a mingling of ash colour, white, and obscure brown; the second increases the ratio of white; the third completes the transition from the dull greyish brown of the first year.

Sir W. Jardine observes, that in captivity from three to five years are required to effect a thorough change.

This beautiful Eagle is a native of the temperate and northern regions of both continents, but is much more common in America, where it is adopted as the national standard of the United States. "Formed," says Wilson, "by nature for braving the severest cold; feeding equally on the produce of the sea and of the land; possessing powers of flight capable of outstripping even the tempests themselves; unawed by anything but man; and from the ethereal heights to which he soars looking abroad at one glance on an immeasurable expanse of forests, fields, lakes, and ocean deep below him; he appears indifferent to the little localities of change of seasons, as in a few minutes he can pass from summer to winter, from the lower to the higher regions of the atmosphere, the abode of eternal cold, and thence descend at will to the torrid or the arctic regions of the earth."

Though preying indiscriminately on every kind of animal, especially small mammalia, and not even refusing carrion when pressed by hunger, the White-headed Eagle gives the decided preference to fish. Not that he obtains his prey by his own exertions as a fisher, or at least very seldom, and then only in the shallows; he watches the labours of the Osprey, and forces that industrious fisher to give up his booty. Wilson's spirited description of the contest has been often quoted; nor is the sketch by Audubon of this bird's ferocious attack upon the Wild Swan less replete with descriptive energy. The favourite localities of the White-headed Eagle are the borders of lakes, the rocky margins of the larger rivers, and especially the precipitous shores of the ocean.

The nest is generally placed in the topmost branches of lofty trees, often in the centre of a morass or swamp, and is formed of a mass of sticks, sods, grass, &c. It is increased by fresh layers annually, being repaired and used year after year until it becomes of such magnitude as to be observable at a great distance. The young are fed with fish, which often lie scattered in a putrid state round the tree, infecting the air for a considerable distance. The young are at first covered with a cream-coloured cottony down, which gradually gives place to the greyish brown feathers of the first year.

The adult plumage is as follows: head, upper part of the neck, and the tail, pure white; body of a deep chocolate brown; beak, cere, and tarsi, whitish yellow; irides almost white.

We have figured an adult male and an immature bird one third the size of nature.

OSPREY.

Pandion haliæetus; *(Savig.)*

Genus PANDION.

Gen. Char. *Beak* short, strong, rounded and broad, cutting edge nearly straight. *Nostrils* oblong-oval, placed obliquely. *Wings* long; the second and third quill-feathers the longest. *Legs* strong and muscular; tarsi short, covered with scales. *Toes* free, nearly equal in length; outer toe reversible; all armed with strong curved and sharp claws; under surface of the toes rough and covered with small pointed scales.

OSPREY.

Pandion haliæetus, *Sav.*

Le Balbuzard.

There is no species of the great family of *Falconidæ* whose range of habitat is so universal as that of the Osprey; and there certainly is none to whose habits attaches more interest than to those of this noble bird. While some of its race prey upon quadrupeds, and others upon the feathered tribes, the Osprey gains his subsistence almost exclusively from the waters, the scaly tenants of that element constituting its food: hence it is observed, that the countries in which he takes up his abode must be at least temperate, since it is evident that if the waters be frozen, he would be compelled by necessity to seek a more congenial climate. Such is evidently the case; and hence the Osprey is everywhere migratory, visiting the northern latitudes only during the months of spring and summer. In Europe this bird is but thinly dispersed; but to counterbalance this it is found in every portion of it, at least where wide rivers, lakes and arms of the sea offer it the necessary supply of food. The British Isles are not so much frequented as other parts of Europe; and when one of these birds does make its appearance, its magnitude and peculiar actions call forth the attacks of so many assailants, that it is either quickly destroyed or driven to seek a safer asylum elsewhere. Indeed it can hardly be said to be a welcome visiter, since it makes the greatest havoc among the stock of fish-ponds and rivers, not readily leaving if once established where its prey abounds.

In some parts of America the Osprey is very common, especially in the United States, where it makes its appearance on the return of spring: hence it is a welcome visiter, since its arrival betokens the opening of the rivers and the return of the hordes of fish. Here along the borders of mighty streams, undisturbed and unmolested save by the Bald Eagle, its professed enemy, it builds its nest in tall trees, constructing it of sticks and turf, so as to form a large mass, on the edges of which other small birds congregate and nidify without the slightest injury: in fact, the Osprey, or Fish-Hawk, is a quiet bird, with little ferocity or daring in his temperament. His manner of taking his prey is very remarkable: hovering for a time on wide-spread wings over the water, he then sails about, intently gazing on the element beneath. The moment a fish appears, down he plunges like an arrow, almost disappearing beneath the water, but rising in a moment, with the victim grasped in his strong and incurved talons: throwing the spray from his burnished plumage, he soars aloft, and hastens to his nest to share the spoil with his young, or feast upon it at leisure. Often, however, is the Osprey robbed of his prize. We have alluded to the Bald or White-headed Eagle as his foe, who frequently chases him when loaded with his booty, which he is forced to relinquish to his stronger opponent. The spirited narrative of the contest between these two birds—the one to retain, the other to obtain, the booty—in Wilson's American Ornithology, is probably familiar to all our readers; if not, we recommend them to peruse it. The eggs are generally three, of a dull white, blotched with dark red or yellow brown.

In the adult state of plumage, the whole of the upper surface is of a rich glossy brown; the top and sides of the head are mingled white and brown, and a brown line passes from behind the eye to the shoulder; the throat, chest and underparts are white slightly dashed with a few lines of rusty brown; tail barred; cere and nostrils light brown; tarsi blueish lead colour; irides yellowish orange.

The young are distinguished by the feathers of the upper surface being edged with whitish, and the chest being almost wholly of a pale brown.

We have figured an adult and a young bird about half the natural size.

SHORT-TOED EAGLE.
Circaëtus brachydactylus; *(Vieill)*

E. Lear del et lith. Printed by C. Hullmandel.

Genus CIRCAËTUS, *Vieill.*

Gen. Char. *Beak* robust, convex, compressed laterally; the upper mandible with its edges straight, and the point crooked; the inferior blunt at its tip. *Nostrils* oval and transverse. *Tarsi* naked, reticulated, elongated, and thick. *Toes* short, the outer two united at their base by a web; the lateral and hind toes nearly equal; nails short and strongly curved. *Wings* long, the third quill-feather the longest; the first shorter than the sixth.

SHORT-TOED EAGLE.

Circaëtus brachydactylus, *Vieill.*

L'Aigle jean le blanc.

In the present Eagle is exhibited one of those links which in the family *Falconidæ* are so numerous and so clearly appreciable, uniting group to group by intermediate forms so nicely balanced as to embody in themselves the main characters of the more typical genera between which they are interposed. In the "Règne Animal" of the Baron Cuvier, that great naturalist judiciously observes that the genus *Circaëtus* holds an intermediate place between the fishing-eagles (*Haliæëtus*), the ospreys (*Pandion*), and the true buzzards (*Buteo*); and he adds that it has the wings of the eagles and buzzards, with the reticulated tarsi of the ospreys.

Of all the eagles and buzzards none appear to have a wider range than the Short-toed Eagle. Its European localities, according to M. Temminck, are principally Germany and Switzerland: in France it occurs occasionally; but in Holland and the British Islands it has never been seen. It is also dispersed nearly through the whole of Africa and India, countries peculiarly favourable to it, in as much as its food consists principally of snakes and reptiles, which especially abound in the hotter portions of the globe. In the nature of its food and in the elongation of its tarsi we cannot fail to trace a marked approximation to the true Harriers, which, it is well known, are inveterate destroyers of every kind of reptile.

Like most of the *Falconidæ*, the Short-toed Eagle undergoes a succession of changes before it attains a permanent state of plumage. The colouring is so well detailed by M. Temminck that it is useless to attempt any addition. He states that the young have the upper parts of a deep brown, but that the base of each feather is of a pure white, the throat, breast, and belly being of a reddish brown, little or not at all blotched with white; the bars on the tail almost imperceptible; the beak bluish; the tarsi greyish white.

The colouring of the adult male is as follows:

The head is very large; below the eye is a space clothed with white downy feathers; the top of the head, cheeks, throat, breast, and belly are white, variegated by a few blotches of light brown; shoulders and wing-coverts brown, the base of every feather being white; tail square at the end, of a greyish brown barred with brown of a deeper tint, and white underneath; tarsi long, and, as well as the toes, of a light bluish grey; beak black; cere bluish; irides yellow.

The female is distinguished by having less white in her plumage, and by having the head, neck, and breast more thickly blotched.

Our Plate represents a bird in a state intermediate between youth and maturity, in which, as may be observed, the flanks and thighs are transversely barred with brown: the figure is about one third less than the natural size.

COMMON BUZZARD.
Buteo vulgaris; *(Bechst).*

E. Lear del. et lithog. Printed by C. Hullmandel.

Genus BUTEO.

GEN. CHAR. *Beak* rather weak, bending from the base, sides compressed, widening from the base, where the culmen is broad and flat; under mandible shallow, with the tip obliquely truncated. *Cere* large. *Nostrils* pyriform. *Wings* long and ample, the third or fourth quill-feather being the longest, the first four having their inner webs deeply notched, the third, fourth, and fifth having their outer webs deeply notched. *Tarsi* short, naked or feathered to the toes. *Toes* rather short, the front ones united at the base. *Claws* strong but not much hooked.

COMMON BUZZARD.

Buteo vulgaris, *Bechst.*

La Buse.

UNLIKE the true Falcons, whose vigorous flight and aërial disposition place them at the head of the Raptorial birds, or the spirited and bold short-winged birds of the genera *Astur* and *Accipiter*, the species of the present genus, though possessed of considerable bodily powers, are sluggish, timid, and inactive; still they are admirably adapted by nature to fill the office for which they are designed in the œconomy of the creation. Slowly soaring on buoyant wings, the Common Buzzard surveys the earth beneath in search of the smaller mammalia and reptiles which constitute its food, and upon which they pounce with a rapid and noiseless descent; nor does it disdain, when pressed by hunger, to partake of carrion, or such offal as chance throws in its way. Such may be regarded as the character of the Buzzard, which is an inhabitant of all the wooded districts of the British Islands, more particularly those of the southern districts. It is still more abundant in France, Holland, and all the temperate parts of Europe, being everywhere stationary.

To illustrate all the changes which this bird undergoes, and which are, indeed, characteristic of the Buzzards in general, would far exceed the space allotted to each subject of the present work; we have therefore given a figure of the bird in that state which is most common to the species: it is these changes that have led to a great multiplication of the species, and to no little confusion, in the works of the older ornithologists.

Its nest is constructed of sticks in the densest part of the wood, and it sometimes takes up with the deserted nests of Crows, Pies, &c. The eggs are two or three in number, of a dirty white colour, slightly spotted with reddish brown.

From our own experience, we are enabled to say, that the birds of one year old are much lighter in their plumage, particularly on the under surface, than those of the succeeding year, and may be easily recognised by their having the upper portions of their plumage, which is of a very dark brown with violet reflections, edged on each feather with a light yellowish white margin. The next year they become still darker, the back and breast assuming an almost uniform tinge of the same colour, being irregularly broken with transverse bars of yellowish white: the tail is also darker, particularly towards its base, which is generally white or whitish in the bird of the year. In the very advanced stage the colouring is still more uniform, of a pale cinereous brown, with faint indications of an occipital crest, which is represented by two or three feathers more elongated than the rest, and of a darker colour; the cere and legs lemon yellow; irides hazel.

The Plate represents an adult bird about two thirds of the natural size.

ROUGH-LEGGED BUZZARD.

Buteo Lagopus, *(Flem.)*

E. Lear del. et lith. Printed by C. Hullmandel.

ROUGH-LEGGED BUZZARD.

Buteo Lagopus, *Flem.*

La Buse pattué.

THE Rough-legged Buzzard enjoys a much more extensive range of habitat than the preceding species (*Buteo vulgaris*), which is strictly confined to the Old World, while the bird here represented is dispersed over nearly the whole of the Arctic Circle. A beautiful figure of this bird will be found in the 'Fauna Boreali-Americana' of Messrs. Swainson and Richardson, which upon examination will prove, beyond a doubt, its identity with the specimens killed in Europe. Its residence in the northernmost parts of America does not appear to be permanent, for Dr. Richardson informs us that it retires southwards in October to winter upon the banks of the Delaware and Schuylkill, returning again to the north early in spring. " A pair of these birds," says this gentleman, " were seen at their nest, built of sticks, in a lofty tree, standing on a low, moist, alluvial point of land almost encircled by a bend of the Saskatchewan. They sailed round the spot in a wide circle, occasionally settling on the top of the tree, but were too wary to allow us to come within gunshot."

The Rough-legged Buzzard is abundant over the whole of the North of Europe, but is more thinly dispersed over its temperate and warmer parts. It is not a permanent resident in any of the British Isles, but visits them periodically, being in some seasons tolerably abundant, while in others it is scarce. During its stay it commits great depredation in the rabbit warrens, in the neighbourhood of which it may generally be fouud; it also preys upon rats, hamsters, moles, lizards, frogs, and, according to Mr. Selby, wild ducks and other birds. " In the winter of 1815," says this gentleman, " Northumberland was visited by some of these birds, and several opportunities were afforded me of inspecting both living and dead specimens. Those which came under examination closely resembled each other as to colour and markings, though some individuals were darker along the belly than others; and the quantity of white upon the upper half of the tail was not always of equal breadth. Two of these birds, from having attached themselves to a neighbouring marsh, passed under my frequent observation. Their flight was smooth but slow, and not unlike that of the Common Buzzard, and they seldom continued for any length of time on the wing. They preyed upon wild ducks and other birds, which they pounced upon the ground; and it would appear that mice and frogs must have constituted a great part of their food, as the remains of both were found in the stomachs of those that were killed."

The plumed tarsi of this species at once distinguish it from its near ally the Common Buzzard, to which it assimilates in its general contour, as well as in many of its actions, and its general economy.

The nest, according to M. Temminck, is built in lofty trees; the eggs, which are four in number, being white spotted with reddish brown.

Like the common species this bird undergoes a variety of changes between youth and maturity. The sexes are alike in plumage.

Adults have the head, neck, and throat yellowish white, with narrow streaks of brown; back and wing-coverts brown, with paler edges; lower part of the inner webs of the quills white; upper tail-coverts and base of the tail white, the remainder being brown crossed with bands of the same colour, but of a darker tint; breast yellowish white with large spots of brown; under surface brown; thighs yellowish white, with brown arrow-shaped spots; tarsi clothed with feathers of a yellowish white, with a few small brown specks; bill bluish black, darkest at the tip; cere and irides bright yellow; toes reddish yellow; claws black.

The Plate represents an adult male about two thirds of the natural size.

HONEY BUZZARD.

Pernis apivorus; *(Cuv.)*

Drawn from Nature & on stone by J & E Gould. *Printed by C Hullmandel.*

Genus **PERNIS,** *Cuv.*

Gen. Char. *Bill* slender, weak, bending gradually from the base to the tip; cutting margin nearly straight; cere occupying half the length of the bill; under mandible sloping gradually to the tip. *Nostrils* long, narrow, placed very obliquely in the cere and opening forwards: lores thickly covered with small soft tiled feathers. *Wings* long and ample; first feather shorter than the sixth, and the third and fourth the longest in the wing; inner webs of the first four notched, and the outer webs of the third, fourth, and fifth sinuated. *Tail* long and slightly rounded. *Legs,* tarsal half feathered, lower or naked part reticulated. *Toes* rather slender, the inner and outer ones of nearly equal length, the anterior joints of all scutellated. *Claws* weak, slightly hooked, with the inner edge of the middle one dilated.

HONEY BUZZARD.

Pernis apivorus, *Cuv.*

La Buse Bondrée.

The Honey Buzzard, which is the type of Cuvier's genus *Pernis*, is much more sparingly diffused over the continent of Europe than the Common Buzzard (*Buteo vulgaris*), from which it differs in possessing a more feeble and softer bill, which is wider in the gape, and in having shorter and less powerful tarsi and toes, the claws of which are straighter and less retractile: it may also be easily distinguished from the members of the genus *Buteo* by the small and closely set feathers which cover the space between the bill and the eye, which space in all the rest of the *Falconidæ* is either bare or thinly covered with fine hairs or bristles.

We have good reason to believe that the Honey Buzzard is far more abundant in the British Islands than is generally suspected, several instances having come to our knowledge, not only of its capture, but also of its breeding in this country. Its flight is easy and graceful, and, like its near ally the *Buteo vulgaris*, its great size readily attracts the notice of the keeper and sportsman, to whom it soon becomes a prey when it takes up its abode in our woods or parks. The range of this bird is not confined to Europe alone, as is proved by our having frequently observed it in collections from India.

Its favourite food appears to be insects, wasps, bees, and their larvæ, to which are added lizards, small birds, mice, and moles.

It is subject to a number of changes in the colouring of its plumage, some individuals being of a uniform dark bronzy brown, while others have the head, neck, and under surface almost white with broad transverse bars of brown: this latter state is considered to characterize the young bird.

It builds in lofty trees, constructing a nest of twigs lined with wool and other soft materials; the eggs are small, of a yellowish white marked with numerous spots of reddish brown.

Crown of the head brown tinged with bluish; upper surface brown of various tints edged with yellowish brown; throat yellowish white with a few brown streaks on the shafts of the feathers; under surface yellowish white, with triangular spots and bars of chestnut; tail dark brown, with three bars of blackish brown; bill bluish black; cere greenish; irides yellow; tarsi and feet yellow; claws black.

There being no difference in the colouring of the sexes, we have figured a bird in the plumage of its first year.

GOSHAWK.

Astur Palumbarius. *(Bechst)*.

Drawn from Nature & on Stone by J. & E. Gould. Printed by C. Hullmandel.

Genus ASTUR.

Gen. Char. *Beak* short, bending from the base, compressed; upper mandible festooned on its cutting margin. *Nostrils* oval, opening obliquely forwards. *Wings* short, when closed reaching only one half the length of the tail; fourth quill-feather the longest; inner webs of the first five deeply notched. *Legs* covered in front with broad scales; middle toe much longer than the lateral ones, which are equal; hind toe strong; claws curved, strong and sharp.

GOSHAWK.

Astur palumbarius, *Bechst.*

L'Autour.

The *Falco palumbarius* of Linneus, the *Astur palumbarius* of the present day, may be regarded as the most noble and typical species of its genus,—a genus separated from the Falcons by the absence of the true dentation of the mandibles, and by possessing a short and more rounded form of wings, together with a slender and less robust body; and distinguished from the genus *Accipiter* by its short and powerful tarsus, and by the diminished length of the middle toe, which, from its length, in the latter genus forms so conspicuous a character.

The genus seems somewhat extensively distributed, both in the Old and New World: from India in particular we know of several interesting examples; while at the same time America is not deficient in birds of this form, the well-known *Astur atricapillus* of the northern portion of that country being the nearest representative of our species, and until lately confounded with it.

The *Astur palumbarius* is found in considerable abundance in all the wooded districts of Central Europe, though in the present day of very rare occurrence in our own island. M. Temminck informs us that it is also equally scarce in Holland.

This elegant and noble bird minutely resembles in its general habits our well-known Sparrow Hawk, and is not excelled in spirit or daring by the noblest of the Falcons. Its manner of taking its prey, however, appears to us exceedingly different. Pursuing it with assiduity, undaunted courage ,and perseverance, it does not stoop upon it like a Falcon, but glides after its victim, in a line, with the utmost velocity. It was anciently much esteemed in falconry, and its mode of taking its prey is more successful than that of the Falcon, although it does not exhibit those aërial evolutions which are so much admired in the Jerfalcon. The Goshawk was especially used for taking hares and partridges,—game which do not call into play the Falcon's peculiar mode of flight.

The male and female offer the same disproportionate difference in size as the Sparrow Hawk, and the former has the transverse markings finer and more distinct. The colouring of the two sexes is otherwise closely similar. The young, in the first and second year, possess, instead of the transverse bars on the breast, large oblong dashes of brown, upon a ground of white tinged with rufous.

In the adult, the whole of the upper surface is of a dull blueish grey, the under surface white with transverse somewhat zigzag bars of black, and wavy lines of the same colour across the shaft of each feather; the tail ash-coloured above, with four or five bars of blackish brown; irides and feet fine yellow.

The Plate represents a female in full plumage, and a young bird in its immature stage about three fourths of the natural size.

SPARROW HAWK.

Accipiter fringillarius, *(Ray)*.

Falco nisus; *(Linn.)*

Drawn from Life & on Stone by J. & E. Gould. *Printed by C. Hullmandel.*

SPARROW HAWK.

Accipiter fringillarius, *Ray*.

L'Epervier.

Of the smaller European birds of prey the Sparrow Hawk is one of the most bold and intrepid, and, unlike many of the true Falcons of its own size, which live in a great measure upon insects, it preys almost exclusively upon the birds of the Passerine order, but it does not hesitate to attack those of a larger size, and proves a destructive enemy to Pigeons, Partridges, and young poultry;—hence it is one of those predatory tyrants which are peculiarly obnoxious among the preserves of game, especially during the breeding season. It is often seen (pressed no doubt by the necessity of providing for its young,) hovering about the borders of the wood, or lurking in the hedge-row, and ever and anon pouncing upon some unfortunate victim which has arrested its attention. Quick-eyed and rapid, it darts upon its quarry like an arrow, and pursues it with unrelenting pertinacity, undaunted even by the presence of man, in whom the terrified fugitive has been often known to trust for a chance of safety in the desperate emergency; and many instances are on record of the Lark and Pigeon rushing into houses through open windows, followed by the intrepid foe. The flight of the Sparrow Hawk, though distinguishable for celerity, is not of that soaring character which we observe in the true Falcons; instead of descending upon its prey from aloft, and striking it down, or if missing the stroke, mounting again and repeating a similar assault, it darts at it without rising to any altitude, and follows up the chase till enabled to effect its capture. This peculiarity in its flight will at once distinguish it from the Kestrel,—a bird more common, and in a state of nature often mistaken for it.

The Sparrow Hawk is universally, although but moderately distributed throughout the whole of Europe as well as in the adjacent continents of Asia and Africa. The great disparity in size and dissimilarity in colouring between the male and female are among the most remarkable peculiarities connected with the present species. The young also exhibit a decided contrast in their plumage to the adults. These differences we shall endeavour to explain in our descriptions of each.

The adult female is in length fourteen inches; the whole of the upper surface is of a dark greyish brown; but the feathers of the shoulders, if examined, are found to be barred with broad dashes of white, the end only being of the colour which appears generally; on the back of the neck there is a large white patch, each feather being slightly tipped with brown; an obscure stripe of white surmounts the eye; the throat is white with small longitudinal specks; the breast and underparts are also white with beautiful transverse bars; the tail is brown like the back, and crossed with four bands of a darker colour; cere yellowish green; irides and tarsi yellow.

With markings like those of the female, the male has the upper surface of a dark blueish ash colour, but the throat and under parts are rufous, exhibiting the longitudinal specks and transverse bars as in the female, but more obscure; the cere, irides and tarsi as in the female; in length scarcely twelve inches.

The young male has the head and back of the neck, which is destitute of the white patch, of a reddish colour blotched with brown; the feathers of the back and wings are edged with reddish; the scapulars are marked with large spots of white; the under surface yellowish white, transversely barred with reddish; cere greenish yellow; irides greyish ash; tarsi livid.

The Sparrow Hawk frequents wooded and mountainous districts, where it makes great havoc among quails, larks, and small birds in general; small quadrupeds and lizards also form part of its diet.

It builds its nest in trees; the eggs being generally four in number, of a dull blueish white marked with angular red blotches.

Our Plate represents a male and female of this elegant little Hawk in their adult plumage, somewhat less than the natural size.

JER FALCON.

Falco Islandicus, *(Lath)*

Drawn from Life & on Stone by J. & E. Gould. *Printed by C. Hullmandel.*

Genus FALCO.

GEN. CHAR. *Beak* short, thick, strong, curved from the base; upper *mandible* with a prominent acute tooth. *Nostrils* rounded. *Tarsi* stout, short. *Toes* long, strong, armed with curved and sharp claws. *Wings* long, pointed, the first and third feathers long and equal, but shorter than the second feather, which is the longest.

JER-FALCON.

Falco Islandicus, *Lath.*

Le Faucon gerfaut.

THE Jer-falcon may be considered the type of the true Falcons, pre-eminent as it is in all the characters and attributes which distinguish the most noble of the birds of prey. It is a native of most of the Northern parts of Europe, and occasionally visits the Orkney and Shetland Isles. It was seen by Captain Sabine on the west coast of Greenland, and according to Dr. Richardson is a constant resident in the Hudson's Bay territories, where it is known by the name of the speckled Partridge Hawk, and Wanderer, and where it subsists by destroying Plovers, Ptarmigan, Ducks and Geese.

The falconers who visit this country almost every season with their trained Peregrine Falcons for sale, all agree in declaring that the Jer-falcon which they obtain constantly from Norway, is a different bird from that which they consider the true Falcon of Iceland. They say that these two Falcons differ in the comparative length of their wings in reference to the tail; the Iceland Falcon is, to them, a much more valuable, as well as a much more rare species; that they require a different system of training, as well as of general management. They describe the Iceland Falcon as a bird of higher courage than the Jer-falcon, of a more rapid and bolder flight, and that he can be flown successfully at larger game. His gyrations are said to be wider, his mount higher, and his stoop to the quarry more impetuous, grand, and imposing; and a well-trained specimen commands in consequence a much larger price. One of these falconers observed, that the Iceland Falcons he had trained, were, to the number of Norway Jer-falcons, but as one to twenty; another, from his own experience, considered them as still more rare. The question, Are there two species? has occurred to systematic writers in Ornithology; but we doubt whether the specimens contained in our various collections will afford sufficient data to make the separation.

In the adult bird, the prevailing colour of the plumage is white, barred over the upper parts of the body, wings and tail with narrow dark bands; top of the head streaked with dusky lines; all the under parts pure white; beak blueish black; cere yellow; irides dark hazel; tarsus and toes bright yellow. In very old males, the plumage is almost entirely white. Females have much more brown colour disposed over the upper parts, and young birds of the year have scarcely any white; the prevailing colour of their plumage is a uniform brownish ash; some of the feathers of the upper parts of the body, wings and tail varied by being tipped or barred with dingy white; top of the head, and under surface of the body marked with longitudinal patches of brown; cheeks light brown; throat white.

The Jer-falcon breeds in the highest aud most inaccessible rocks, and, according to Dr. Fleming, lays from three to five spotted eggs, of the size of those of a Ptarmigan. The old birds defend their nest and young with great courage.

The figure in the forepart of our Plate was taken from a fine example of this bird, presented to the Zoological Society by the Earl of Cawdor, which was shot on His Lordship's estate, Stackpole Court, Pembrokeshire, and was strongly suspected of having carried on successful warfare among some pheasants. The figure behind is that of a young bird.

In his memoir on the Birds of Greenland, Captain Sabine observes, that "the progress of this bird from youth, when it is quite brown, to the almost perfect whiteness of its maturity, forms a succession of changes, in which each individual feather gradually loses a portion of its brown, as the white edging on the margin increases in breadth from year to year; this has been the cause of the variety of synonyms authors in general refer to;" and, we may add, will also explain the various changes that occur during the life of this bird, between the two periods which we have represented by the subjects chosen for our Plate.

LANNER FALCON.

Falco lanarius; *(Linn)*

E. Lear del et lith. Printed by C. Hullmandel.

LANNER FALCON.

Falco lanarius, *Linn.*

Le Faucon lanier.

The native habitat of this rare bird are the eastern portions of Europe and the adjacent parts of Asia and Africa. It rarely passes further westward than the central parts of the European continent; it is scarcely ever seen in France or Holland; and never visits Great Britain. So extremely rare it is that we are unable to refer our readers to any collection in this country, either public or private, in which an example of this fine Falcon may be seen. We are much indebted to our highly valued friend M. Temminck for the loan of the two fine specimens from which our figures are taken.

In point of affinity the Lanner is directly intermediate between the Gyr-falcon and the Peregrine, the adult female being nearly, if not quite, equal in size to the male Gyr-falcon, while the male is of the usually diminutive size common to the true Falcons. In colouring it differs very considerably from either of the above-mentioned species, never possessing the strongly barred plumage of black and white which characterizes the Gyr-falcon, nor the transverse markings which are found on the breast of the Peregrine. While at Vienna we had an opportunity of observing a fine living specimen in the Royal Menagerie at Schoenbrunn: as far as we could perceive it did not evince any peculiarity of manner to distinguish it from the Gyr-falcon or the Peregrine; it appeared perfectly content and docile in captivity, although it is known to be extremely bold and daring in capturing its prey when in a state of nature. If we may judge from the general appearance of this bird, its strong bill, powerful body, and pointed wings, we should say that no one of its congeners is better adapted for the purpose of Falconry.

The adult has the crown of the head reddish brown, longitudinally marked with streaks of dark brown; over the eye a mark of yellowish white, which extends to the occiput; all the upper surface of a deep brown tinged with ash, each feather being bordered with reddish brown; a narrow line of brown from the base of the bill beneath the ear-coverts, which are yellowish white; all the under surface yellowish white, each feather having lanceolate marks of dark brown; the tail brown transversely barred with a darker colour; the cere and legs yellow; and the irides brown.

The young of the year differs from the adult in having the cere and legs blue instead of yellow; in having the breast much more strongly marked with brown; and in having the whole of the upper surface of a darker tint.

The Plate represents an adult male and a young bird rather less than the natural size.

PEREGRINE FALCON.
Falco peregrinus. *(Linn.)*

E Lear del.

Printed by C Hullmandel

PEREGRINE FALCON.

Falco Peregrinus, *Linn.*

Le Faucon pélerin.

Equally typical with the Jerfalcon and Lanner, the Peregrine, although less in size, possesses the characteristic boldness and ferocity of the genus. Being plentiful throughout the northern and middle portions of the Old World, it has at all times been abundantly employed in falconry, and still continues to be used by the few who continue a practice now almost obsolete. Whether we are to consider the Peregrine of North America and the extreme southern point of that vast continent, as well as that which is met with in New Holland, and other islands of the Pacific, as specifically identical with our European bird, is a point on which naturalists are not unanimously agreed; for ourselves, we consider that there exists the same difference, at least, between the Magellanic birds and those killed in Europe as between the Barn Owls of these two portions of the world, or as between the Goshawk of North America and that of Germany: the same observation equally applies to the New Holland species. But whether these differences, which are always appreciable by the experienced naturalists, are to be regarded as indicative of specific distinctions, or as varieties only dependent upon climate or other causes, is a subject which admits of much controversy; we ourselves are inclined to consider that these differences in birds closely allied are not at all times dependent upon extraneous causes, more especially where the differences are not in the tint of the colouring, but consists of a diversity in the shape and disposition of the markings; still, however, if it could be ascertained that birds, differing as do the Magellanic and European Peregrines, would breed with each other, and produce a fertile offspring, we should then be constrained to regard them as simple but permanent varieties.

In England this beautiful Falcon remains the whole year round: it appears to give preference to the bold rocky cliffs that border the sea, in the most inaccessible parts of which it builds its eyrie, generally laying four eggs, of a uniform dark red colour. The young, from the time of being fledged to their full maturity, which is not attained until four or five years have elapsed, undergo a series of changes, so remarkable as to have caused a list of numerous synonyms and no little degree of confusion: the persevering observations of modern naturalists have, however, cleared up the confusion, and rectified the mistaken views with which the works of the older writers abound; still one circumstance has attended this modern investigation, which shows how difficult it is to avoid error, even in the closest scrutiny; we allude to the fact of several writers having contended that the Lanner, a species perfectly distinct from the Peregrine, was in fact nothing but the young of the latter: this also is now found to be a mistake, and we trust that our Plates of the two species will still more clearly illustrate the subject. We need scarcely comment on the rapidity of flight in which this species so much excels, nor upon its destruction of various kinds of game, water fowl, particularly ducks, teal, &c.

The sexes differ considerably in size, the male being much the smallest, and in general more blue on the upper surface.

The adults have the bill lead colour, becoming black at the tip; cere, naked skin round the eyes, and the feet yellow; whole of the upper surface bluish lead colour, approaching to black on the head and cheeks, the feathers of the back and wings being barred with a deeper tint; quills brownish black, the inner webs barred with white; tail barred with bands of black and grey, the tips white; throat and breast yellowish white; under surface white with a tinge of rufous, and regularly barred with transverse lines of black.

The young of the year differ in having the whole of the upper surface brown, each feather being margined with a lighter colouring; the breast and under surface light fawn brown, with oblong longitudinal dashes of blackish brown; tail brown, with bands of a darker colour; cere and legs greenish yellow; irides the same as in the adult, deep hazel approaching to black.

The Plate represents an adult and a young bird, of the natural size.

HOBBY.

Falco subbuteo; *(Linn:)*

Drawn from life and on Stone by J & E Gould. Printed by C. Hullmandel.

HOBBY.

Falco subbuteo, *Linn.*

Le Faucon hobereau.

THE Hobby, although possessing all the typical characteristics of the genus *Falco,* is nevertheless wanting in that determined spirit and energy which distinguishes, not only the large, but also many of the smaller species of its race,—for example, the Merlin, which boldly attacks and kills birds far larger than itself, while the Hobby, which is a miniature representation of the Peregrine Falcon, (a bird noted for its daring and rapacious habits,) subsists in a great measure on insects, which it takes on the wing, and for the capture of which its rapid flight gives it great facility: nevertheless, it also attacks the smaller kinds of birds, especially Larks, among which it makes great havoc; and has been even trained, though not without difficulty, to fly at Quails and Partridges. The Hobby has a wide range throughout Europe, where it appears to be universally migratory, passing southwards with the approach of winter. In our island it arrives in spring, and departs in the month of October, and, with the exception of the Kestrel and Sparrowhawk, is one of the commonest of our smaller birds of prey. It frequents in preference wooded districts, near the margins of rivers, along which it may be observed to glide, rapidly darting from its perch in pursuit of dragon flies and the larger coleoptera. Its nest is built in trees, and, according to M. Temminck, sometimes even in bushes of moderate size: it is said occasionally to usurp the nest of the Crow,—a circumstance in which it agrees with many of its congeners. The eggs are three in number, of a dull white mottled with reddish brown.

Among the true Falcons, no bird presents less sexual difference either in size or colour. The male, in its adult plumage, has the upper surface of a deep blackish blue; the throat white with a black moustache passing from beneath the eye and stretching downwards to the sides of the neck; the lower parts whitish with longitudinal dashes; the thighs and lower tail-coverts reddish; the tail obscurely barred with black; the beak lead colour; the cere, eyelids and feet yellow; and the irides brown. Length fourteen inches.

The female is but little larger than the male, and the young soon assume the markings of the old birds; but the tints are duller, and the feathers are strongly edged with rufous, which prevails especially over the head; the longitudinal dashes of the under parts are brown, the ground colour inclining to light reddish yellow; the cere and tarsi yellowish green.

The Plate represents an adult male and female of the natural size.

RED FOOTED FALCON.

Falco rufipes. *(Bechst)*

Drawn from Life & on Stone by J & E Gould. Printed by C. Hullmandel

RED-FOOTED FALCON.

Falco rufipes, *Bechstein.*

Le Faucon à pieds rouges.

THIS small but true Falcon is one of the most elegant of the European species, and has lately become an object of still greater interest to the British ornithologist, from the circumstance of five or six examples having been recently taken in this country.

In the fourth volume of Loudon's Magazine of Natural History, page 116, Mr. Yarrell has recorded, that in the month of May 1830, three specimens of this Falcon were observed together at Horning in Norfolk.— Fortunately all three birds were obtained, and proved to be an adult female and two young males, in different states of plumage. A fourth specimen, a female, has been shot in Holkham Park.

A notice has since been read at the Linnean Society from Mr. Foljambe, of the capture of a male in Yorkshire; and a female lived nearly two years in the Gardens of the Zoological Society in the Regent's Park. From some of these examples, and from others in the collections of private friends, to which we have constant access, we have had ample opportunities of examining the many very interesting changes of plumage which occur in both sexes during their progress from youth to maturity.

The upper figure in our Plate represents an adult female. M. Temminck in his *Manuel*, page 33, describes this bird as having the upper part of the head marked with dark longitudinal streaks. Our specimen, from which the figure in the Plate was coloured, has the head of one uniform tint, without streak, but with a dark circle round the eye; and the female killed in Norfolk, of which we have seen a drawing, resembles our own bird exactly. Both these examples are considered to be adult. The immature female has the head streaked with dusky lines, which it retains through the second year; but it appears certain, from specimens before us, that these markings are lost at an advanced age. The feathers of the back and wing-coverts are then blueish-black, edged with lighter blue. The plumage of the other parts of the adult female is sufficiently portrayed in our figure. The young female has the top of the head brown, with dusky streaks; throat and ear-coverts white; eyes encircled with black: it has also a small black moustache extending from the eye downwards; the sides of the neck, breast, and all the under parts yellowish-white, with brown longitudinal streaks on the breast and abdomen; upper parts brown, the feathers edged with reddish-brown; tail with numerous alternate bars of brown and reddish-white, the tips white. Young male birds appear first in plumage similar to that of the female, changing at their moult to a light blueish-grey, and subsequently assuming the dark lead-colour so conspicuous on the head, back, and wings of the adult male bird represented by our lower figure. The thighs, vent and under tail-coverts are deep ferruginous; cere, orbits and feet orange-red; claws yellow-brown, darker at the tips. The fine adult male specimen from which our figure was coloured is in Mr. Yarrell's collection. The general uniformity in the colour of the males, contrasted with the pleasing variety of the females, is one of the most striking characteristics of this species, which is common over the greater part of the North of Europe; but of its habits or nidification little is recorded. Meyer, who has examined the stomachs of these birds, found in them only the remains of large coleoptera.

Our bird is the Orange-legged Hobby and Ingrian Falcon of Dr. Latham, so named from its inhabiting the province of Ingria in Russia, where it is called *Kobez;* it is also the *Falco vespertinus* of Gmelin. The adult male appears to have been unknown to Buffon as a distinct species, and is figured in the *Planches enluminées* of that Naturalist, No. 431, under the name of "a singular variety of the Hobby."

MERLIN.

Falco æsalon: *(Temm.)*

Drawn from Life & on Stone by J & E. Gould. *Printed by C. Hullmandel.*

MERLIN.

Falco Æsalon, *Temm.*

Le Faucon Émérillon.

ALTHOUGH the Merlin is the least of the European birds of prey, still it possesses all the features which characterize the most typical of its genus. Its undaunted courage and power of rapid flight embolden it to attack birds far superior to itself in weight and magnitude ; hence, when hawking was a favourite pastime with our ancestors, the Merlin was trained to the pursuit of partridges, woodcocks, snipes and larks ; and so determined is its spirit, and so certain its aim, that it has not unfrequently been known to strike a partridge dead, from a covey, with a single blow. Its flight is so low, that while skimming across large fallow or barren grounds, it often appears to touch the earth with its wings. In the southern parts of the British Isles, it is only a winter visiter, arriving at the departure of the Hobby ; but Mr. Selby has fully proved that in the northern parts it is stationary, and, unlike the Falcons in general, incubates on the ground, constructing a nest among the heather. "The number of the eggs," says Mr. Selby, who has discovered their nests in these situations in Northumberland, "is from three to five, of a blueish white, marked with brown spots, principally disposed at the larger end."

The advanced state of ornithological science, as it regards the changes in plumage of our native birds, enables us to affirm that the Stone Falcon (*Falco Lithofalco*, Auct.) is none other than the male Merlin in its advanced stage of plumage, the bird undergoing changes in this particular which characterize more or less the whole of the *Falconidæ*. The uniform dark tints of the adult are not fully attained before the third year.

The Merlin is extensively spread over the countries of Europe ; but M. Temminck informs us that it is scarce in Holland, though it appears, from the accounts of other authors, to be met with in Germany in winter. As regards its nidification, the above-mentioned naturalist differs materially from Mr. Selby in the situation he assigns to it for the purpose of breeding, which he states to be trees, or the clefts of rocks : the truth perhaps may be, that in different countries it may choose different localities, according as opportunities may favour it.

In the adult male the bill is blueish ; the crown of the head, back and wing-coverts blueish grey, the stems of each feather being black ; primaries black ; tail blueish grey with four bars of black, and a broad band of the same colour near the end ; tip white ; throat and upper part of the chest white ; cheeks and all the under parts buff orange, with broad oblong blackish spots ; cere, legs and orbits yellow ; irides brown.

The female somewhat exceeds her mate in size ; and although she never attains the rich colouring of the male as figured in the accompanying Plate, approximates very closely to it at a very advanced age. The generality of individuals taken have the plumage similar in colour and markings to the upper bird, which represents a male in immature plumage.

The female and young birds have the top of the head of an obscure brown marked with oblong spots of black ; stripe over the eye white ; upper surface and scapulars brown, tinged with grey, each feather being spotted and edged with brown ; quills blackish brown, obscurely spotted with brown ; under wing-coverts rufous with white spots ; throat white ; breast and under surface pale brown marked with longitudinal spots as in the male, but broader and less distinct ; tail obscure brown with five or six rufous bars and tipped with white ; cere, orbits and tarsi yellow ; irides brown.

The Plate represents two males, one the old bird, the other a young bird of the first year, with which the female, except when very old, agrees in plumage.

LEAD-COLOURED FALCON.

Falco concolor: *(Temm.)*

Drawn from Nature & on stone by J & E Gould.

Printed by C Hullmandel.

LEAD-COLOURED FALCON.

Falco concolor, *Temm.*

Le Faucon concolore.

This species appears to be dispersed over the whole of Northern Africa, being abundant in Abyssinia and on the banks of the Nile; it is also said to occur on the western portions of that continent, at least M. Lesson in his 'Traité d'Ornithologie' states its habitat to be "le Sénégal, la Barbarie, l'Egypte, l'Arabie." That a bird of this kind should cross the Mediterranean and visit the continent of Europe is not surprising, and we learn from M. Temminck that such is really the case; it is therefore entitled to a place in our work.

The accompanying figure was taken from what we conceive to be an adult male; its structure is in every respect similar to that of the typical Falcons, and from the lengthened form of its wing and the general tone of its colouring, it is nearly allied to the Hobby (*Falco Subbuteo*); it may however be at once distinguished not only from that, but from every other species of the true Falcons, by its uniform lead-coloured plumage, whence its specific appellation.

Although no facts are on record as to its mode of life, we may reasonably conclude that insects and small birds constitute its principal subsistence, and that in its general economy it closely assimilates to the Hobby.

With the exception of the primaries, which are blackish brown, the entire plumage of the *Falco concolor* is of a uniform leaden grey, with the shaft of each feather darker; cere and feet yellow; bill and claws black.

Our figure is of the natural size.

KESTREL.

Falco tinnunculus; *(Linn)*

Drawn on stone by E. Lear. Printed by C. Hullmandel.

KESTREL.

Falco tinnunculus, *Linn.*

Le Faucon cresserelle.

This indigenous Falcon is by far the most common species of those inhabiting Europe, over the whole of which continent it is universally diffused, as well as in those portions of Asia and Africa which are either immediately connected with or otherwise opposed to its shores; the whole of the northern parts of the latter country affording it a natural habitat.

Although we believe that the *Falco tinnunculus* has not been discovered in America, still that extensive continent has produced several species whose form and colouring unite them to the Kestrels of the Old World, and, as we before stated in the description of the Lesser Kestrel (*Falco tinnunculoïdes*), they appear to form one of the most natural groups in the family of *Falconidæ*.

The Kestrel may be daily observed making its graceful flights over fields and barren grounds in search of its natural food, which consists of mice, frogs, small birds and insects, while in pursuit of which its attention is often suddenly arrested, and poising itself in the air, which it fans with its long and pointed wings, it suddenly pounces down upon its victim with the utmost impetuosity, and may be frequently seen rising from the ground with its prey firmly fixed in its talons, and flying off to some retired situation to devour it, or, if in the season of incubation, conveying it to its young.

The male at the age of three years, when it is in full plumage, is adorned with the most delicate and sober colours, added to which it possesses a perfect symmetrical contour of body,—circumstances unquestionably ranking it as one of the most beautiful species of its genus. The female after the first moult undergoes no change; and the young males until after the age of two years are not distinguishable from her: this is the cause that so large a proportion of the birds bear the plumage just referred to, since but comparatively few survive the second year of their existence.

The birds of this division are of a more feeble character and less courageous disposition than the nobler groups of the *Falconidæ*; and, though easily tamed, cannot be used in the chase with sufficient certainty, notwithstanding the assertion of authors that they were formerly trained to the capture of Snipes and Partridges. They frequently take possession of the deserted nest of a Crow or Magpie for the purpose of incubation, yet it is far from uncommon for them to deposit their eggs on the bare surface of a ledge of rocks:—these eggs are from four to six in number, of a reddish brown colour with darker speckles or blotches, varying considerably in intensity. The young, like most of the nestlings of the hawks, are for the first month entirely clothed with a white down.

In the adult male, the bill, the tail (with the exception of a bar of black near the extremity of its feathers which terminate in white), the rump, and the fore-part of the head are of a fine blueish grey; the back and wing-coverts of a reddish fawn colour, each feather having at its extremity an arrow-shaped spot of black; primaries dark brown, their edges lighter; breast, belly and thighs of a pale cream-colour tinged with brown, and sprinkled on the breast with brown spots of a linear form, but assuming a rounder shape on the lower part of the body.

In the female, the whole of the upper parts and tail are of a browner hue than those of the male, each feather having several bars of a dark brown, and the tail likewise barred with brown, but terminating with a black band and white tips as in the male; the primaries are also brown with paler edges; the whole of the other parts resemble the male.

The Plate represents a male and female about three fourths of their natural size.

LESSER KESTRIL.

Falco tinnunculoides. *(Natter).*

Drawn from Nature & on Stone by J. & E. Gould.

Printed by C. Hullmandel.

LESSER KESTREL.

Falco Tinnunculoïdes, *Natter.*

La Cresserellette.

THIS elegant little Falcon, although closely allied to the Common Kestrel, is to be distinguished from that species by its smaller size, its greater length of wing, the white colour of the nails, and the entire absence of markings on the back :—the female, however, agrees so closely in plumage with the female of *Falco Tinnunculus*, that we have not considered it necessary to introduce a figure of her into our Plate, which represents the adult male. Notwithstanding, she still retains the characteristics of the species : viz. inferiority of size, length of wing, and white nails ; circumstances by which she may at once be identified. As far as we have been able to ascertain, this bird has not been discovered in the British Islands, but is common in the southern parts of continental Europe, especially Spain, Italy, and the South of France, frequenting rocky and mountainous districts as well as lofty spires, church steeples, and ruins, selecting such situations for its breeding places ;—the female (which rather exceeds the male in size) generally laying four eggs, very much resembling those of the Kestrel.

Although we cannot doubt that small mammalia and birds form part of the food of this species, still we have reason to know that it subsists in a great measure on the larger coleopterous and hymenopterous insects, which it takes on the wing, darting at them with great quickness and precision of aim.

In some of its characters, and especially its lengthened wing, which reaches the extremity of the tail, the *Falco Tinnunculoïdes* approaches the typical form of the genus more nearly than our Kestrel ; yet in both species we perceive a departure from those strongly marked features which pre-eminently distinguish the more noble of the group ;—*i. e.* a less muscular form of body, a beak the tooth of which is more rounded and less acute, tarsi less robust, talons less curved and weaker, in union with a disposition more timid and an appetite less blood-thirsty ; characters which proclaim a grade below that of their more daring congeners.

It is not the intention of the Author of this work to enter into an analysis of existing genera, or to establish new ones ; nevertheless, he may be allowed to suggest an inquiry to those who are more particularly engaged in systematic arrangements ;—viz. whether there be not room for a further removal of this bird, and those in evident relationship to it, from the more typical species which compose the Genus *Falco;* naturalists having availed themselves of less prominent characters in the formation of genera, (*Astur* and *Accipiter*, for instance,) between which there is the closest affinity. On the contrary, there is between the group which we now refer to, and the Falcons *par excellence*, a well-marked distinction in habits, disposition, style of colouring, and food,—sufficient, we think, to constitute a clear ground of separation. This proposed group would contain at least three well-marked species of the Old Continent ; viz. the present bird, the Common Kestrel, and the *Falco rupicolus;* to which may be added the *Falco sparverius* of Latham, and several other species of America and its adjacent islands.

In size the *F. Tinnunculoïdes* is inferior to the Kestrel, the total length of the male being eleven inches.

The wings reach to the extreme tip of the tail, which is rounded ; the top of the head, occiput, and sides of the neck are of a fine uniform ash-colour ; the whole of the upper surface, with the exception of some of the larger wing-coverts, the secondaries, quills and rump of a brownish red without any markings, the latter being of a blueish ash-colour, as is the tail also, which is crossed with a black band, and at its extremity tipped with white. The inferior surface is of a clear brown red, thinly sprinkled with small black dashes and longitudinal marks.

Beak blueish ; cere and space round the eyes yellow ; feet yellow ; nails white.

Young males of the year differ little from the adult female.

KITE.

Milvus vulgaris, *(Flem.)*

E. Lear del. et lithog.

Printed by C. Hullmandel.

Genus MILVUS.

Gen. Char. *Bill* of moderate strength, nearly straight at the base, rapidly incurved in front of the cere to the tip, which forms an acute hook; *cere* short. *Nostrils* oval, rather obliquely placed in the cere. *Wings* very long; the first feather short; the fourth the longest; the first five having their inner webs notched. *Tail* long and forked. *Legs* with the tarsi very short, feathered below the joint; the naked frontal part scutellated. *Toes* rather short, the outer united at its base to the middle one. *Claws* long and strong, moderately incurved, with the inner edge of the middle one thin and dilated.

KITE.

Milvus vulgaris, *Flem.*

Le Milan royal.

This elegant species, although generally diffused over the British Islands, is much less common than formerly; indeed the destroying hand of the gamekeeper has completely extirpated it in many of the inland counties, particularly such as are but thinly wooded. The only retreats wherein the Kite now finds an asylum are larger woods and forests of denser growth, in whose impervious recesses itself and its brood are effectually shrouded from observation. The districts where the Kite may be most frequently observed at the present period, are the more uncultivated portions of Wales and the adjoining counties, as well as the wild tracts of rocky moorlands in both the northern and southern parts of the island. Throughout the continent of Europe it appears to have a wider range,—except in Holland, in consequence of a scarcity of large woods and uninhabited wilds. While on the wing performing its aërial evolutions, nothing can excel the ease and grace with which the Kite sails along surveying the earth below, its flight generally consisting of widely extended circles, during the performance of which the wings appear to be entirely motionless, the tail acting as a rudder to guide its course; in this manner it ofttimes soars to so great a height as to be almost imperceptible. Its prey, which consists of mice, rats, leverets, young gallinaceous birds, ducks, reptiles, fishes, and insects, is sought for while it is soaring in the air at a moderate distance from the ground, and is taken by a woop so noiseless and rapid, that little or no warning is given of its approach; in this way it sometimes commits great havoc among the young broods of poultry, pheasants, partridges, &c.

In general form and colouring of plumage the sexes bear a close resemblance, nor do the young birds undergo any very decided change from youth to maturity.

The process of incubation is commenced early in the spring. The nest is constructed of sticks lined with wool and hair; and is situated in the thickest part of the forest: the eggs are generally three or four in number, of a greyish white, more or less distinctly speckled with reddish brown.

The head and neck are clothed with narrow pointed feathers of greyish white, each having a central dash of dark brown; the whole of the upper surface is bright ferruginous brown, each feather having its centre blackish brown; tail and thighs rich rufous brown; under surface brownish white, with dark longitudinal blotches; bill dark brown; cere and tarsi bright yellow; irides straw yellow.

The Plate represents an adult in full plumage about three fourths of the natural size.

BLACK KITE.
Milvus ater

Printed by C. Hullmandel.

BLACK KITE.

Milvus ater.

Le Milan noir ou parasite.

THIS species, which may at all times be distinguished from the Common Kite of England (*Milvus vulgaris*, Flem.) by the darker colour of its plumage and by the numerous longitudinal stripes on the head and neck, is dispersed in considerable numbers over the southern portions of Germany, the whole of France, Switzerland, and the European countries bordering the Mediterranean Sea. No instance is on record of its having paid a migratory visit to the British Islands; still, judging from its extraordinary powers of flight, and from the wandering habits of the generality of the *Falconidæ*, it is not improbable that it may have penetrated so far west as our island, and have been mistaken for the common species: we throw out this hint in order to induce British ornithologists generally, and particularly those who reside in the southern parts of England, to investigate this subject whenever an opportunity offers.

The range of the Black Kite eastward appears to be very great, as we have seen examples of it in several collections from India, particularly the Himalaya mountains; it is also equally abundant in Northern Africa.

We have observed this species in a state of nature and in confinement both in Germany and France; it bore so strict a resemblance in its manners to the common species, that the addition of a second description is totally unnecessary.

It incubates on trees, and lays three or four yellowish white eggs very thickly spotted with brown.

The sexes are alike in plumage, and the young when a year old resemble the parents.

Bill black; cere, feet, and legs yellow; irides silvery yellow; head and neck longitudinally striped with brown and greyish white; all the upper parts deep brown; under surface reddish brown with a longitudinal stripe of a darker tint down the centre of each feather; primaries blackish brown; tail slightly forked and of a dark grey brown, transversely rayed with darker brown.

The Plate represents an adult bird about two thirds of the natural size.

SWALLOW TAILED KITE.

Nauclerus furcatus. *(Vig.)*

Drawn from Nature & on Stone by J & E Gould. *Printed by C. Hullmandel.*

Genus NAUCLERUS, *Vig.*

Gen. Char. *Bill* small, weak, considerably hooked, with a small and nearly obsolete festoon in the middle. *Orbits* and sides of the head thinly provided with feathers. *Wings* very long; the first and second quill-feathers internally emarginate towards the tip. *Tail* very long, and deeply forked. *Tarsi* very short, not longer than the hind toe and claw; plumed half way in front, the remaining portions covered with angulated scales. *Toes* short; the two lateral almost equal, the hinder nearly equal to the inner. *Claws* grooved beneath.

SWALLOW-TAILED KITE.

Nauclerus furcatus, *Vigors.*

La Milan de la Caroline.

Two examples of this elegant bird having been taken in this country, the first in Argyleshire, the second in Yorkshire, we have considered that it is entitled to be included among the Birds of Europe, and have accordingly given it a place here. We also agree with Mr. Vigors and Mr. Swainson that this bird requires to be separated generically from those of the genus Elanus of Savigny.

For a correct knowledge of the habits and manners of this handsome bird we are indebted to the ornithologists of the United States of America, in different parts of which at particular seasons of the year it appears to be very abundant. In the history of this species by Wilson and Mr. Audubon, many interesting details will be found, and as one or the other of these works are in the hands of every lover of nature and ornithology, we shall avail ourselves of the less perfectly known History of the Birds of the United States and Canada by Mr. Nuttall, who says, "This beautiful Kite breeds and passes the summer in the warmer parts of the United States, and is also probably resident in all tropical and temperate America, migrating into the southern as well as the northern hemisphere. In the former, according to Vieillot, it is found in Peru, and as far as Buenos Ayres; and though it is extremely rare to meet with this species as far as the latitude of 40 degrees in the Atlantic States, yet, tempted by the abundance of the fruitful valley of the Mississippi, individuals have been seen along that river as far as the Falls of St. Anthony, in the 44th degree of north latitude."

"They appear in the United States about the close of April or beginning of May, and are very numerous in the Mississippi territory, twenty or thirty being sometimes visible at the same time, often collecting locusts and other large insects, which they are said to feed on from their claws while flying; at other times also seizing upon the nests of locusts and wasps, and, like the Honey Buzzard, devouring both the insects and their larvæ. Snakes and lizards are their common food in all parts of America. In the month of October they begin to retire to the south, at which season Mr. Bartram observed them in great numbers assembled in Florida, soaring steadily at great elevations for several days in succession, and slowly passing towards their winter quarters along the Gulf of Mexico."

The flight of this bird is described as being smooth and graceful in the extreme, and it remains on wing nearly the whole of the day, roosting at night in high trees. The nest is usually placed among the top branches of the tallest oak or pine, and is formed of sticks, intermixed with moss and grass, lined with a few feathers. The eggs are from four to six in number, of a greenish white, with a few irregular blotches of dark brown at the large end. The young birds are at first covered with white down.

In the adult bird the beak is bluish black, the cere of a lighter blue, the irides dark; the whole of the head, neck, breast, and under surface of the wings, sides of the body, thighs and under tail-coverts pure white; the back, wings, primaries, secondaries, upper tail-coverts and tail-feathers black, with a purple metallic lustre, the tertials black on the outer webs, but patched with pure white on the inner; tail very deeply forked; legs and toes greenish blue; claws faded orange brown.

We have figured the bird of the natural size.

BLACK-WINGED KITE.

Elanus melanopterus *(Steph.)*

Drawn from Nature & on Stone by J. & E. Gould. Printed by C. Hullmandel.

Genus ELANUS, *Savig.*

GEN. CHAR. *Bill* weak, of mean length, compressed, nearly straight at the base, the tip hooked. *Wings* long, with the second feathers generally the longest, the first and second having their inner web strongly notched. *Tail* long, more or less forked. *Tarsi* short, feathered for half their length, and the naked part reticulated. *Claws* strong and incurved; the under surface in some species partly rounded.

BLACK-WINGED KITE.

Elanus melanopterus, *Leach.*

L'Elanion blanc.

WHEN we consider the wide range of this beautiful species, scattered as it is over all the temperate and warmer portions of the Old World, it is a matter of no surprise that its capture has of late years been so frequent in Europe. It is abundantly dispersed along the banks of the Nile, and in fact the whole of Africa and India is inhabited by it; neither do specimens from Java and New Holland present any specific differences from those taken in Europe. In all probability no part of Europe affords it a permanent residence. Spain, Italy, and the Grecian Islands are the portions of our quarter of the globe most frequented by the Black-winged Kite; instances are, however, on record of its having been captured in the middle of Germany: it must therefore, like many other species, be regarded merely as an irregular visitor which has crossed the Mediterranean from the opposite shores of Africa. From the great length of its wings, together with its short and feathered tarsi, we are led to infer that it is capable of rapid and powerful flight, and that like its allies in America it possesses the power of remaining suspended in the air for a great length of time.

Its food consists principally of insects, chiefly captured in the air, to which are sparingly added lizards, frogs, snakes, and birds.

The sexes are very much alike in colour, but the female is said to be rather larger than her mate. The young of the first autumn may be distinguished from the adults by their having the back strongly tinged with brown, and the end of each feather encircled with buffy white; the sides of the chest brown, and the feathers on the breast streaked down the centre with dark brown.

The adult has the head and the whole of the back of a fine grey; the centre of the wings black; the primaries and secondaries greyish brown, with lighter grey edges; the shoulders of the wings, throat, all the under surface, and tail pure white; cere and toes yellow; bill and claws black; irides orange.

We have figured an adult and a young bird of the natural size.

E. Lear del. et lith.

MARSH HARRIER.
Circus rufus (Briss.)

Printed by C. Hullmandel.

Genus CIRCUS, *Briss.*

GEN. CHAR. *Bill* bending from the base, weak, much compressed, and forming a narrow rounded culmen ; tomia of the upper mandible exhibiting a very small sinuation near the middle of the bill; under mandible shallow and rounded at the point. *Nostrils* rather large; broadly oval; nearly concealed by the reflected and upward curving hairs of the lores. *Head* surrounded by a ruff of stiffish tiled feathers. *Wings* long, the fourth feather barely exceeding the third, but being the longest in the wing; first four having their inner webs notched; the third, fourth, and fifth having the outer webs sinuated. *Tail* long, slightly rounded. *Tarsi* long, slender, feathered in front for a short distance below the joints, the naked part scutellated. *Toes* of mean length, rather slender, middle toe the longest, outer toe rather exceeding the inner, and joined at the base to the middle one by a membrane, third toe shortest. *Claws* moderately incurved and very short, those of the inner and hind toes the largest.

MARSH HARRIER.

Circus rufus, *Briss.*

Le Busard Harpaye ou de Marais.

THE size of this bird renders it so conspicuous that it cannot fail to attract attention wherever it appears; it is consequently most probable that the greater number of those which are seen in our island are not native-bred specimens, but have wandered from the adjacent continent; and we are confirmed in this opinion by the circumstance of most of the birds which have been shot being in the youthful or immature state of plumage: we know also that young birds are in the habit of wandering greater distances from their birthplace than adults. So great, indeed, are the chances against their attaining a state of mature plumage in our island, that we do not recollect a single instance of a specimen in the plumage of the bird figured in our Plate, having been killed here: that it is many years in attaining this plumage is very evident, and it is equally certain that it breeds while yet in the deep chocolate-coloured plumage by which it is distinguished during the first and several succeeding years. It will be seen that when it has attained the perfect livery, the wings and tail have assumed that delicate grey so characteristic of the Harriers in general, while the feathers of the remaining parts of the body are not only of a different tint, but are also of a different form, being more or less lanceolate instead of round. Although we are not able to state it as a fact, yet we are inclined to believe that it is the male only which possesses the beautiful grey colouring alluded to above. Even in its youthful state the young of this bird exhibit considerable differences of colouring, some being of a uniform chocolate brown, while others have the crown of the head, cheeks, and shoulders of a rich buff.

The Marsh Harrier appears to enjoy a wide extent of habitat, being found in the low marshy districts of Europe, Africa, and a great portion of Asia; as is proved by our having received it in collections from the Himalaya mountains. Like the rest of the Harriers its flight is buoyant and sweeping, but generally at a low elevation: it traverses over the moors and marshes in search of its prey, which consists of frogs, lizards, mice, insects, and even fish.

The nest is placed on the ground among low bushes or reeds, generally near the edge of the water: the eggs are four in number, white and rounded.

We take our description of the adult bird from the "Manuel" of M. Temminck.

Head, neck, and breast of a yellowish white, with numerous longitudinal dashes of brown occupying the centre of each feather; scapularies and wing-feathers reddish brown; quills white at the base, and black for the remainder of their length ; secondaries and tail-feathers of an ashy grey ; whole of the under surface light rufous marked with yellowish blotches; beak black ; cere greenish yellow ; irides reddish yellow ; tarsi yellow.

The young of the year has the plumage of a very strong chocolate brown ; the wing-coverts, the quills, and the tail-feathers tipped with brownish yellow; the top of the head, occiput, and throat more or less pale ; irides blackish brown.

The Plate represents an adult and a young bird about three fourths of the natural size.

HEN HARRIER.
Circus cyaneus, (Flem.)

HEN-HARRIER.

Circus cyaneus, *Meyer*.

Le Busard St. Martin.

It is to be regretted that this delicately plumaged Hawk, which a few years ago was common in our island, is now so scarce as rarely to admit of its being observed in a state of nature. Like many of its congeners, much mischief has been laid to its charge ; and without even for a moment attempting to balance the good which it effects by destroying hundreds of snakes, lizards, and mice in the course of a single year, with the injury it does by preying on a limited number of leverets and other young game, which are only open to its attacks for the period of a few weeks, its ruthless destruction is diligently persevered in by the gamekeeper and sportsman without the least consideration ; in fact, so rapidly have many of our native *Falconidæ* decreased within these few years, that there is but little doubt many species once numerous will ere long be entirely extirpated.

This fine Harrier enjoys an extensive range of habitat independently of Europe, over the whole of which, wherever situations favourable to its residence occur, it is found in greater or less abundance : it also inhabits similar situations over the greater part of Africa and India. A species nearly allied, if not absolutely identical, exists in the northern portions of the American continent.

The flight of the Hen-Harrier while in quest of its prey is strikingly peculiar, and is altogether different from that of the birds of every other group of the *Falconidæ* ; it is light and buoyant, but performed at no great elevation from the ground, which it quarters with the utmost regularity, traversing a certain extent of country and returning nearly to the same place at a given time for many days together. Thus skimming along with noiseless wings, it strongly reminds us of one of the Owls, and it pounces down upon its prey with unerring precision ; this, as we have before stated, consists principally of mice, leverets, lizards, snakes, frogs, and unfledged birds, never daring to contend with large birds, or quadrupeds of even moderate size.

In this country the localities to which the Hen-Harrier is almost exclusively limited, are wide heathy moorlands, extensive wastes, and furze-covered commons, to which may be added low marshes, flat lands, bordering lakes, and morasses. In these wild and solitary situations it incubates and rears its young, its nest being placed on the ground, among the tufted herbage most prevalent on the spot ; the eggs resembling those of the Owl, but larger, only being of a dull dirty white without any spots.

The difference between the male and female is so remarkable, as at no distant date to have led to the supposition that each sex was a distinct species ; an error, the correction of which is due to our talented ornithologist Colonel Montagu. This is now so clearly understood as not to need any especial remark ; we would, however, observe that this extraordinary feature is exhibited in most of the species of the genus *Circus*, a genus almost universally dispersed over the globe.

The young birds of both sexes for the first two years are precisely alike in their colouring, which differs but little from that of the adult female, and it is this circumstance which militated against the idea of the Hen-Harrier and the Ringtail being identically the same.

It is only after the second year that the male begins to assume the delicate silvery grey which in the state of maturity pervades the whole of the upper surface.

We give the details of the colouring as follow :

The adult male has the head, neck, chest, and whole of the upper surface, with the exception of the rump and the two outer tail-feathers on each side,—which are white, the latter having a fine transverse band of greyish brown,—of a fine blueish silvery grey ; quills black ; under surface white, with a few faint blotches of brown disposed in the centre of a great part of the feathers ; legs, upper part of the cere, and irides brown.

The female has the whole of the upper surface chocolate brown, the feathers of the head, and back of the neck bordered with reddish sandy yellow ; the ear-coverts deep brown ; the marginal feathers of the face short and stiff, of a sandy yellow with deep brown shafts ; whole of the under surface reddish yellow, with longitudinal dashes of brown ; the tail barred alternately with bands of light and deep umbre brown ; legs and upper part of the cere yellow ; irides hazel.

The Plate represents a male and female of the natural size.

PALLID HARRIER.
Circus pallidus; *(Sykes.)*

Drawn from Nature & on Stone by J. & E. Gould. Printed by C. Hullmandel.

PALLID HARRIER.

Circus pallidus, *Sykes.*

For the knowledge of the occurrence of this species of Harrier in Europe we are indebted to M. Temminck, who has transmitted for our use a fine male, which he states was killed on the banks of the Rhine. This bird, the *Circus pallidus* of Colonel Sykes, is abundantly dispersed over a great portion of India, but up to the year 1832 remained uncharacterized, in consequence of its having been considered as identical with the *Circus cyaneus.* The differences, however, which exist between those closely allied species were then clearly pointed out by Colonel Sykes in his Catalogue of the Birds of the Dukhun; and as these differences have been well defined by that gentleman in the Proceedings of the Zoological Society of London for April 1832, we prefer to make use of his own words: "This bird has usually been considered the *Circ. cyaneus* of Europe; but it differs in the shade of its plumage (male and female); in the back-head of the male not being white spotted with pale brown; in the absence of dusky streaks on the breast; in the rump and upper tail-coverts being white barred with brown ash; in the inner webs of four of the tail-feathers not being white; and in the bars of the under tail being seven instead of four. The female resembles the female of *Circ. cyaneus,* but the plumage is two shades lighter, the tail is barred with six broad fuscous bars instead of four, and the tail-feathers are much more pointed. The remains of six lizards were found in the stomach of one bird. I never saw these birds perch on trees. They frequent the open stony plains only. The sexes were never seen together." To this we may add that we have compared the specimen sent to us by M. Temminck with others from India, that not the slightest difference exists between them, and that the barred upper tail-coverts and paler colouring of this species will readily distinguish it from the *Circ. cyaneus.*

Since the above was written we have seen as many as eight or ten specimens in one collection, the whole of which were killed in Spain, from which we are led to infer that it is there a common species, and that in all probability it also abounds throughout the northern portions of Africa.

Head and all the upper surface pale grey very slightly tinged with brown; upper tail-coverts white, spotted or rather barred with pale brown; quills white at the base passing into deep brown at their extremities, and margined externally with greyish; tail pale grey, the outer feathers becoming nearly white; the whole crossed with six fuscous bars, which are most conspicuous on the outer feathers, where they assume a rufous tint; bill blue; cere and legs yellow; irides greenish yellow.

We have figured the bird of the natural size.

ASH-COLOURED HARRIER.
Circus cineraceus; (Mayer).

ASH-COLOURED HARRIER.

Circus cineraceus, *Meyer*.

Le Busard Montagu.

THE present elegant bird excels its congener the Hen-Harrier in the relative admeasurement of its wings and tail, and though less robust, is even more elegant in its proportions. In habits and manners, and the localities it frequents, there is little difference between them, a circumstance which, together with its colouring, was the cause of its being so long considered as identical with that species. Its distinguishing characters consist in the elongated wings, across which extends a conspicuous band of black in the male; the rich chestnut dashes on the under surface, and bars of the same colour on the outer tail-feathers. The female is scarcely to be distinguished from the female of the other species, except by the elongated wings and the general slenderness of the body. The discovery of this bird as a distinct species is due to Colonel Montagu, in whose writings we have a detailed account of its specific differences, together with considerable information respecting its general manners and history.

The Ash-coloured Harrier a few years back was deemed a bird of great rarity, but is now fully as common as its relative the Hen-Harrier. Mr. Selby informs us that he has taken it in Northumberland, where it breeds upon the moors and open lands: the southern districts of England, however, appear to be its favourite residence. We have ourselves received numerous examples from the fens of Cambridgeshire and Lincolnshire. On the Continent it appears almost universally distributed, especially in the eastern and southern provinces.

Its food consists of small mammalia, such as moles, rats, mice, and young hares, to which are added snakes, lizards, frogs, &c.

Its place of nidification is on the ground, among rushes, furze, or any low brushwood suited to its purpose.

Its flight is peculiarly buoyant, and perhaps exceeds in rapidity and lightness that of any other European Harrier.

The sexes offer the same distinctions of colouring that we see in the Hen-Harrier, but we find the young for the first six months of their existence to be more uniform in their colouring, the plumage being less variegated by spots or dashes.

The male has the head, neck, whole of the upper surface, and middle tail-feathers blueish grey; a distinct band of black crosses the middle of the wing; quill-feathers black; outer tail-feathers white, barred with chestnut and tipped with grey; under surface white, with regular longitudinal dashes of rich chestnut; bill black; cere, irides, and tarsi fine yellow.

The female has the whole of the upper surface of a deep chocolate brown; the top of the head lighter than the rest of the body; each feather with its centre of a deeper tint, so as to give it a spotted appearance; around the eye is an obscure circle of dull white; ear-coverts rich brown; under surface light reddish brown, with longitudinal dashes of a deeper colour: these in the young of both sexes are scarcely to be discerned; tail brown, the outer feathers lighter, and exhibiting bars of deep umber; cere, irides, and tarsi as in the male.

The Plate represents a male and female of the natural size.

BARN OWL.

Strix flammea; *(Linn.)*

Drawn from Life & on Stone by J. & E. Gould.

Printed by C. Hullmandel.

Genus STRIX.

Gen. Char. *Beak* straight at the base, the tip arched, and hooked; cutting margin of the upper mandible nearly straight; under mandible sloping to the point, and doubly notched. *Nostrils* oval, obliquely placed on the anterior ridge of the cere; facial disk large, complete. *Wings* long and ample; the second quill-feather the longest in the wing, the first but little shorter, equal to the third, and slightly notched on its inner web near the tip. *Legs* with tarsi long and slender, clothed with downy feathers; toes thinly covered with hairs; claws long, sharp, moderately curved, and all more or less grooved beneath.

BARN OWL.

Strix flammea, *Linn.*

La Chouette effraie.

Whether our well-known Barn Owl be identical with those found in almost every portion of the globe, notwithstanding their slight variations or differences of plumage, we have not been able satisfactorily to determine; and it yet remains a question whether the Owls so nearly resembling the present, from the United States, South America and its adjacent islands, together with others from Africa, India and New Holland, be merely varieties depending upon climate, food and a combination of circumstances, or, on the contrary, radically distinct, each constituting a different species.

The genus *Strix* as limited by modern authors, and taking this bird as its type, possesses, besides the varieties above alluded to, many which must certainly be considered as truly distinct species. Of these we have seen several from New Holland, one from India, and one from the West Indian Islands. It is a genus at once distinguished from all other genera of the family of *Strigidæ* by the elongated bill, the loose and downy texture of the plumage, and by the beautiful style of colouring which pervades the upper surface of the body.

The *Strix flammea* is spread over the whole of Europe, and appears to be everywhere stationary, at least such is the case in our own island, where they inhabit barns, ruins, church-towers and hollow trees, remaining concealed all day, but issuing at the approach of evening, when they prowl, on light and noiseless wing, in search of their prey, night being the time when the species of this genus exert their powers and display their destructive energies. Dazzled by the light of day, for which their powers of vision are not adapted, they remain motionless and inanimate in their retreats, shading their eyes with the thin membranous veil which they possess for the purpose of drawing over the pupils. To observe them in this state, we should not suppose them endowed with that energy and quickness of action which they display at night, when, intent upon their search, they skim over the meadows with every sense alive to the object of their pursuit: so rapidly, indeed, do they pounce upon their victims, that even the little active mouse is seized before aware of its approaching fate. Although mice form the principal part of their subsistence, it is nevertheless certain that they sometimes prey upon young birds, rats and leverets; and instances have been known of their committing depredations among the finned inhabitants of lakes and ponds.

In the plumage of these interesting birds there exists considerable variety, some individuals being fawn-coloured on the upper and under surface, spotted and dashed with dark grey, while others are purely white on the under surface; and others again white on the same part, with minute spots of grey. So far as we have been able to judge from dissection, the individuals killed in this country with pure white breasts, as represented in the Plate, are invariably adult males, the females and young males having the breast more or less speckled, and the edge of the facial disk tinged with fawn colour.

The *Strix flammea* incubates in holes of trees, old buildings, and similar situations, generally laying three or four nearly round white eggs.

The young, for a considerable period, are covered with a thick coating of white down, and their retreat is always found to contain vast quantities of pellets or castings, consisting of the indigestible parts of their food.

The Plate represents an adult male of the natural size.

EAGLE OWL.

Bubo maximus *(Sibbald)*.

Drawn on Stone by E. Lear.

Printed by C. Hullmandel.

Genus BUBO.

Gen. Char. *Bill* short, strong, curved and compressed at the point. *Nostrils* pierced in the cere, large, oval, or rounded. *Head* furnished with tufts of feathers. *Wings* rather short, concave, third and fourth quill-feathers generally the longest. *Legs* and *toes* covered with feathers, outer toe reversible; *claws* long, curved and sharp.

GREAT-HORNED OR EAGLE OWL.

Bubo maximus, *Sibbald.*

Le Hibou Grand-duc.

Among the species of this singular race belonging to Europe, the Great-horned or Eagle Owl holds the first place in point of size and in majesty of appearance; nor is it inferior, or if so, but slightly, to any known species from other parts of the world. It forms a typical example of the genus *Bubo*, a group distinguished by a tuft of elongated feathers above each eye, usually denominated ears, though, as may be seen, these plumes have no connexion whatever with the true organs of hearing.

The present division, as well as one or two others, the species of which possess bright yellow-coloured irides, appears to enjoy the power of vision to a greater extent either in dull daylight or by the bright light of the moon; and even during sunshine they are by no means so confused and distressed as their allies contained in the restricted genus *Strix*, possessing eyes, the pupils of which, capable of prodigious enlargement, adapt them more exclusively to the dusk of evening or the sombre darkness of night. The true habitat of this noble species is the more northern portions of Europe: M. Temminck states, that it is so extensively spread as to occur at the Cape of Good Hope. We have ourselves seen it in collections from China; and Dr. Latham adds, that it is found at Kamschatka as well as in the northernmost parts of America. Granting, however, that it is diffused thus extensively, still its true habitat appears to be among the large forests of the wild and desolate regions of Norway, and the parallel latitudes of Sweden and Russia. It is less common in Germany and Switzerland, and of rare occurrence in France and England; still, from the frequent captures of it that have been made in the British Islands, it may be classed among our birds, especially as it seems to be deterred from settling among us more by the want of secluded and unmolested retreats than by an uncongeniality of climate. The Great-horned Owl may likewise be considered as one of the most powerful of its race, boldly preying upon the largest game. Perched upon some branch, and obscured by the shadows of evening, it marks its ill-fated quarry,—the fawn reposing among the fern,—the hare nibbling the grass,—the grouse couching among the heath;—silently and rapidly down it pounces, strikes its talons into its victim, and commences the work of destruction. Less noble game, such as moles, rats, and lizards, may be also ranked among its articles of food.

This fine bird chooses the clefts of rocks, or the hollows of decayed trees of antique growth, for the purpose of nidification, laying three eggs of a rounded shape and white colour.

The female is larger than her mate, and her colours are more bright.

The upper surface of the body is a mingled blending of brown and yellow, with zigzag lines and bars; below the ground colour is yellow, with black longitudinal dashes on the chest, and fine transverse irregular bars over the whole of the rest of the plumage; irides bright fiery orange; beak and nails black. Length nearly two feet.

We are indebted to the Hon. Daniel Finch for an example of this fine Owl for illustration in this Work.

Our figure is about three fourths of the natural size.

EASTERN GREAT HORNED OWL.

Bubo Ascalaphus.

E. Lear del et lith

Printed by C. Hullmandel.

EASTERN GREAT HORNED OWL.

Bubo Ascalaphus.

Le Hibou Ascalaphus.

This fine species of horned Owl would appear to represent in the temperate portions of Asia and Africa, the *Bubo maximus* of Norway, Russia, &c.; and if we mistake not the *Otus* (*Bubo*) *Bengalensis* of our "Century of Birds" must rank as synonymous with the present species. In Europe the eastern and southern portions appear to be the only parts visited by the *Bubo Ascalaphus*, M. Temminck giving Sicily and Sardinia as places in which it has been observed.

Of the habits and manners of this species nothing has been placed on record, nor are we able to afford any information on the subject. Specimens have been sent to the Zoological Society by Sir Thomas Reade from Tunis, and a single individual formed a part of the collection made during the late expedition to the Euphrates.

Feathers of the facial disk buffy white terminated with black; crown of the head dark brown, each feather irregularly edged and terminated with buffy white; feathers surrounding the neck deep buff, becoming paler at the tip, and with an irregular mark of dark brown down the centre; the remainder of the upper surface, wings and tail dark brown, irregularly blotched with reddish buff, pale buff and lighter brown; all the under surface deep buff, the feathers on the centre of the abdomen being much paler and crossed with several fine and irregular bars of brown at their extremities; feathers of the legs deep buff, becoming much paler on the front of the tarsi and on the toes; bill and claws black.

We have figured an adult male nearly of the natural size, from a specimen forwarded to us by M. Temminck.

LONG EARED OWL.

Strix otus, *(Linn.)*

Otus vulgaris: *(Flem.)*

Drawn from Nature & on Stone by J. & E. Gould *Printed by C. Hullmandel*

Genus OTUS.

Gen. Char. *Bill* bending, and forming an elliptic curve, the cere covering the basal ridge for nearly half the length of the bill; cutting margin of the upper mandible straight, the under one having the tip obliquely truncated and notched. *Nostrils* oval, obliquely placed. *Fascial disc* of moderate size and complete. *Conch* of the ear extending from the outer angle of the eye to behind the limb of the lower jaw, the opening defended by a flap or *operculum*. *Head* furnished with egrets. *Wings* long; the second quill-feather the longest. *Tail* even, and scarcely showing any concavity beneath. *Legs* and *toes* feathered to the insertion of the claws. *Toes* rather short; the outer one reversible. *Claws* moderately curved, long, and very sharp; rounded beneath, except the middle one, which is grooved, and with a sharp inner edge.

LONG-EARED OWL.

Strix otus, *Linn.*

Otus vulgaris, *Flem.*

Le Hibou moyenduc.

The habits of this Owl lead it to frequent thick woods, in the depths of which it lives retired from observation, concealing itself during the day amongst the foliage of the holly and ivy-clad trees, whence it emerges at the approach of evening in quest of food; and dissection confirms the opinion that small mammalia, such as mice, moles, and rats form its principal subsistence; in fact, as regards food, it agrees closely with the others of its tribe. Most of the woody districts, especially in the northern portions of England, and in Scotland, are the chief localities wherein it may be found in our own islands; it is, however, by no means so common as the Tawny Owl, which generally frequents the same situations. It has a wide range over the continent of Europe. The same species, and one so closely allied to it as to render it difficult to say whether it can be fairly separated, is found in the United States and the northern regions of America as far as the sixtieth degree of latitude.

Unlike the Tawny and Barn Owls, which breed in hollow trees, the present species evinces a partiality for the deserted nests of the Crow, Raven and Magpie for the purpose of nidification: it lays about four white eggs; the young are covered with a full coat of white down, which lasts for a considerable time, and disappears gradually as the feathers advance.

The sexes offer little or no external difference either in size or in the colour of the plumage.

The head is ornamented with two egrets, consisting each of several feathers of a pale yellow colour at the edges, with broad central dashes of black; the whole of the upper surface is of a tawny yellow clouded with grey, each feather having its centre black, and the whole being checquered with zigzag bars and dots of brown and black; the under surface is pale tawny with longitudinal dashes of black; tail barred; facial disc light grey, clouded with tawny brown; irides orange; beak black.

Our Plate represents a male of the natural size.

SHORT-EARED OWL.

Strix Brachyotos; *(Linn.)*
Otus Brachyotos; *(Cuv.)*

Drawn from Nature & on Stone by J & E Gould. *Printed by C. Hullmandel.*

SHORT-EARED OWL.

Strix brachyotos, *Lath.*

Otus Brachyotos, *Cuv.*

La Hibou brachyote.

The Short-eared Owl is so universally dispersed as to render it probable that it may be observed over the whole of the four continents, with the exception of the high northern regions. We have ourselves been enabled to compare specimens of this species from the Straits of Magellan, Brazil, and North America, with others from every part of Africa and India, all of which were so strictly similar in their markings and size that it was impossible to distinguish them.

Unlike the rest of its tribe, which habitually reside among trees and rocks, the Short-eared Owl reposes on the ground, and prefers extensive moors and marshes to thickly wooded districts. Although it is sparingly dispersed during summer over the northern parts of England and Scotland, in which localities it is known to breed, still it must be regarded as a migratory bird both in the British Islands and the greater portion of the Continent. In Holland it is particularly abundant during the months of September and October, about which period it makes its annual migration to England, where it arrives in companies of from five to twenty or thirty in number, and gradually disperses over the marshes and extensive fields of turnips which border the whole of our eastern coast. Its flight is strong and vigorous, and from its diurnal habits it may be frequently observed, particularly in gloomy weather, on the wing at midday, hunting for small birds, mice, frogs, &c., which constitute its principal food.

When in a state of repose, it secretes itself on the ground, either in a tuft of long grass, heath, or among the thickest part of the turnips, and it is seldom roused from this retreat until closely approached. It is to be regretted that these secluded and retiring habits tend much to its own destruction by the facility with which it is discovered by the gunner with the assistance of his pointer, which will generally point on scenting it.

In a note in his valuable edition of Wilson's American Ornithology, Sir William Jardine informs us that he has found the nest of this bird on the extensive moors at the head of Dryfe (a small rivulet in Dumfriesshire), that the eggs are five in number, and that the "nest is formed upon the ground among the heath, the bottom of the nest scraped until the fresh earth appears, on which the eggs are placed, without any lining or other accessory covering. When approaching the nest or young, the old birds fly and hover round, uttering a small shrill cry, and snappiug with their bills. The young are barely able to fly by the 12th of August, and appear to leave the nest some time before they are able to rise from the ground. I have taken them, on that great day to sportsmen, squatted on the heath like young black game, at no great distance from each other, and always attended by the parent birds."

Feathers covering the nostrils brownish white, with black shafts; circle immediately around the eyes blackish brown; remainder of the facial disk yellowish brown, mottled with blackish brown; circle of small feathers behind the facial disk mottled with tawny white, blackish brown, and white, except opposite to the orifice of the ear, where they are wholly blackish brown; on each side of the forehead four or five feathers somewhat longer than the rest, which are erected and depressed at pleasure; head, back, and wing-coverts dark brown, deeply edged with tawny brown; quills pale reddish brown, with several broad bars of dark brown on their outer webs; the inner webs are also barred, but not so numerously or so regularly as the outer; the tips of all ending in ashy grey; fore part of the neck and breast buff orange, each feather streaked down the centre with dark brown; under surface pale yellowish brown, with dark brown shafts; tarsi and toes dull yellowish white free from spots, the feathers assuming a hairy appearance on the toes; claws blackish grey; bill blueish black; irides gamboge yellow.

The Plate represents an adult male of the natural size.

SCOPS-EARED OWL.

Scops Aldrovandi; *(Will and Ray.)*

Drawn from Nature & on Stone by J. & E. Gould.

Printed by C. Hullmandel.

Genus SCOPS.

Gen. Char. *Bill* curved from the base; the upper ridge of the culmen flattened; the cere short. *Nostrils* round, placed in front of the cere. *Facial disc* small, and incomplete above the eyes; auditory conch small, and without an operculum; forehead with egrets or tufts. *Wings* long, the third feather the longest. *Tail* even or slightly rounded, concave beneath. *Legs* rather long. *Tarsi* feathered to the toes, which have their upper joints reticulated, and the anterior ones scutellated. *Claws* sharp, moderately curved, and partially grooved beneath. Plumage soft and downy.

SCOPS-EARED OWL.

Scops Aldrovandi, *Will.* and *Ray*.

Le Petit Duc.

The range of this beautiful little Owl is so extensive, that few of the larger species are more widely distributed. Independently of its existence throughout most of the countries of Europe, it is found both in Africa and Asia; and individuals from China have come under our notice, differing in no respect from specimens killed in our own island. Its occurrence here is, however, extremely rare in comparison to adjacent parts. It is abundant in France, Switzerland, and all the southern and eastern portions of Europe: in Holland and the north-western portion of the Continent it is almost as rare as it is in England.

In Europe it appears to be strictly migratory, arriving late in spring, when moths and the larger coleopterous insects, upon which it principally subsists, abound; but in the hotter portions of the Old World, where such insects are always abundant, numbers of these birds are stationary throughout the whole of the year: to these, its most common food, are added birds, mice, and other small animals. In its manners it is principally nocturnal, issuing forth from its hiding-place on the approach of twilight, in chase of those insects which are also roused from their state of repose to activity at the same time. In confinement it is docile and contented, and especially interesting from its minute size and the elegance of its markings.

It breeds in the holes of decayed trees, clefts of rocks, and old buildings, the eggs being four or five in number, of a pure white.

The sexes offer little or no variations of colour; indeed the female so exactly resembles the male as not to be distinguished except by dissection.

The general colour is grey blended with brown; and freckled with minute markings of black, relieved by bold longitudinal dashes down the centre of most of the feathers; the head is ornamented with egrets capable of being elevated and depressed at will; a few black dashes encircle the disc of the face; the quills are barred alternately with rich brown and yellowish grey; irides and feet brilliant yellow.

The Plate represents an adult bird of the natural size.

GREAT CINEREOUS OWL.

Strix Lapponica; *(Retz.)*
______ cinerea; *(Gmel.)*
Surnia cinerea.

E. Lear del. Printed by C. Hullmandel.

Genus SURNIA, *Dum.*

GEN. CHAR. *Beak* short, arched. *Disc* of the *head* small and incomplete. *Ears* small, oval. *Egrets* none. *Legs* very plumose. *Tail* elongated wedge-shaped.

GREAT CINEREOUS OWL.

Strix Lapponica, *Retz.*

Strix cinerea, *Gm.*

Surnia cinerea.

THOSE who would wish to visit the haunts of this noble species of Owl, one of the very finest of its race, must leave the abodes of civilization and penetrate into the dreary regions of the arctic circle, where nature wears her rudest and wildest dress, for it is an inhabitant of that portion of both continents; and although solitary individuals now and then make their appearance in Lapland, Norway, and Russia, yet it would appear that the northern parts of America are its true habitat, as in that truly scientific work the "Fauna Boreali-Americana," by Messrs. Swainson and Richardson, the latter gentleman informs us that "It is by no means a rare bird in the fur-countries, being an inhabitant of all the woody districts lying between lake Superior and latitudes 67° or 68°, and between Hudson's Bay and the Pacific. It is common on the borders of the Great Bear Lake; and there and in the higher parallels of latitude it must pursue its prey, during the summer months, by daylight. It keeps, however, within the woods, and does not frequent the barren grounds like the Snowy Owl, nor is it so often met with in broad daylight as the Hawk Owl, but hunts principally when the sun is low; indeed it is only at such times, when the recesses of the woods are deeply shadowed, that the American hare and the murine animals, on which the Cinereous Owl chiefly preys, come forth to feed."

Through the great intercourse which the Hudson's Bay Company has with the polar countries of America, this bird is more common, perhaps, in the cabinets of London than in those of any part of the Continent; we are not, however, aware that there is in London any other European specimen than the one from which our figure is taken, and which was kindly entrusted to our care for that purpose by our obliging friend the Baron de Feldegg of Frankfort.

To the countries above mentioned as the habitat of this species, we may add the extensive region reaching across the north of Siberia from Russia to Kamtchatka, which we may regard as the nursery from whence those individuals have strayed which have been killed in various parts of Europe. Of these instances M. Temminck mentions a specimen in the cabinet of Vienna and one in his own collection, both of which are females, and a male in the Museum of Paris, which was placed there by M. Paikul, a Swede; the latter, he states, measures twenty inches, and the one in his own collection two feet eight inches, being an admeasurement larger than that of the female of *Bubo maximus*.

Of its nidification we have no further information than that communicated by Dr. Richardson in the work above quoted; in which he informs us that he discovered a nest "on the top of a lofty balsam poplar, built of sticks, and lined with feathers. It contained three young, which were covered with a whitish down."

The sexes differ in size considerably, but in their markings are so similar that the description of one will serve for both.

The face is grey barred with concentric circles of brown; the whole of the upper surface, wings, and tail are grey, marked with bars and zigzag interlineations of blackish brown; the under parts are lighter than the upper, with longitudinal dashes and obscure bars of brown, especially on the thighs and flanks; beak yellow at the tip; tarsi feathered to the claws, and of the same colour as the under surface; claws black; irides bright yellow.

The Plate represents an adult male about three fourths of the natural size.

SNOWY OWL.
Strix Nyctea, (*Linn.*)
Surnia Nyctea, (*Dum.*)

SNOWY OWL.

Strix Nyctea, *Linn.*

Surnia Nyctea, *Dum.*

La Chouette Harfang.

The recorded instances of the capture of this noble Owl within the British Islands no longer leave a doubt as to the propriety of giving it a place in our Fauna; its visits are, however, extremely uncertain, and generally occur at very lengthened intervals. Mr. Selby informs us that he has in his possession two very fine specimens, male and female, which were killed near Rothbury, in Northumberland, in the latter part of January 1823, during the severe snow-storm that was so generally felt throughout the North of England and Scotland at that period.

The arctic regions constitute the true habitat and native place of abode of the Snowy Owl, from the severities of which climate it retreats when, on the approach of extraordinarily severe weather, the various small animals upon which it preys have either removed southward or sought shelter beneath the encrusted frozen snow. It would appear that its migrations are extended further south on the American continent than in the Old World, where it is seldom observed so far as Holland and France: it is sometimes found in the North of Germany, more frequently in Russia, Sweden, and Norway, and occasionally in the Feroe, Shetland, and Orkney Islands. It is one of the most robust and powerful of its race: its food consists of alpine hares, rabbits, rats, lemmings, and grouse; and even the wary fox has been known to fall a victim to its attacks. The indefatigable Wilson informs us that it is a dexterous fisher, pouncing upon its finny prey and securing it by an instantaneous stroke of its foot; and Dr. Richardson states in the second volume of the Fauna Boreali-Americana, that he has seen it pursue the American hare, making repeated strokes at the animal with its foot. It hunts in the day; and, indeed, unless it could do so, it would be unfit to pass the summer within the arctic circle. When seen on the barren grounds it was generally squatting on the earth, and if put up, it alighted again after a short flight; but was always so wary as to be approached with great difficulty. In the woody districts it showed less caution, and, according to Hearne, has been known to watch the Grouse-shooters a whole day for the purpose of sharing in the spoil. "On such occasions it perches on a high tree, and when a bird is shot, skims down and carries it off before the sportsman can get near it."

It appears to affect different situations for the purpose of nidification, sometimes choosing the ledges of precipitous rocks, and at others, according to Dr. Richardson, making "its nest on the ground and laying three or four white eggs, of which two only are in general hatched. In winter, when this Owl is fat, the Indians and White residents in the fur-countries esteem it to be good eating. Its flesh is delicately white."

The Snowy Owl is subject to considerable variations of plumage in the first three or four years of its existence, and during this period it is characterized by a plumage more or less strongly barred with brown, which markings become more indistinct as the bird advances in age, and they disappear entirely in old males, leaving them of a pure white. As is the case with most of the Raptorial birds, the female is considerably larger than her mate, but in other respects is not distinguishable.

In the adult male the plumage is wholly white; the irides fine yellow; the bill and claws black, the former being nearly covered by bristly feathers projecting from its base, and the latter, which are long and very sharp, being nearly concealed by the long hairy feathers that clothe the legs and toes. The head, compared with those of other Owls, is small in proportion to the size of the bird.

Our Plate represents an adult and a bird of the second year, about one third less than the natural size.

URAL OWL.
Surnia Uralensis.

Drawn from Nature & on Stone by J. & E. Gould.   Printed by C. Hullmandel.

URAL OWL.

Surnia Uralensis, *Dum.*

Strix Uralensis, *Linn.*

Le Hibou de l'Oural.

The obscure and almost untraversed regions which this fine Owl habitually frequents must be deemed not only the cause of its great scarcity, but also of the little that is known respecting its habits and manners. Although the large size and the remarkable development of the facial disk of this bird readily distinguish it from the more typical species of the present genus, nevertheless we have inserted it in this place, believing that in general habits and manners it will be found to agree in a great measure with the other species of the genus.

The Ural Owl is a native of the northern regions of Lapland and Siberia; it is also found, but very sparingly, in the North of Sweden and Norway. In Hungary and Livonia, according to M. Temminck, it is somewhat more abundant; we must, however, regard it as one of the rarest of the European Owls, our own specimen being, we believe, the only one in England, nor does it occur in many of the largest collections on the Continent. Though a native of the arctic circle in the Old World, it does not appear to have been hitherto discovered in the parallel latitudes of America, and we have every reason to believe it to be a stranger to that continent.

Like the rest of the larger owls its food consists of small mammalia, such as leverets, rats, and mice, and not unfrequently the Ptarmigan and other birds.

It is said to construct its nest in the holes of trees, and to lay two white eggs.

The sexes are alike in plumage, but the young of the year differ in having the ground colouring of a pale greyish brown; the upper parts irregularly spotted with brown and light red, varied with blotches of white; the wings and tail barred with grey, and the whole of the under parts longitudinally streaked and blotched with brown.

The adults have the whole of the face greyish white, the rim of the facial disk consisting of white feathers spotted with black; the whole of the upper surface longitudinally blotched with brown and white; the under surface dusky white, every feather having a brown streak down the centre; the wings and tail barred with brown and yellowish white, the latter being of considerable length and remarkably graduated; beak yellow; tarsi covered with greyish white hairs; nails brown; irides brownish yellow.

The Plate represents a male rather less than the natural size, the adult bird being two feet in length.

HAWK OWL.

Surnia funerea; *(Dumeril)*.

Strix funerea; *(Gmel:)*.

Drawn from Nature & on Stone by J. & E. Gould. *Printed by C. Hullmandel.*

HAWK OWL.

Surnia funerea, *Dumeril.*

Strix funerea, *Gmel.*

Le Chouette caparacoch.

Of the European examples of the genus *Surnia*, a genus established by M. Dumeril for the reception of such of the Owls as approach the *Falconidæ* in habits, manners, and general structure, the Hawk Owl, although the least, is nevertheless one of the most typical. It possesses an almost unlimited range of habitat throughout the northern and arctic regions of both continents, and is not unfrequently seen in Germany and even France. No example, however, is on record of its having been seen in the British Islands, which is rather remarkable, considering that the Snowy Owl, its most nearly allied relative, has been so frequently captured within the British dominions. Like that fine species, the Hawk Owl is endowed with the faculty of seeing its prey, if not in the bright light of day, at least during dull weather and long before sun-set in the evening, and from this circumstance, which has led to its being considered as a feeder by day, in connexion with its structure, it may be regarded, together with the rest of its genus, as forming the passage between the Harriers on the one hand, and the true nocturnal Owls on the other.

Its food consists of rats, mice, birds, and insects.

According to the best information we can obtain, it builds in trees, and lays two white eggs.

The sexes differ in no respect except a trifle in size, and in the intensity of the markings.

The forehead is thickly dotted with white and brown, the facial disc is greyish white, partly encircled by a crescent-shaped band of black, which passes over the ears; the upper surface is irregularly blotched with brown and white, the latter colour predominating on the shoulders; the wings are brown, irregularly barred with white; the whole of the under surface is greyish white, barred with transverse rays of brown, the shaft of each feather being also brown; tail brown, barred with white; tarsi greyish white; toes yellow; irides bright yellow.

The Plate represents an adult male of the natural size.

BARRED OWL.

Strix nebulosa, *(Linn.)*

Ulula ———. *(Cuv.)*

E. Lear del et lith.

Printed by C. Hullmandel.

Genus ULULA.

Gen. Char. *Bill* nearly straight at the base, the tip hooked, with a rounded culmen, cutting margin of the upper mandible having a small lobe or sinuation near the middle. Facial disk large and complete; auditory conch rather large, and defended by an operculum. *Wings* short, rounded, concave; the first quill-feather very short; the fourth the longest in the wing, with the third and fifth nearly equal to it. *Tail* reaching beyond the closed wings, rounded, bent, and concave beneath. *Legs* having the tarsi plumed, and the toes more or less so. *Claws* moderately curved, long, short, all more or less grooved beneath.

BARRED OWL.

Strix nebulosa, *Linn.*

Ulula nebulosa, *Cuv.*

La Chouette nébuleuse.

In the regions of the Old World the Barred Owl scarcely ever extends its migrations further south than Norway, Sweden, and Russia, in which countries it is so sparingly distributed as rather to be regarded as an accidental visitor than a native species. The northern and temperate portions of America appear to be its true habitat, for it is dispersed over the whole of the United States, where, Mr. Audubon informs us, its peculiar cry of *Whah, whah, whah-aa,* may be heard towards evening proceeding from every part of the forest. According to this diligent observer of nature, the flight of the Barred Owl is smooth, light, and noiseless, and capable of being greatly protracted. Mr. Audubon further remarks that its powers of vision during the day seem to be of an equivocal character, he having seen one alight on the back of a cow, which it left so suddenly, on the animal moving, as to leave no doubt in his mind that the Owl had mistaken the object upon which it had perched for something else: at other times he has observed that the approach of the Grey Squirrel intimidated it, if one of these animals accidentally jumped on the branch close to where it was sitting, although the Barred Owl destroys numbers of this species of Squirrel during the twilight. It is a well-known fact that the eyes of those Owls whose habits are strictly nocturnal differ both in colour and construction from those which feed partially by day, or rather whose greatest powers of vision are developed in the twilight and during dark and gloomy days. Had we not been acquainted with the habits of this bird and the colour of its eyes, we should probably have assigned it a place among the Owls forming the genus *Surnia,* to which division it bears a strong resemblance both in the colour of its plumage and in its general contour. The noiseless flight of the Barred Owl may be attributed to the peculiar nature of its plumage, which, like that of all other nocturnal species, is extremely soft and yielding, enabling it to steal quickly upon its victim without exciting observation or alarm.

Its food consists of young hares and rabbits, mice, small birds, frogs, lizards, &c.

Its eggs are deposited in the holes of decayed trees or the deserted nests of Crows and Hawks; they are round, of a pure white, and from four to six in number.

The male and female differ somewhat in size, the males being the smallest; and they are also subject to considerable varieties of plumage, some specimens, particularly those found in Europe, being of a very dark colour, while others are very light.

The plumage of the generality of specimens may be thus described:

The face light ash encircled with lines of brown; the upper part of the plumage, together with the quills and tail, is of a brownish grey, transversely rayed with white and yellowish bars; the front of the neck and chest transversely barred with greyish ash and yellowish white markings; the lower part of the breast and flanks yellowish grey with longitudinal stripes of brown; feet and toes covered with short grey feathers; beak yellow; irides blackish brown.

The Plate represents an adult male, rather less than the natural size.

TAWNY OR WOOD OWL.

Strix aluco. *(Linn.)*
Surnia ___ *(Dum.)*

Drawn from Life & on Stone by J. & E. Gould. *Printed by C. Hullmandel.*

TAWNY OR WOOD OWL.

Strix Aluco, *Linn.*

Syrnium Aluco, *Savigny.*

La Chouette hulotte.

We have followed Baron Cuvier in adopting, or at least in adding to our names of this bird, the generic appellation of Savigny, who separated this species from the more typical Owls on account of the short and curved beak, the large size of the facial disk, and the toes feathered to the claws.

This bird measures from fourteen to fifteen inches in length, and with the exception of the Barn Owl is the most common of the British species. It is to be found generally throughout most of the well-wooded districts of Great Britain, and inhabits in abundance the large forests of the European continent. According to M. Temminck it is rather a rare bird in Holland.

In this country the Tawny Owl takes up its abode in woods and old plantations, preferring such as are thickly set with holly and firs, and well grown over with ivy. Here it remains quiet and secluded during the day, but at nightfall becomes clamorous and hoots aloud. In the breeding season it searches for a hole in a tree, or in default of finding such a convenience takes possession of the deserted nest of a Hawk or Crow, in which its eggs are deposited. These are of large size, measuring $1\frac{7}{8}$ inch in length by $1\frac{1}{2}$ inch in width, equally rounded at both ends and perfectly white. The females begin to sit as soon as they have laid their first egg, and the young for a considerable time after exclusion are a shapeless mass of grey down. The parent birds attend their young brood with great assiduity, and supply them plentifully with mice, shrews, moles, and the young of various other mammalia of larger size.

The beak in this species of Owl is yellowish white, short and curved; irides dark blue; the feathers forming the facial disk light brown; the feathers surrounding the disk marked with numerous dark spots; head, neck and back reddish yellow brown, spotted and streaked with dark brown in the direction of the shaft of each feather; on the scapulars and wing-coverts are large white spots forming conspicuous rows; under surface reddish white with brown bars; wing- and tail-feathers reddish brown, barred with very dark brown, under sides reddish ash, with lighter-coloured bars, outer edges of the quill primaries beautifully serrated; legs and toes covered with short downy feathers of reddish grey, with brown specks. Claws nearly black, long, curved and sharp. The females when compared with the males are larger in size and darker in colour, approaching to deep red brown. By mistake the word Surnia instead of Syrnium was printed on our Plate.

We have figured a bird of the natural size.

LITTLE OWL.

Strix nudipes; *(Nilsson)*.

Noctua nudipes; *(Mihi)*.

E. Lear del. et lith.

Printed by C. Hullmandel.

LITTLE OWL.

Strix nudipes, *Nilsson.*

Noctua nudipes, *Mihi.*

La Chouette chevêche.

THE *Strix passerina* of Linnæus, of which *Acadica* is a synonym, is the title of a very different bird from the one here figured; and, although most modern naturalists from some unaccountable cause have assigned the term *passerina* to the present species, as if it were that which Linnæus so designated, we have thought it necessary, in justice to truth, to correct this misnomer by restoring the old name of *nudipes* given to it by Nilsson.

The Little Owl must be considered one of the rarest of our occasional visitors; its presence appearing to depend entirely upon accidental circumstances. It is plentifully distributed over the whole of the temperate portions of Europe. M. Temminck states that it is abundant in Holland and Germany, but that it is never seen in high northern latitudes.

We cannot undertake to say whether the Little Owl is to be classed among the migratory birds of its race, as we are not in possession of any details of its habits and manners. It appears, however, to display all the characteristics of the genus to which it belongs, preying in the dusk of the evening and during twilight on mice, moles, small birds, and large insects. Having had an opportunity of observing it in captivity, we are enabled to state that its conduct under such circumstances is precisely similar to that of other species when in a similar situation. During the day it sits in almost motionless repose, occasionally snapping with its bill, when disturbed, but resuming its quiet position as soon as the annoyance ceases: on the approach of evening it becomes lively and alert, and by its animated manners betrays its anxiety for food and liberty. Its eggs, which are four or five in number, are deposited sometimes in the holes of trees, but more frequently in old walls and ruined towers.

The sexes are alike in plumage, and the young attain at an early period the adult colouring.

The upper parts are of a brownish grey marked with large irregular blotches of white, the feathers on the top of the head being regularly spotted with yellowish white; throat white, separated by a brown belt from the chest; a white circle surrounds the eye; the whole of the under parts dusky white, irregularly clouded and blotched with brown; tail brown, barred with yellowish brown; bill and feet yellowish straw colour; irides straw yellow.

The Plate represents an adult of the natural size.

TENGMALM'S OWL.

Strix Tengmalmi; *(Gmel.)*
Noctua Tengmalmi; *(Selb.)*

E. Lear del et lith.

Printed by C. Hullmandel

TENGMALM'S OWL.

Strix Tengmalmi, *Gmel.*

Noctua Tengmalmi, *Selby.*

La Chouette Tengmalm.

In all probability this little Owl extends its range over the whole of the Arctic Circle, in which inhospitable region it appears to represent the *Noctua nudipes*, a species inhabiting more temperate parts, and with which it has more than once been confounded. The *Noctua Tengmalmi* is abundant in Russia and Norway; it is also found, but more rarely, in Germany and France, and it has been captured two or three times in the British Islands. Mr. Selby mentions one example in particular, which was killed near Morpeth in Northumberland in 1812, and forms a part of that gentleman's collection. In the 'Fauna Boreali-Americana' Dr. Richardson states his belief "that it inhabits all the woody country from Great Slave Lake to the United States. On the banks of the Saskatchewan it is so common that its voice is heard almost every night by the traveller wherever he selects his bivouac. Its cry in the night is a single melancholy note, repeated at intervals of a minute or two; and it is one of the superstitious practices of the Indians to whistle when they hear it. If the bird is silent when thus challenged, the speedy death of the inquirer is augured; hence its Cree appellation of Death-bird.

When it is disturbed or accidentally wanders abroad by day, it is so dazzled by the sun that it becomes stupid, and may be easily taken with the hand.

It is said to build a nest of grass, in holes or clefts about half way up a pine-tree, and to lay two eggs, in the month of May.

The sexes are alike in plumage.

Facial disk greyish white mingled with black, except that portion immediately before and behind the eye, where it is wholly black; crown, nape, and back part of the neck pale brown spotted with white, those on the latter part being the largest, and surrounded with darker brown; back, wing-coverts, and scapularies pale brown spotted with white, the spots on the mantle being nearly concealed by the tips of the feathers; quills pale brown, having on their exterior webs a few oval spots of white forming imperfect bars; the extremities of the outer web of the first quill reverted, of the second for half its length, and of the third only a small portion near the tip; tail pale brown crossed by five rows of white spots, giving it the appearance of being barred; under surface white slightly tinged with buff; tarsi and toes thickly clothed with soft hair-like feathers of a buff colour; bill and irides bright yellow; claws black.

The Plate represents an adult male of the natural size.

SPARROW OWL.

Strix pafserina; *(Linn.)*

Noctua ______ *(Mihi)*.

Drawn from Life & on Stone by J. & E. Gould. *Printed by C. Hullmandel.*

SPARROW OWL.

Strix passerina, *Linn.*

S. acadica, *Gmel.*

Noctua passerina, *Mihi.*

La Chouette chevêchette.

We only follow the opinions and example of several of the best naturalists of the European continent in considering this very minute Owl, the *Chevêchette* of M. Temminck's *Manuel d'Ornithologie*, p. 96, as the true *passerina* of Linneus, who, in the 12th edition of the *Systema Naturæ*, p. 133, says of this bird, "*magnitudo passeris.*" It measures but little more than six inches in length; we have therefore called it, in reference to its diminutive size, the Sparrow Owl, intending to distinguish it from the *Strix passerina* of authors, the *Chevêche* of M. Temminck, p. 92, which measures nine inches in length, and to which another specific name has been given.

The Sparrow Owl is an inhabitant of Livonia and of the northern regions generally, seldom venturing farther south than the colder parts of Germany, where it is only seen in large forests, and has not, that we are aware, been taken in any part of the British Islands; nor is it, we believe, ever found in America.

All the upper parts of the head and body are of dark greyish brown, varied with spots of white; the under parts white, with longitudinal patches of brown; on the flanks, the brown spots have a direction across the feathers; throat and sides of the neck almost white; the tail, the feathers of which are rather long, exhibits four narrow white bands; the feet are feathered to the extremities of the toes; the beak and irides yellow.

The female is rather darker in the general tone of her colour, inclining to chocolate brown, and the white spots are less brilliant. She lays two white eggs; the nest is made in a hole of a tree in the forest, or occupies an aperture of a rock.

The ordinary food of this species consists of mice, coleopterous insects, and large moths.

We have figured a bird of the natural size.

THE

BIRDS OF EUROPE.

BY

JOHN GOULD, F.L.S., &c.

IN FIVE VOLUMES.

VOL. II.

INSESSORES.

LONDON:

PRINTED BY RICHARD AND JOHN E. TAYLOR, RED LION COURT, FLEET STREET.

PUBLISHED BY THE AUTHOR, 20 BROAD STREET, GOLDEN SQUARE.

1837.

LIST OF PLATES.

VOLUME II.

NOTE.—As the arrangement of the Plates during the course of publication was found to be impracticable, the Numbers here given will refer to the Plates when arranged, and the work may be quoted by them.

INSESSORES.

LIST OF PLATES.

* Named erroneously Sylvia hippolais.

† Named erroneously Motacilla alba. As I have every reason to believe that this species, one of the most elegant and familiar of our native birds, will prove to be quite distinct from either Motacilla lugubris or Motacilla alba, I would beg leave to name it after my valued friend William Yarrell, Esq.

EUROPEAN GOATSUCKER.
Caprimulgus Europæus; *(Linn.)*

Drawn from Nature & on Stone by J. & E. Gould.

Printed by C. Hullmandel.

Genus CAPRIMULGUS, *Linn.*

Gen. Char. *Bill* very short, weak, curved at the tip, broad and depressed at the base; the upper mandible deflected at the point; gape very large, and extending to or beyond the posterior angle of the eyes; basal edge of the upper mandible bordered with strong moveable bristles, directed forwards. *Nostrils* basal, tubular, or with a large prominent rim, clothed with very small feathers. *Wings* long, the first quill shorter than the second, which is the longest of all. *Tail* rounded or forked, of ten feathers. *Tarsi* short. *Toes* three before and one behind, the anterior ones united as far as the first joint by a membrane; the claw of the middle toe broad, and serrated on the inner edge.

EUROPEAN GOATSUCKER.

Caprimulgus europæus, *Linn.*

L'Engoulevent ordinaire.

The *Caprimulgus europæus* was until within the last few years the only species of this curious and interesting race of birds known to inhabit Europe; a second has, however, been discovered by M. Natterer in the South of Spain, and described by him under the name of *Caprimulgus ruficollis*, from the conspicuous red band which crosses the back of the neck, certainly a far more appropriate term than the one (*europæus*) given to the present bird, which is now applicable to both species, or to any others that may hereafter be discovered.

The European Goatsucker is a migratory bird, inhabiting all the temperate portions of Europe during summer, and retiring southward beyond the Mediterranean on the approach of winter. It arrives in the British Islands from the middle to the end of May, and departs again about the latter end of September or beginning of October. While here it is distributed over the whole of the kingdom, residing in woods, plantations, thick beds of fern, (whence its provincial name of Fern Owl,) and districts clothed with tall grasses. Being strictly nocturnal in its habits, it avoids as much as possible the bright light of day, but on the approach of twilight it may be seen hawking for *Melolonthæ*, *Phalænæ*, and other nocturnal insects.

The flight of the Goatsucker is rapid in the extreme while in pursuit of its prey, and is accompanied by a number of evolutions, similar to those of the Swallow, but which are, if possible, performed with still greater ease and facility.

It makes no nest, but lays two eggs on the bare ground, amongst fern, heath, or long grass, sometimes in woods or furze, but always near woods, in which it may conceal itself by day. The eggs are white, marbled with light brown and grey.

The Goatsucker reposes mostly on the ground; and when it perches on the limb of a tree, it is commonly along the branch, and not across it like other birds.

"The male," says Montagu, "makes a very singular noise during the period of incubation, not unlike the sound of a large spinning-wheel, and which it is observed to utter perched, with the head downwards; besides which it emits a sharp squeak, repeated as it flies."

The whole of the upper surface and the throat are of an ashy grey, numerously spotted and streaked with dark brown, and tinged with pale or yellowish brown; the head and back streaked longitudinally with black; beneath the base of the under mandible runs a stripe of white, which extends along each side of the head; in the centre of the throat a patch of white; under surface yellowish brown, transversely barred with black; outer webs of the quills blotched with red brown, the three exterior feathers having a large white patch on the inner webs near their tips; tail irregularly marked with black, grey, and yellowish brown, two outer tail-feathers on each side deeply tipped with white; bill and irides dark brown; tarsi pale brown.

The female differs from the male only in being destitute of the white spots on the quill-feathers, and of the white tips of the lateral tail-feathers.

We have figured a male of the natural size.

RED-COLLARED GOATSUCKER.

Caprimulgus ruficollis; *(Temm.)*

Drawn from Nature & on stone by J & E Gould. Printed by C. Hullmandel.

RED-COLLARED GOATSUCKER.

Caprimulgus ruficollis, *Temm.*

L'Engoulevent à collier roux.

THROUGH the kindness of Mr. John Natterer and the Directors of the Imperial Cabinet at Vienna, who have liberally forwarded the original specimen for our use, we are enabled to give a figure of this interesting species of Goatsucker, which is so extremely rare, that we know of no examples in the museums of this country, nor in any of those on the Continent, with the exception of Vienna. Northern Africa is, we doubt not, its natural habitat, whence it may occasionally pass into Europe, but so rarely that no other examples are on record than those referred to in the following notes, which accompanied the above-mentioned specimen, and which we prefer giving in M. Natterer's own words.

" *Caprimulgus ruficollis*: male. Shot the 14th of July, 1817, some miles distant from Algeziras, in the oak-woods by daylight. Iris dark brown; legs reddish grey; nostrils oval with their borders much elevated; length, 12 inches 8 lines; extent of the wings, 22½ inches; the tail exceeding the wings by 1½ inch.

" Another male was shot on the 20th of July in the valley of the Rio del Miel near Algeziras, flying very low, an hour after sunset. Length, 13 inches and 3 lines; extent of the wings, 23¼ inches; the tail exceeding the wings by 1 inch and 8 lines.

" The female I shot while flying near the same spot on the 21st of July. Length, 12 inches and 8 lines; extent of the wings, 22½ inches; the tail exceeding the wings by 2 inches. The female closely resembles the male, differing only in having less white on the throat; the white tips of the two outer tail-feathers only 8 lines long, and tinged with brown on the outer web; the white spots on the three first primary quills smaller and tinged with ochre, and without any corresponding mark on the outer web as in the male; the remainder of the plumage is exactly the same as that of the male.

" The name of the bird in the part of Spain where it was killed is *Samala*. It seems to be a very rare bird, for I passed several nights in the adjacent woods without discovering any more examples."

The male may be thus described:

Forehead, sides, and back of the head, back, and six centre tail-feathers ashy brown, with numerous extremely fine freckles of dark brown, which are most decided on the tail-feathers, where they assume the form of irregular bars; feathers on the centre of the head dark brown with paler edges freckled with a darker tint; throat, sides of the face, ear-coverts, and back of the neck rufous, which is very rich on the latter, where it forms a decided collar; from the angle of the mouth to the back of the neck passes a narrow line of white; on each side of the neck an irregular patch of white feathers with a crescent mark of deep brown at the tip; wing-coverts the same as the back, but having the tip of each feather rufous; scapularies dark brown, with a broad margin of buff on their outer edge, and with grey freckled with dark brown on the inner; primaries dull brown; with a broad white spot on the inner web, forming a bar on the under side of the wings and a faint indication of it on the outer web; the secondaries and the tips of the primaries, particularly on their inner webs, having irregular and faint markings of grey; two outer tail-feathers on each side dark brown, irregularly blotched on the outer web with reddish brown and largely tipped with white; all the under-surface pale rufous, with irregular transverse arrow-shaped markings of deep brown, which are most numerous on the breast; bill and feet reddish grey.

We have figured the male of the natural size.

SWIFT.
1. Cypselus murarius, *(Temm.)*.

WHITE-BELLIED SWIFT.
2. Cypselus Alpinus, *(Temm.)*.

Drawn from Life and on Stone by J. & E. Gould.

Printed by C. Hullmandel.

Genus CYPSELUS, *Illig.*

GEN. CHAR. *Beak* very short, triangular, large, all its base concealed, depressed; *gape* extending beyond the eyes; *upper mandible* hooked at the point; *nostrils* longitudinal, near the ridge of the beak, open, the edges raised and furnished with small feathers. *Tarsi* very short. *Toes* four, all directed forwards and entirely divided; *nails* short, strong and hooked. *Tail* composed of ten feathers. *Wings* very long; the first *quill-feather* a little shorter than the second.

SWIFT.

Cypselus murarius, *Temm.*

Le Martinet.

WE know of no birds, and certainly none in Europe, possessed of equal power of flight with the species of the present genus: in fact, their natural habitat appears to be the air, their short feet and strong claws serving more especially to cling to the rough surface of rocks, towers, and high buildings, or firmly securing them while in a state of repose. Their extraordinary length of wing, combined with the shortness of their tarsi, prevents them walking on, or rising from, any level situation, unless by repeated exertions, or taking advantage of some slight elevation; they are therefore seldom, if ever, to be seen on the ground. These birds are distinguished not only by the velocity of their flight, but by the smooth and graceful sweeps they take during their aërial career in pursuit of the various insects constituting their food. In fine and serene weather, when these insect tribes ascend to an almost incredible elevation, the Swifts occupy the highest regions of the atmosphere, persevering in the chase almost beyond the reach of sight. Their degree of elevation, however, doubtless depends on that of the insects, and the latter on the influence of the weather; so that the low or high flight of these birds may be taken as a barometrical index of the state of the air, and the consequent probability of rain or the contrary.

The Swift, which is spread over Europe generally, is essentially migratory. It arrives in England at the beginning of May, and leaves us again as early as August or September. It breeds in old buildings, steeples, ruins, towers, and rocks. The eggs are white.

The colour of the plumage, with the exception of the throat, which is white, is of a uniform sooty-black with bronze-coloured reflections. The sexes offer no external marks of distinction.

WHITE-BELLIED SWIFT.

Cypselus alpinus, *Temm.*

Le Martinet à ventre blanc.

THE claims of this fine Swift to a place in the Fauna of Great Britain rest on the circumstance of one noticed by Mr. Selby in the "Transactions of the Northumberland, Newcastle, and Durham Natural History Society;" and one we have personally inspected, which was killed by the gardener of R. Holford, Esq., on his estate at Kingsgate near Margate, in whose possession it now remains.

The natural habitat of the *Cypselus alpinus* is more exclusively limited to the middle of the southern districts of Europe, particularly its alpine regions, and the shores of the Mediterranean, being very abundant at Gibraltar, Sardinia, Malta, and throughout the whole of the Archipelago; and to these may also be added the northern parts of Africa. In its manners it closely resembles our well-known Swift, but possesses, if possible, still greater powers of flight.

It would appear that the clefts of rocks and high buildings are the sites which this bird chooses for the purpose of nidification; the female laying three or four eggs of a uniform ivory white.

The sexes of this species present but little differences, the colour of the female being rather less decided; in the male a uniform greyish brown is spread over the whole of the upper surface, which descends across the breast in the form of a band, along the flanks and over the inferior tail-coverts; the throat and the middle of the belly are of a pure white, the tarsi covered with brown feathers, and the irides brown. Length from nine to ten inches.

CHIMNEY SWALLOW.

Hirundo rustica; (Linn.)

Drawn from Nature & on stone by J. & E. Gould. *Printed by C. Hullmandel.*

Genus HIRUNDO, *Linn.*

Gen. Char. *Bill* short, much depressed, wide at the base; upper mandible bent at the tip and carinated; the gape extending as far back as the eyes. *Nostrils* basal, oblong, partly covered by a membrane. *Tarsi* short. *Toes* slender, three before and one behind; the outer toe united to the middle one as far as the first joint. *Tail* of twelve feathers, generally forked. *Wings* long, acuminate, the first quill-feather the longest.

CHIMNEY SWALLOW.

Hirundo rustica, *Linn.*

L'Hirondelle de Cheminée ou domestique.

The migration of the Swallow and the laws which regulate its movements are now so well understood that it will scarcely be necessary to advert to them here. In the British Islands and in all other portions of the European Continent, the period of its arrival may be calculated upon with tolerable certainty, a scattered few generally appearing in all parts of the same latitude from the fifth to the tenth of April, after which period their numbers become suddenly augmented, and the work of reproduction is almost immediately proceeded with. Two broods are generally produced in the course of the season, the first being mostly able to fly before midsummer, and the second in the month of August. The young on leaving the nest are assiduously supplied with food, and carefully attended by their parents, until they are sufficiently strong to provide for themselves. The task of incubation being accomplished, the Swallows congregate in extensive flocks, and in obedience to the laws of nature retrace their steps, and pass the remainder of the year in more southern countries, where the insect food so essential to their existence is ever abundant. In this migratory movement, the adults, we are inclined to believe, always precede their progeny, which remain with us as long as the weather continues open and a sufficient supply of insects can be obtained for their subsistence.

The members of this aërial tribe are only excelled in their power of flight by the *Cypseli*, or Swifts, and are seen to the greatest advantage in the air, where their dexterity in securing their prey, the manner in which they drink while passing over the stream, and the celerity with which they feed their young while on the wing, cannot fail to call forth our admiration.

It is now generally admitted that the Barn Swallow of America is quite distinct from the British species; consequently the range of our bird is limited to the Old World. In the summer months the *Hirundo rustica* is universally dispersed over the whole of Europe, whence, as before stated, it migrates periodically into tropical regions; and as all migratory animals move from north to south, and *vice versâ*, Africa constitutes its winter residence.

In the British Islands the sites chosen for its nests are the interiors of chimneys and of coal-pits; but in many parts of the Continent, where these situations are not frequently met with, it builds on church towers, old ruins, the eaves of houses, barns, and other outbuildings. The eggs are four or five in number, their colour white, speckled with reddish brown and pale blue.

Forehead and throat rich chestnut; the remainder of the head, a band across the breast, and the whole of the upper surface black with blue reflections; tail very deeply forked, the two outer feathers extending far beyond the others; a large white spot on the inner webs of all the feathers, except the two middle ones; all the under surface white tinged with reddish brown, which is deepest on the vent and under tail-coverts; bill and feet black.

The female has rather less of the rich chestnut on the forehead, the black is less brilliant, and the outer tail-feathers much shorter than in the male.

The young is entirely destitute of the chestnut on the forehead, the throat is merely tinged with rufous; the band across the breast is but faintly indicated; all the upper surface resembles that of the adult, but the tints are much more dull; the wings are also shorter, and it is destitute of the long tail-feathers, which are not acquired till the first moult.

Our Plate represents an adult and a young bird of the natural size.

RUFOUS SWALLOW.

Hirundo rufula; *(Temm.)*

Drawn from Nature & on Stone by J & E. Gould.

Printed by C. Hullmandel.

RUFOUS SWALLOW.

Hirundo rufula, *Temm.*

L'Hirondelle rousseline.

Although Africa, particularly the southern and western portions of that continent, constitutes the native habitat of this beautiful species of Swallow, the contiguous portions of Europe are not without its occasional presence, and notwithstanding it is now considered a portion of the Fauna of this quarter of the globe, still we do not believe it occurs at regular periods, as is the case with the *Hirundo rustica*, but that it occasionally strays across the Mediterranean from the northern coasts of Africa, where it is also a rare species.

In Le Vaillant's ' Oiseaux d'Afrique' will be found an interesting account of this species, from which we learn that it is so familiar that it readily enters the houses of the inhabitants, particularly those in the interior, frequently building its nest in the sleeping-room of the family; that the nest differs from those of the other species of the genus, being a hollow ball, the entrance to which is constructed in the form of a long tube, through which the female passes into the interior, which is lined with any loose and soft materials the bird may find at hand. The eggs, which are from four to six in number, are white sprinkled with small brown spots.

Like the other members of the genus, its food consists of insects and their larvæ.

The top of the head, occiput, back, and wing-coverts are black with steel blue reflexions; the remainder of the wings and tail brownish black, with a faint indication of steel blue colour; back of the neck, rump, and upper tail-coverts deep rufous; sides of the face and throat pale rufous, which colour gradually becomes deeper and richer on the under-surface, flanks, and under tail-coverts; beak, irides, and feet black.

The female resembles the male, except that the crown of the head is rust red instead of black, and the tail-feathers are not so long.

We have figured an adult male of the natural size.

ROCK MARTIN.

Hirundo rupestris, (*Linn.*)

Drawn from Nature & on Stone by J & E Gould.

Printed by C Hullmandel.

ROCK MARTIN.

Hirundo rupestris, *Linn.*

L'Hirundelle de Rocher.

This bird is very abundant along the shores of the Mediterranean, and occurs also in such portions of Southern Europe as abound in rocky and precipitous places: it is an inhabitant of Savoy and Piedmont, but is more scarce in Switzerland, Germany, and the middle of France. As far as we are aware, it has never been found in the northern parts of Europe, nor has it as yet been observed in the British Isles. It is larger than the Common Sand Martin, to which species it bears a close resèmblance both in the colouring of its plumage and also in its general economy. It builds its nest and rears its young in the holes of rocks; the eggs being five or six in number, white, marked with minute dots.

We need hardly observe that its general habits and manners are in strict accordance with those of its family. Its food consists of insects, which it takes during flight.

The sexes offer no distinguishable difference in the markings of their plumage.

The whole of the upper surface is a uniform light brown; the quills and tail-feathers being darker, the inner webs of all the feathers of the latter, except the two middle ones, having in their centre a large oval blotch of white; the under surface is of a dull sandy white, slightly tinged with rufous.

We have figured a male and female of the natural size.

MARTIN.

Hirundo urbica; *(Linn.)*

Drawn from Nature & on stone by J. & E. Gould. Printed by C. Hullmandel.

MARTIN.

Hirundo urbica, *Linn.*

L'Hirondelle de fenêtre.

LIKE the Swallow this little fairy-like bird is strictly migratory, resorting during our winters to climes far to the south of the British Islands, and indeed to any other portion of Europe; whence it does not return till the spring, generally making its appearance about the middle of April: but in this respect the Martin, as well as most other insectivorous birds, is influenced in a great measure by the state of the season, a certain degree of temperature being necessary for bringing forth the insects upon which its existence depends.

The flight of the Martin is not so rapid, nor attended with such sudden evolutions as that of the graceful Swallow, but it is nevertheless performed with great ease and buoyancy; and although it does not possess so long a wing in proportion to its size as any other of its European brethren, and is consequently less adapted for continued flight, still it is seldom if ever seen resting either on trees or on the ground, but is continually traversing the air with apparently untiring wings, except during the period of nidification, when it descends to the earth for the purpose of collecting mud employed for the construction of its nest, which is erected under the eaves of houses and windows, the sides of rocks, under the arches of bridges, &c.: the nest when complete is a most compact and solid structure, firmly cemented together, the labour of one day being allowed to remain until the substance has got hard and dry, before the little mason proceeds to heap on more wet materials; the only means of ingress or egress is a small hole on the most sheltered side of the nest: the interior is well lined with straw, hay, and feathers. The eggs are five in number, of a pinkish white. At first the young birds are fed in the nest, afterwards the parents cling to the outside by means of their claws, and feed them at the entrance; when able to fly, they are fed on the wing for a considerable time, like the Swallow, and they occasionally resort to the house-top, or to the branches of some neighbouring tree, where they are also supplied with food by their parents, until they have acquired sufficient strength and confidence to launch forth and provide for themselves.

Like the Swallow, the Martin produces two broods in the year, the first of which are able to fly in July, and the second in August or the beginning of September. Early in October the Martins assemble in large flocks, frequently so numerous as almost to cover the roofs of houses, particularly in the villages situated on the borders of the Thames. About the middle of the month they commence their migration, continuing to depart in flocks till the early part of November, after the sixth or eighth of which month few are to be seen.

The notes of its song, which is frequently uttered during the period of incubation, are guttural, but soft and pleasing.

The female differs but little from the male: the young during the first autumn are readily distinguished by the less degree of brilliancy in their colouring.

The head, back of the neck, and back glossy bluish black; wings and tail brownish black; rump and all the under surface pure white; bill black; tarsi and toes clothed with white downy feathers.

We have figured an adult male and female of the natural size.

SAND MARTIN.

Hirundo riparia, *(Linn)*

Drawn from Nature & on stone by J. & E. Gould. *Printed by C. Hullmandel.*

SAND MARTIN.

Hirundo riparia, *Linn.*

L'Hirondelle de rivage.

WITH the exception of the place chosen for the purpose of incubation, the Sand Martin resembles its congeners; but in this respect the whole tribe of which this delicate species forms a part, are singularly different, both as regards the form of their nests and the situations chosen for their reception: for instance, if we examine the four species which take up their summer residence in the British Islands, we find that the Swift is directed to ruins, particularly inaccessible towers, and large public buildings; the Swallow gives preference to the entrances of chimneys and the mouths of pits; while the little Fairy Martin adheres its hard clay nest to the sides of our dwellings, as if to court our protection and care; and the delicate little bird which forms the subject of the present plate is directed by the impulse of nature to nidify in places remarkably different from either of these, viz., steep and precipitous sand-banks, pits of chalk, &c., more particularly the former, appearing to give a preference to banks overhanging water, though we have occasionally observed their breeding-places far remote from any water. In the most inaccessible parts of these situations this little excavator digs a horizontal hole of considerable depth, sometimes even to three or four feet, at the far end of which it places a nest, loosely constructed of dried grass and feathers, in which are deposited four or five delicate eggs, of a clear white. As soon as the young gain sufficient strength, they reach the edge of their subterraneous passage, and at an early period wing their way after their parents, who soon teach them to capture insects for themselves and become independent of their assistance.

The task of reproduction being performed, the Sand Martin congregates in flocks, which are greatly augmented by the end of autumn, at which period the multitude assembled almost surpasses belief. The naturalist cannot look upon the vast herd of these little birds which collect to roost in the osier-beds on the banks of the Thames before their final departure, without admiring the design of an all-wise Creator in protecting these humble creatures during the period above alluded to, as by their means alone are kept in subjection the vast myriads of insects, which would otherwise become so numerous as to defy the power of man, with all his ingenuity, either to annihilate, or to work any apparent diminution of their numbers.

We have often observed the Sand Martin as early as the month of March, from which circumstance it must be ranked among the earliest of our summer visitors. There are few birds that have a more extensive range, being common throughout Europe, the continent of Africa, some portions of India, and, if we mistake not, it is spread over a great portion of the American continent.

It feeds exclusively on flies, which it captures on the wing, being endowed with powers of flight equal to any other species of the genus.

The sexes are alike in plumage, and the young of the year resemble the adults, except that the feathers of the upper parts are edged with a lighter margin.

The top of the head, upper surface, the body, and tail dark brown; throat, belly, and under tail-coverts white; primaries, bill, and feet blackish brown.

The young of the year have the same colouring as the adult, except that each feather on the upper surface is bordered with yellowish white.

We have figured an adult, and a young bird of the year, of the natural size.

BEE EATER.

Merops Apiaster, *(Linn)*.

Drawn from Life & on Stone by J & E. Gould.

Printed by C. Hullmandel.

Genus MEROPS.

GEN. CHAR. *Beak* elongated, pointed, quadrangular, slightly arched, the ridge elevated. *Nostrils* basal, lateral, ovoid, open, slightly covered by hair directed forward. *Tarsi* short. *Toes* three before and one behind; the second toe united to the middle one as far as the second articulation, and the middle toe to the inner one as far as the first articulation; the hind toe large at its base, with a very small claw; the second *quill-feather* the longest, the first very short.

BEE-EATER.

Merops apiaster, *Linn.*

Le Guepier vulgaire.

THE station which the species of this genus occupy appears to be intermediate between the Kingsfishers on the one hand, and the Swallows on the other: to the former they are allied by their elongated form of beak, shortness of tarsi, brilliancy of plumage, similarity of places of nidification, and the white colour of the eggs; and to the latter by their gregarious habits, their lengthened wing, their great and continuous powers of flight, and their manner of capturing while on wing the insects which constitute their food.

The present beautiful species is we believe the only one of its genus which Europe can claim as its own. It is a bird of migratory habits, visiting, in the greatest abundance, the warmer portions of the Continent, especially Italy, Spain, Sicily, the Archipelago, and Turkey; and, not unfrequently, France, Germany and Switzerland, straying at uncertain intervals across the Channel to the shores of England, sometimes singly, and sometimes in small flocks of eight, ten, or even twenty, but never remaining with us or attempting to breed, our climate being in all respects uncongenial to its habits. Montague informs us that it is nowhere so plentiful as in the southern parts of Russia, particularly about the rivers Don and Volga.

In the situation it chooses for a place of nidification, it greatly resembles our Sand-Martins, preferring precipitous sand-banks and the edges of rivers, in which it scoops out deep holes, generally in an oblique direction.

The eggs are from five to seven in number, of a pure white; but whether deposited on the bare ground or in a nest we are not able to say with any degree of certainty, as it is a point on which different authors hold contradictory opinions.

In its manners it very much resembles the Swallow tribec, ontinuing like them for a length of time on the wing, and traversing backwards and forwards in pursuit of its food, which consists of flies, gnats, and small coleoptera, as well as bees and wasps, to which it is peculiarly partial, and from which it derives its name. Although its flight is, as above stated, like that of the Swallow, still we are informed its allied species in India are frequently in the habit of taking their food like the Flycatchers, whose manners they closely imitate, sitting motionless on a branch, darting at the insects as they pass, and returning again to their station. We have some grounds for suspecting that this peculiarity obtains more or less with all the species of this genus.

The sexes of the Bee-eater in general offer no material differences of plumage, except perhaps that the colours of the female are rather more obscure,—a circumstance which, as in the Kingsfisher, extends also to the young, the adult colouring of the plumage being assumed at an early age.

The beak is black, and one inch and three quarters in length; the irides red; the forehead yellowish white merging into blueish green; the occiput, back of the neck, and upper part of the back rich chestnut, fading off on the rump into a brownish amber; from the base of the beak proceeds a black mark which passes beneath the eye, and spreads over the coverts of the ear; the wings, except a large middle stripe of brown, are greenish, with something of an olive tinge; the quill-feathers inclining to blue and ending in black; the tail greenish; the chin and throat bright yellow, bounded by a black line which ascending reaches the ear-coverts; the breast and whole of the under surface blue, intermingled with reflections of green; the first quill-feather rudimentary, the second the longest.

The tail is square with the exception of the two middle feathers, which are an inch longer than the others; feet and tarsi reddish brown. Total length from ten to eleven inches.

We have represented an adult male in its finest state of plumage.

ROLLER.

Coracias garrula: *(Linn.)*

Drawn from Life and on Stone by J. & E. Gould.

Printed by C. Hullmandel.

Genus CORACIAS.

Gen. Char. *Beak* moderate, compressed, higher than broad, straight, cutting; *upper mandible* curved towards the point. *Nostrils* basal, lateral, linear, pierced diagonally, partly closed by a membrane furnished with feathers. *Feet, tarsus* shorter than the middle toe, three toes before and one behind, entirely divided. *Wings* long, first primary a little shorter than the second which is the longest.

ROLLER.

Coracias garrulus, *Linn.*

Le Rollier vulgaire.

The Roller is one of the most beautiful of birds, and although the extraordinary brilliancy of its plumage, varying in an assemblage of the finest shades of blue and green, as well as its great rarity, might render its claim to a place in our Fauna doubtful in the minds of some of our readers to whom the species is but little known, so many instances of its occurrence in this country will be found recorded by various authors, that it is our pleasant duty to consider it British as well as European.

This handsome bird is said to be common in the oak forests of Germany, and also in many of those of Denmark and Sweden. It is less plentiful in France, and according to M. Temminck is never seen in Holland. Its capture in this country has generally happened along the extended line of our eastern coast, from Norfolk northwards. Frequenting large woods generally, it builds in the holes of decayed trees, and lays from four to seven eggs of a smooth and shining white, in form a short oval almost round, very like those of our Kingfisher, but much larger.

The whole length of this bird is about twelve inches; the bill is black towards the point, becoming brown at the base, with a few bristles; irides of two circles, yellow and brown; head, neck, breast and belly various shades of verditer blue, changing to pale green; shoulders azure blue; back reddish brown; rump purple; wing primaries dark blueish black, edged lighter; tail-feathers pale greenish blue, the outer ones tipped with black, those in the middle are also much darker in colour; legs reddish brown. In old males the outer tail-feathers are somewhat elongated. Adult females differ but little from the males, but young birds do not attain their brilliant colours till the second year.

Their food consists of worms, snails and insects generally, and in their habits these birds are remarkably noisy and restless.

Our figure is of the natural size, and was taken from a fine adult male.

KINGFISHER.
Alcedo ispida, *(Linn)*.

Drawn from Life and on Stone by J & E Gould.

Printed by C Hullmandel.

Genus ALCEDO.

Gen. Char. *Beak* long, straight, quadrangular, and acute. *Nostrils* placed at the base of the beak, oblique, and nearly closed by a naked membrane. *Feet* small; *tarsi* short, naked. *Toes* three before, of which the external toe is united to the middle one as far as the second articulation, and the middle toe to the inner one as far as the first; *hind toe* large at its junction with the tarsus. Third *quill-feather* the longest.

KINGSFISHER.

Alcedo ispida, *Linn.*

Martin pêcheur.

When we behold the brilliant colours of this bird as it darts by us like a meteor, displaying the metallic lustre of its plumage, we cannot help fancying for the moment that we behold some erratic native of a tropical clime.

The appetite of the Kingsfisher is voracious, and his manners shy and retiring: dwelling near lonely and sequestered brooks and rivers, he sits for hours together motionless and solitary on some branch overhanging the stream, patiently watching the motions of the smaller fishes which constitute his food; waiting for a favourable moment to dart with the velocity of an arrow upon the first that is near enough the surface or within the reach of his aim, seldom failing in the attempt. He then returns to his former station on some large stone or branch, where he commences the destruction of his captive, which is effected by shifting its position in his bill, so as to grasp it firmly near the tail, and then striking its head smartly against the object on which he rests: he now reverses its position, and swallows it head foremost; the indigestible parts are afterwards ejected in a manner analogous to that of the Owls and birds of prey.

The Kingsfisher, however, does not confine himself entirely to this mode of watching in motionless solitude; but should the stream be broad, or no favourable station for espionage present itself, he may be seen poising himself over it at an altitude of ten or fifteen feet, scrutinizing the element below for his food,—plunging upon it with a velocity which often carries him considerably below the surface. For these habits his muscular wedge-shaped body, increasing gradually from a long pointed bill, aided by the sleek metal-like surface of the plumage, which at the same time freely passes through and throws off the water, seems expressly to adapt him.

The wing of the Kingsfisher is short but powerful; hence its flight is smooth, even, and exceedingly rapid.

Silent except during the pairing and breeding season, (when he occasionally utters a sharp piercing cry, indicative perhaps of attachment,) and equally solitary and unsocial in his habits, the Kingsfisher dwells alone, seldom consorting with others, or even with his mate, except in the period of incubation and during the rearing of the young, when their joint labours are necessary, and both unite with great assiduity in the office of procuring the requisite supplies of food. The places selected for this purpose are steep and secluded banks overhanging ponds or rivers, where in a hole, generally at a considerable distance above the surface of the water, and extending to the depth of two or three feet into the bank, the female, without making a nest, lays five or six eggs of a beautiful pinky white. As soon as the young are hatched, the parent birds may be seen incessantly passing to and from the hole with food, the ejected exuviæ of which in a short time form around the unfledged brood a putrid and offensive mass.

The young do not leave the hole until fully fledged and capable of flight; when, seated on some neighbouring branch, they may be known by their clamorous twittering, greeting their parents as they pass, from whom they impatiently expect their supplies. In a short time, however, they commence plundering for themselves, assuming at that early age nearly the adult plumage. The *Alcedo ispida* is the only species of the genus found in Europe, the western parts of which, including the British Isles, seem to be its proper habitat. The young appear to possess habits of partial migration, at least in our British Islands, wandering from the interior parts along the courses of rivers to the coast, frequenting, in the autumnal and winter months, the mouths of small rivulets and dykes near the sea; but more particularly those along the line of the southern coast and the shores of adjacent inlets.

The annexed Plate represents a male, between which and the female there is no distinguishing difference of plumage. The bill is black; irides dark; the crown of the head, cheeks, and wing-coverts, of a deep shining green, each feather tipt with a lighter metallic hue; the rest of the upper surface, a brilliant azure; the ear-feathers rufous, behind which a white spot extends to the nape of the neck. The throat white, the under surface fine rufous; the legs bright orange.

Length seven inches; weight from two ounces to two and a half.

BLACK AND WHITE KINGFISHER.
Alcedo rudis; *(Linn.)*

BLACK AND WHITE KINGSFISHER.

Alcedo rudis, *Linn.*

Le Martin Pêcheur Pie.

ALTHOUGH the continents of Africa and Asia constitute the natural habitat of this species, still from its occurrence in some of the islands of the Grecian Archipelago, it is necessarily added to the Fauna of Europe, although it is but a rare and accidental visitor. It is abundantly dispersed over the whole of Africa, particularly on the banks of the Nile and the other rivers of Egypt, as also those of Syria and the adjacent countries. The discovery of this Kingsfisher in a quarter of the globe where until lately only a single species had been found, renders it a bird of great interest, and we regret that we cannot lay before our readers any authentic information relative to its peculiar habits and manners. In the third part of his "Manuel" M. Temminck merely states that it feeds on fishes, and lays white eggs. In the form of its bill and in its general structure, it so nearly resembles the *Alcedo ispida*, that although differing from it in size and in the character of its plumage, we may reasonably suppose its general economy to be in strict accordance with that species.

As is the case with most species of its tribe, the plumage of the sexes is very similar; the female and young, however, appear to have the chest crossed with a somewhat narrow and single band of white, while the male has two bands, the upper one of which becomes very broad towards the shoulders, and gradually diminishes towards the middle of the chest; in other respects they are so similar that one description will be sufficient.

Crown of the head and occiput black; the whole of the upper surface varied with numerous bars of black and white, the latter colour terminating and bordering all the feathers; a black band extends from the angle of the beak and spreads over the ears; all the under parts white except the bands of black, which extend across the chest; primaries black; tail white at the base, the remainder barred with black and white; bill black; feet reddish brown.

The Plate represents a male and female of the natural size.

1. PIED FLYCATCHER.
Muscicapa luctuosa; *(Temm.)*

2. WHITE-COLLARED FLYCATCHER.
Muscicapa albicollis *(Temm.)*

Drawn from Nature & on Stone by J & E Gould.

Printed by C. Hullmandel.

PIED FLYCATCHER.

Muscicapa luctuosa, *Temm.*

Le Gobe-mouche becfigue.

In the British Isles this interesting little bird is exceedingly local in the districts it chooses for its periodical visits. Arriving, on the return of spring, from the more congenial and warmer portions of the Old Continent, it takes up its abode, not, as might be expected, in the southern parts of our island, but in the northern and midland counties, especially Lancashire, Yorkshire and Derbyshire, finding probably either food or some other inducement, of which we have no knowledge, that is suited to its wants. In France and Germany it is far from being scarce; but its most favourite tract is along the European shores of the Mediterranean, and over the whole of Italy. In Holland, and, we believe, in Denmark, Sweden and Russia, it is never seen.

The Pied Flycatcher is a most active and unwearied pursuer of the insect tribes, being continually in motion, darting at them as they pass, or searching after them among leaves and flowers. Whether it be for the sake of the fruit itself, or for the insects which abound near the figs, certain it is, that from the circumstance of the bird being constantly found in the neighbourhood of fig trees, it has obtained the name of *Beccafico*. It constructs a nest in the holes of trees, and lays from four to six eggs, of a uniform pale blue colour.

WHITE-COLLARED FLYCATCHER.

Muscicapa albicollis, *Temm.*

Le Gobe-mouche à collier.

Although we have seen this species in a collection of British birds, and were informed that it was supposed to have been killed in England, still we have every reason to believe that it seldom advances so far north as England, or even the adjacent provinces of France or Germany; it is in fact more strictly confined to the central portion and Asiatic confines of Europe.

As regards the distinguishing characteristics of these two closely allied species, we may observe, that the adults in the plumage of summer may be easily distinguished from each other by the absence in the former of the entire collar round the neck; but to render the description of both these species more complete, we beg leave to translate the following passage from the valuable *Manuel d'Ornithologie* of M. Temminck: "The female of the two species, the males in their winter clothing, and the young, all resemble each other so closely as frequently to deceive. They may be easily distinguished, except in the first species, by the little white speculum which occupies the centre of the wing in *M. albicollis*, whilst the wing is of one colour in *M. luctuosa*; secondly, by the lateral feathers of the tail, of which the two external have the edge whitish, more or less spread, according to age, in *M. albicollis*, whilst in the *M. luctuosa* the three lateral tail-feathers are slightly bordered with white edges. The manner of living, the note or call, and the song of the male offer very marked differences: the eggs also differ in colour. It is to M. Lotinguer that we are indebted for the knowledge of the double moult which annually takes place in these birds."

In winter, according to the above-quoted author, both these species lose the black plumage of summer, and assume a uniform brown livery over all the upper parts, at which season the young, female and males, resemble each other: on the return of spring they moult again, the males assuming their black livery.

In the month of October, and during the greater part of the winter, both these species collect in countless flocks in Italy, where they are known by the common term of *Beccafici*, and at which time they are taken in immense numbers for the table, being considered an especial luxury.

The eggs of *M. albicollis* are greenish blue spotted with brown.

The Plate represents a male and female of *M. luctuosa*, and a male of *M. albicollis*, of the natural size, in the plumage of summer.

RED-BREASTED FLY-CATCHER.

Muscicapa parva; *(Bechst.)*

Drawn from Nature & on stone by J & E. Gould. *Printed by C. Hullmandel.*

RED-BREASTED FLYCATCHER.

Muscicapa parva, *Bechst.*

Le Gobe-mouche rougeatre.

So rare is this species of Flycatcher in the collections of Europe, and so little has been recorded of its history, that we are led to hope that the accompanying Plate illustrating the old and young bird, and the present notice of its habits and manners, will prove a trifling addition to our knowledge of European ornithology. During a recent visit to Vienna we had opportunities of observing it in a living state, both in its immature and adult plumage. Its actions and manners are strikingly peculiar, and appear to partake of those appertaining to the species of more than one genus; it resembles the Robin not only in the colour of its plumage but in several of its actions, being sprightly and animated, constantly jerking its tail and depressing its head in the manner our Redbreast is observed to do; it also imitates the action of the Whinchat in the depressed oscillating movement of the tail: thus it appears to form an intermediate link between the *Muscicapidæ* on the one hand, and the *Saxicolinæ* on the other. In the comparative length and robust form of its legs this intermediate station is also further evinced; for though the tarsi have not the strength which we see in the true *Saxicolæ*, still they are more developed than in the genuine Flycatchers. It is a bird of migratory habits, and in Europe its habitat appears to be limited almost exclusively to the eastern portions of the continent. It is tolerably abundant in the neighbourhood of Vienna, and is known to breed annually in the woods of that district. From the circumstance of our having seen it in collections from the East Indies, particularly from that portion adjacent to Persia, it is doubtless widely diffused over the intermediate regions.

The sexes are alike in their colouring, but the female is less brilliant than the male. The upper figure in our Plate represents an adult male, and the lower one that of the young bird of the year in its second plumage, the first having been spotted like that of the Robin. M. Temminck states that the moult is simple, but that the colours of the plumage, particularly on the under surface, change periodically. Like the *Muscicapidæ* in general, the Red-breasted Flycatcher is quick and active, taking its prey on the wing with great dexterity. Its food consists of soft-winged insects, to which in all probability berries are occasionally added.

The nest, according to M. Temminck, is placed among the interwoven twigs of trees or in the forks of the branches, but of the number or colour of its eggs no information has yet reached us.

The male has the whole of the upper surface brown; the four middle tail-feathers and the extremities of the outer ones blackish brown; the base of the latter being white; the throat and breast of a bright rufous; the under surface white tinged with rufous brown on the flanks; the beak, legs, and irides brown.

The young have the breast, which is so richly tinted in the adult, white with a slight tinge of yellow.

The figures are of the natural size.

SPOTTED FLYCATCHER.
Muscicapa grisola, *(Linn.)*

Drawn from Nature & on stone by J & E Gould. *Printed by C Hullmandel.*

SPOTTED FLYCATCHER.

Muscicapa grisola, *Linn.*

Le Gobe-mouche gris.

This species, like its congener the Pied Flycatcher, is one of the summer visitants that enliven our woods and gardens during this most pleasant season of the year. It is one of the latest of the spring birds, scarcely if ever arriving before the middle of May; but soon after this period it may be found throughout the whole of England and a portion of Scotland, wherever there exists a locality suitable to its economy: after remaining here during the summer, it migrates to more southern and congenial climates during September and October.

In its universal distribution the Spotted Flycatcher differs very considerably from the Pied Flycatcher, which is very local in its habitat; it is also less confined to large woods and plantations, and appears to give a preference to gardens, shrubberies, and orchards. It does not evince the least fear or timidity, but frequently constructs its nest and rears its young over the door of the cottager, or upon the branches of fruit-trees nailed against the walls, sometimes in the decayed holes of trees, and frequently upon the ends of the beams or rafters in the gardener's tool-house and other outbuildings.

The nest is constructed of moss and small twigs, lined with hair and feathers: the eggs are four or five in number, of a greyish white spotted with pale reddish brown. When the young quit the nest, they follow their parents to some neighbouring wood, garden, or plantation, where they are very diligently attended and fed.

The Spotted Flycatcher appears to enjoy a wide range over the continent of Europe, being very generally dispersed from the border of the Arctic Circle to its most southern boundary; and we have also frequently observed it among collections from India.

It is a most active little bird, and is incessantly engaged in capturing the smaller winged insects which pass within the range of its chosen territory. Its favourite perch is generally a decayed branch, from which it sallies forth and " returns after each of these aërial attacks."

Its note is weak and monotonous, being little more than a feeble chirp.

The sexes are precisely alike in the colour and markings of their plumage. The young for a short period after they first begin to fly have the feathers tipped with a spot of yellowish white, giving them a mottled appearance.

The whole of the upper surface is brown, the crown of the head being spotted with a darker brown; throat and belly white; sides of the neck, breast, and flanks streaked with brown; bill and legs dark brown.

We have figured an adult male of the natural size.

GREAT SHRIKE.

Lanius excubitor. *(Linn.)*

Collurio ———— *(Vig.)*

Drawn from life and on stone by J. & E. Gould.

Printed by C. Hullmandel.

Genus COLLURIO, *Vigors.*

GEN. CHAR. *Beak* and *feet* as in the genus *Lanius.* *Wings* somewhat rounded, short; the first *quill-feather* shortest; the second a little shorter than the following ones; the third, fourth and fifth, the longest and nearly equal. *Tail* elongated, graduated.—Type of the genus *Lanius Excubitor*, Linn.

GREAT SHRIKE.

Lanius Excubitor, *Linn.*

Collurio Excubitor, *Vigors.*

La Pie-grieche grise.

THE present species forms the type of the genus *Collurio* as characterized by Mr. Vigors; and the reasons which induced that gentleman to separate certain birds from the genus *Lanius*—of which to form the genus *Collurio,*—were a short time ago explained by him to the Committee of Science and Correspondence of the Zoological Society, in the First Part of whose "Proceedings," page 42, the distinguishing characteristics of these two genera are clearly detailed. The principal feature which constitutes the ground of separation consists in the somewhat rounded form of wing, which in the true *Lanii* is more pointed,—and in the lengthened and graduated tail; to this may be added their general superiority of size:—the Grey Shrikes in this point manifest an ascendancy over the well-known Red-backed Shrike, the more rare Wood-chat, and their allied congeners.

The Great Shrike is one of the migratory birds of Great Britain, appearing however by no means regularly, so as to lead us to expect its annual return, but must be considered an uncertain straggler, being in some seasons very scarce, and in others more abundant, and that only during the months of autumn and winter, so as to make it rather doubtful whether it ever breeds in our island, although we are aware that such is said to be the case. It is extensively spread over Continental Europe, in many parts of which it remains stationary throughout the year; but in others it performs regular periodical migrations, departing and returning with the season.

The *C. Excubitor* is a bold and courageous bird, attacking others much larger than itself, and destroying mice, frogs, and small birds for its food, of which however we believe the hard-winged insects constitute a principal portion. In killing its prey, its chief instrument is its bill, which is thick and strong, and with which it penetrates the cranium of any small animal within its power. It never strikes with its claws in the manner of the Hawks, but uses them merely to assist in grasping and thus securing its victim; for though the legs and toes are slender and apparently weak, they are well armed with claws and have the power of tight compression. The most singular fact, however, respecting its mode of feeding or securing its prey, is its well-authenticated habit of fixing it on a thorn or sharp-pointed stick, which it selects for that purpose, and then proceeding to tear it to pieces, at the same time satisfying its appetite. Of a New Holland bird, (the *Vanga Destructor,*) the same singular habit is recorded.

The specific name of *Excubitor*, or Sentinel, was given by Linneus as one highly appropriate, from the circumstance of this bird being used on the Continent by falconers and persons engaged in procuring falcons for the purpose of hawking, especially the Peregrine during the period of its migrations. The Shrike acts as a sort of monitor, giving warning to the man in attendance of the approach of the sharp-eyed bird of prey, the appearance of which, even at a great distance, immediately elicits its querulous chattering cry. A net trap, artfully contrived, is placed for the capture of the falcon, and a live pigeon secured by a string, over which the man has perfect controul, allures the falcon to the fatal engine. In the mean time the Shrike, having warned the man of the near approach of the expected visitant, retires, as the danger increases, to a hole provided for its safety; there it continues chattering loudly, while the falcon pounces upon the fated pigeon, and is cunningly enticed by the gradual withdrawing of his victim, which he will not quit, within the circle of the net: the check-string once pulled, the capture is achieved. For its peculiar aid in this service, the Shrike is unrivalled, and when hawking was in vogue its merits were duly appreciated.

The favourite resorts of the Great Shrike are high hedges, coppices, and thick trees, among which it breeds, building a nest composed of grasses, moss, and vegetable fibres, and laying from five to seven white eggs marked with ash-grey and brownish blotches. The only difference to be observed in the plumage of the sexes, is that the breast of the female is transversely barred with faint lines of ash colour.

In the adult male the head, neck and back, are of a fine light ash; a band of black passes below the eye, and covers the ear-feathers; wings black, with a white spot in the centre, formed by the white bases of the quill-feathers; the outer tail-feathers white; the remainder black, terminating with white, which becomes more contracted as they approach the two middle feathers,—these are entirely black; beak and feet black. Length nine inches.

We have figured an adult male.

GREAT GREY SHRIKE.

Lanius meridionalis, *Temm.*
Collurio ————— *Vig.*

Drawn from Life & on Stone by J & E Gould.

Printed by C. Hullmandel.

GREAT GREY SHRIKE.

Lanius meridionalis, *Temm.*

Collurio meridionalis, *Vig.*

La Pie-Grieche méridionale.

WE are indebted to the kindness and liberality of the officers of the British Museum, for allowing us to illustrate this fine species of Shrike from a specimen in the collection under their care, to which it was presented by Captain S. E. Cook, who obtained it in the centre of Spain; a country which, with Italy, the southern provinces of France and the districts bordering on the Mediterranean, appears to be its native habitat. Of the European Shrikes, the *Lanius meridionalis* is that which comes least under our notice; it is, however, the largest of its genus, and may be taken as typical of that group comprehended by Mr. Vigors under the generic title *Collurio*,—a group on which we offered some remarks when speaking of that rare British species the *Lanius Excubitor* of Linnæus, with which the present bird closely agrees, but from which it may be distinguished by its somewhat greater size, the darker grey of the upper surface, and especially by the beautiful vinous tinge which prevades the plumage of the breast. In the latter respect, indeed, it resembles the *Lanius minor*; but with this bird it can never be confounded, the disparity of size being at once a distinguishing character between them, to which we may add the long and pointed wings and less graduated tail in *Lanius minor*, as opposed to the short wings and cuneiform tail of the *Lanius meridionalis.*

The habits and manners of this interesting and natural group, ally the species together as closely as their forms and colours; and although we have been able to obtain no precise information respecting our present subject, still we cannot for a moment doubt that it exhibits the same properties and acts the same tyrannical part in its mode of obtaining food as its British congener, preying upon the larger insects, especially those of the Coleopterous order, young or feeble birds, and even small mammalia. The nidification and eggs are unknown.

The top of the head and the whole of the plumage of the back are of a dark ash colour; a large black band passes below the eyes and extends over the ear-coverts; the wings are black, having the origins of the quill-feathers and the tips of the secondaries white; the four middle tail-feathers are black, the exterior ones white; the breast and under part vinous or salmon-colour, fading into grey about the sides and thighs; bill and legs black.

The female differs only in having the tints of the plumage more obscure, with faint transverse bars on the under parts.

We have figured a male in full plumage.

LESSER GREY SHRIKE.

Lanius minor. *(Linn.)*

Collurio ——. *(Vig.)*

Drawn from Life & on Stone by J. & E. Gould. *Printed by C. Hullmandel.*

LESSER GREY SHRIKE.

Lanius minor, *Linn.*

Collurio minor, *Vig.*

La Pie-Grieche à Poitrine rose.

The Lesser Grey Shrike, or Rose-breasted Shrike of Temminck, though belonging to that division of the family to which Mr. Vigors has given the generic title of *Collurio*, (and which is distinguished from the restricted genus *Lanius* by the graduated tail and short rounded wing,) must nevertheless be regarded as forming a link between these two genera; inasmuch as the tail is rather rounded than decidedly graduated, and the wings are more lengthened than in any other species of the genus to which it is now assigned.

Of the European Shrikes, the present is certainly one of the most distinguished for the beauty and delicacy of its colouring: in size, it is inferior to the *Lanius Excubitor* of Linnæus, but possesses a more strong and robust bill than is found in that bird,—to which, however, in manners, and in the general character of its plumage, it closely approximates.

The *Lanius minor* is strictly Continental; no instance, so far as we are aware, being on record of its ever having visited our Island. From M. Temminck, to whose acquaintance with the birds of Europe the scientific world is so much indebted, we learn that its range on the Continent is very extensive, inhabiting, with the exception of Holland (in which it is rarely seen), Turkey, the Archipelago, Italy and Spain, breeding also in some parts of France and Germany, and visiting the northern portions of Europe as far as Russia. Thickets, trees, bushes, and hedge-rows are its favourite resort, among which it constructs its nest: the eggs, six in number, are oblong and of a dull green, having a zone of small spots round the centre, of an olive-grey.

Its food, like that of the other species of the genus, consists principally of insects, such as moths, and coleoptera, to which young or feeble birds are occasionally added.

The beak and legs are black; a black band passes over the forehead, eyes and ears; the top of the head, back and rump of a fine ash; throat white; breast and sides of a delicate rose-colour; wings black, with a white bar across the quill-feathers; the middle tail-feathers black, the two outer feathers quite white, the two next partially white, the succeeding feathers on each side less and less so.

The female differs as little from the male as those of the allied species *Excubitor*, and is only to be distinguished by the more obscure rose-colour of the breast, and the black of the plumage having a tinge of brown.

The young of the year of both sexes after the autumn moult are destitute of the black band on the forehead, which remains grey during the winter, and the plumage of the superior surface is more or less broken with grey, the under parts being obscure; but after the moult of spring, both sexes gain the band over the forehead and ears, the rose-colour of the breast becoming at the same time more lively. Total length eight inches.

Our figure represents an adult male of the natural size.

RED-BACKED SHRIKE.

Lanius collurio; *(Linn.)*

Drawn from Nature & on Stone by J. & E. Gould.

Printed by C. Hullmandel.

Genus LANIUS.

Gen. Char. *Beak* of moderate length, robust, compressed, straight at the base, curved at the tip; edges of the upper mandible emarginated, and exhibiting a conspicuous tooth. *Nostrils* basal, lateral, nearly round, partly covered by a membrane. *Gape* furnished with stiff bristles. *Feet* moderate. *Toes* free. *Acro-tarsia* broadly scutellate. *Wings* somewhat pointed and rather short; the first quill-feather very short; the third the longest; the rest gradually decreasing. *Tail* equal, or somewhat rounded.

RED-BACKED SHRIKE.

Lanius Collurio, *Linn.*

Le Pie-griéche écorcheur.

Among the Shrikes which periodically visit our island, the Red-backed is the best known and most universally spread. Its arrival usually occurs from the middle of April to the beginning of May, the exact period being regulated by the forwardness of the spring, inasmuch as its food consists almost solely of insects, the appearance of which depends upon the temperature of the season. Though found occasionally in the northern counties of England, it is by no means so abundant there as in the middle, and more especially the southern districts. In Scotland it is, we believe, altogether unknown, nor are we aware of its having been discovered in Ireland. It is partial to downs and open pastures, particularly such as are intersected or bordered by thick stunted hedges, where it may be commonly met with singly or in pairs, but never in flocks. Like the rest of the Shrikes, its manners and note are very peculiar, and serve at once to distinguish it from the small birds of other groups. Its chief food, as we have before observed, consists of insects, such as grasshoppers, beetles, and the larger kinds of flies, which it often takes on the wing. It may be generally noticed quietly perched in some commanding situation awaiting the approach of its prey, upon which it darts not unlike a Flycatcher, generally returning to the same perch. Besides insects, it is known to attack young and feeble birds, mice, lizards, slugs, &c., which, as is the case with most of its congeners, it impales on a sharp thorn or spike previously to tearing them to pieces.

Though small in size, the Red-backed Shrike is extremely fierce and courageous, defending itself with great obstinacy when wounded or assailed. On the Continent it is widely distributed, being spread throughout every province of Europe, from the south as far as Russia and Sweden; and we may add that it is also a native of the North of Africa.

The Red-backed Shrike builds its nest in sharp thorny bushes, often at a considerable distance from the ground, constructing it of dried grasses and wool, with a lining of hair. The eggs are five or six in number, of a pinkish white, with spots of wood-brown disposed in zones chiefly at the larger end.

The sexes offer very considerable difference in their colouring; that of the adult male is as follows:

Top of the head, occiput, upper part of the back, and rump, fine grey; a narrow band of black begins above the beak, passes round the eye, and spreads over the ear-coverts; middle of the back and shoulders fine chestnut; quills brown; two middle tail-feathers black; the rest white for more than half their basal length, the extremity being black tipped with white; throat white; under surface pale roseate; bill and tarsi black.

The female wants the beautiful grey hood and mantle, as also the black streak on the face; the whole of the upper plumage is dull reddish brown, with dusky transverse lines more or less obscure; a deeper tint of brown pervades the quills and tail-feathers; the under surface is white, barred on the sides of the neck, the chest, and flanks, with fine semilunar lines of brown.

The young males of the year closely resemble the adult female; in fact, the only difference consists in the feathers of the back being, in the former, more distinctly margined with transverse lines of brown.

Our Plate represents an adult male, and a young male of the year, of the natural size.

WOODCHAT.

Lanius rufus: *(Briss:)*

Drawn from Nature & on Stone by J & E Gould. *Printed by C. Hullmandel*

WOODCHAT.

Lanius rufus, *Briss.*

La Pie-grièche rousse.

ALTHOUGH the Woodchat is abundantly spread over the Continent, particularly the warmer portions, such as Spain, Italy, and all the countries bordering on the Mediterranean, it is so rare in the British Islands as scarcely to merit a place in our Fauna; indeed, two or three instances of its having been seen or captured in England are all that we are acquainted with, and we have ourselves never received it in a recent state. In general habits and manners it strictly resembles its nearly allied species the Red-backed Shrike, *Lanius Collurio*; but a moment's comparison will be sufficient to establish the distinctions of the two species. In one respect, however, the Woodchat appears to be peculiar; we allude to the circumstance of the nearly allied style of colouring of the two sexes, which is contrary to what prevails in the Red-backed Shrike, and agrees more strictly with the rest of the European representatives of the present family. The only point by which the sexes may be distinguished is the less brilliant colouring of the female.

The food of the Woodchat consists of coleopterous and other large insects, and occasionally small and nestling birds. According to M. Temminck, it builds its nest in bushes, selecting the fork of a small branch among foliage sufficiently dense for its concealment. It lays five or six eggs, of a whitish green, irregularly blotched with grey.

The plumage is as follows: A narrow white band borders the margin of the upper mandible, and is followed by a black belt across the forehead; the same colour occupying the ear-coverts passes down the sides of the neck and the middle of the back, where it fades off into grey; the occiput, back of the neck, and upper part of the back, rich chestnut; wings black, the scapularies and a bar across the base of the quills being white; upper tail-coverts white, with a tinge of yellow; tail black, except the outer feather on each side, which is nearly white, and the next, which is tipped with white, and is white at its base; the third is also tipped with white; beak and tarsi black.

In the female the black is tinged with brown, and the chestnut of the back is less vivid.

The Plate represents a male and female of the size of life.

GOLDEN ORIOLE.

Oriolus Galbula, *(Linn)*.

Drawn from Life & on Stone by J & E. Gould. *Printed by C. Hullmandel.*

Genus ORIOLUS.

Gen. Char. *Beak* conical, straight, and sharp-pointed, flattened at its base; upper *mandible* ridged and slightly notched at the point. *Nostrils* basal, lateral, naked, and pierced horizontally in a large membrane. *Feet* three toes before and one behind, the external toe united to the middle one at its base. *Tarsus* not exceeding the middle toe in length; third *quill-feather* the longest.

GOLDEN ORIOLE.

Oriolus galbula, *Linn.*

Le Loriot.

The genus *Oriolus*, as restricted by modern authors, comprises a group of birds pre-eminent for their beauty and the contrast of their colours, rich yellow and deep black dividing the plumage in proportions varying according to the difference of species; and there is no genus the members of which are more naturally united to each other than the present.

The Orioles are strictly confined to the older or longest known portions of the globe; various species being respectively disposed over Asia, Africa, the islands of the Indian Archipelago, and the southern and eastern portions of Europe. Their place in America appears to be supplied by the *Icteri* and the other genera of the family of *Sturnidæ*.

The Golden Oriole is to be considered as merely an occasional sojourner in England, its visits being but few, and only during the months of summer. Although undoubtedly scarce, it is by no means the most rare of those birds which, from their occurrence in England, have been admitted to a place in its Fauna. We are not aware that there are any instances on record of its breeding in this country: still we have reason to suppose that such would be the case, if, when it favours any part of our island with its residence, it were permitted to remain unmolested. In Italy and the whole of the southern provinces of Europe it is very abundant, and is also far from being uncommon in France, Holland, Germany, and some of the districts of Russia; paying all those countries an annual visit, for the purpose of incubation. We have received numerous specimens of this bird from Tripoli and the whole line of the northern coast of Africa, where it is exceedingly common; and to which country, or at least its more northern parts, we have reason to suppose it migrates when absent during the winter months from Europe. We have never received this identical species from India, although there is one, the *O. aureus*, which closely approximates to it, but which may at once be distinguished by its rather smaller size, and by the black line passing through and beyond the eye, while in the European species the black colour reaches only to the anterior edge of the orbit. We have been informed that this latter species has also been found in Europe, which is not unlikely, as the border line of the two continents is merely conventional.

The nest of the Golden Oriole, like that of many of the natives of the tropical climates, is a striking example of ingenuity: it is composed of fibres of hemp or other vegetables ingeniously interwoven together, lined with fine moss or lichen, and suspended at the extremity of the tallest branches of lofty trees; the eggs are four or five in number, the ground-colour being a pure white, marked by a few well-defined purple brown, or black spots. Its food consists of wild berries and fruits, as well as insects and their larvæ.

In the male, the beak is brownish red; irides red; general plumage fine king's yellow, a black streak intervening between the beak and eye; the wings black with a bar of yellow; all the quill-feathers tipped with yellowish white; the two middle tail-feathers black; the rest have their basal half black, and the other portion yellow; tarsi lead-colour; claws black. Length of the bird ten inches.

In the female, the yellow of the upper parts is clouded with an olive tinge, and below with grayish white, each feather having a longitudinal mark of a darker colour; the wings are brownish black, and the tail-feathers, where black in the male, are obscure olive.

The young of the year resemble the female, but have the longitudinal markings of the lower parts stronger; the irides brown, and beak of a dark gray.

Our Plate represents a male and female of the natural size.

BLACK-OUZEL OR BLACKBIRD.
Merula vulgaris; *(Ray).*

Drawn from Nature & on Stone by J. & E. Gould.

Printed by C. Hullmandel.

Genus MERULA.

Gen. Char. *Bill* nearly as long as the head; straight at the base; slightly bending towards the point, which is rather compressed; the upper mandible emarginated; *gape* furnished with a few bristles. *Nostrils* basal, lateral and oval, partly covered by a naked membrane. *Legs* of mean length, muscular. *Toes,* three before and one behind; the outer toe joined at its base to the middle one, which is shorter than the tarsus. *Claws* slightly arcuate; that of the hind toe the largest. Of the wings, the first quill is short, and the third and fourth are the longest.

BLACK OUZEL, OR BLACKBIRD.

Merula vulgaris, *Ray.*

Le merle noir.

This familiar species is very generally distributed over Europe; and although it is stationary with us throughout the year, yet in some parts of the Continent it is a bird of passage; and we may add, that on the approach of severe winters the number in our own island is greatly augmented by temporary visiters from the North.

When we consider the style of colouring, form, and habits of this bird, with those of the Ring Ouzel and some other European species, we cannot but agree with our countryman Ray in considering them fully entitled to rank as a genus distinct and separate from that of *Turdus,* with which Linnæus and his followers blended them.

In their habits these birds are more terrestrial than the true Thrushes, frequenting secluded copses, hedgerows and ravines, as well as gardens and shrubberies, skulking about under the bushes, and retiring from observation with great celerity. Towards evening they may often be seen extremely restless and clamorous, uttering a shrill chatter as they dart from bush to bush in chase of each other before retiring to rest. It is more solitary in its habits than the Thrush, there seldom being more than two or three in the same immediate locality. There is also a slight degree of difference in the general construction of the nest, the colour of the eggs, and the situations chosen for the purpose of nidification, to which we may add the marked difference in colouring between the male and the female.

The song of the Blackbird, though not so melodious as that of the Thrush, is a clear bold strain, which, when heard in the calm mornings and evenings of spring, is very delightful, and renders the bird a general favourite. In its food the Blackbird is perfectly omnivorous, accommodating itself to such as the season offers: in winter it lives on berries, worms, and shelled snails, which it seeks for under hedgerows and other hiding-places; in summer, on worms, insects and their larvæ, as well as all kinds of fruits that the garden and hedgerow afford.

The Blackbird appears to be strictly confined to Europe, but its form is represented in the Himalaya mountains by an interesting and nearly allied species the *Turdus pœcilopterus,* and in the tropical regions of America by two or three other species.

It is an early breeder, often commencing the work of nidification in the months of February and March, building its nest in thick secluded bushes, laurels, ivy, or any densely leaved covert: the nest is constructed externally of moss, small twigs, and fibres plastered with mud, internally of fine dry grass: the eggs are usually five in number, of a blueish green blotched with reddish brown.

The young of both sexes greatly resemble the female, but are generally more spotted.

The male, with the exception of the bill and the orbits of the eyes, which are of a beautiful orange, is entirely of a jet black.

The female is of a deep umber brown inclining to black; the chest, belly and thighs, varied with dashes of a darker colour.

The Plate represents a male and female of the natural size.

RING OUZEL.
Merula torquata; *(Briss.)*

Drawn from Nature & on Stone by J. & E. Gould. Printed by C. Hullmandel.

RING OUZEL.

Merula torquata, *Briss.*

Le Merle à Plastron blanc.

"The periodical visits of this bird to our coast," says Mr. Selby, "are contrary to the others of the genus that migrate, viz. the Fieldfare, the Redwing, and the Common Thrush, as it arrives in spring, and immediately resorts to the mountainous districts of England and Scotland, preferring those that are the most stony and barren." Although it doubtless always breeds in the situations above described, it may not unfrequently be seen traversing the hedgerows of cultivated lands during its passage to and from distant climates.

In general form and appearance it strictly resembles the Blackbird, but in its manners it is much more shy and distrustful, rarely admitting itself to be approached. Unlike that bird, it is not observed skulking among bushes, &c., but affects more open situations, which doubtless renders it habitually cautious, as being more necessary to its safety. Its voice is somewhat harsh and powerful, consisting of a few notes, which, according to Mr. Selby, are not unlike those of the Missel Thrush. On the Continent it is distributed through most of the northern countries, and is very common in Sweden, France, and Germany; indeed, with the exception of Holland, it is universally distributed throughout Europe, as well as the adjacent parts of Asia and Africa. In all these countries it is said to be migratory; and we may easily conceive the cause of this to be the failure of a supply of food in the peculiar situations it frequents, and the consequent necessity of retiring to a more genial climate, where berries, fruits, and insects may be easily obtained.

The male differs from the female in the greater purity and contrast of his colours. The general plumage is black, each feather having a margin of grey; a broad gorget of pure white extends across the chest; the bill is blackish brown at the tip, and yellow at the base; legs blackish.

The plumage of the female is more clouded with brownish grey, the pectoral gorget being less extensive and tinged with dusky brown.

The young males closely resemble the adult female, but in young females the gorget is scarcely perceptible.

The Plate represents an adult bird of each sex of the natural size.

MIGRATORY OUZEL.
Merula migratoria; *(Swains.)*

Drawn from Nature & on Stone by J & E. Gould.　　　　Printed by C. Hullmandel.

MIGRATORY OUZEL.

Merula migratoria, *Swains.*

Le Merle erratique.

This beautiful species of Thrush, if not the theme of poets, has nevertheless called forth many spirited and flowing descriptions, the most animated of which are to be found in the works of Wilson, Audubon, and Dr. Richardson. From the latter author we find that few of the feathered race seek a more northern region for the purpose of breeding than the Migratory Ouzel. "It arrives in the Missouri (in lat. 41½°), from the eastward, on the 11th of April; and in the course of its northerly movement, reaches Severn River, in Hudson's Bay, about a fortnight later. Its first appearance at Carlton House, in lat. 53°, in the year 1827, was on the 22nd of April. In the same season it reached Fort Chepewyan, in lat. 58¾°, on the 7th of May, and Fort Franklin, in lat 65°, on the 20th of that month. Those that build their nests in the fifty-fourth parallel of latitude, begin to hatch in the end of May; but eleven degrees further to the north, that event is deferred till the 11th of June. The snow even then partially covers the ground; but there is, in those high latitudes, abundance of the berries of the *Vaccinium uliginosum* and *Vitis idæa*, *Arbutus alpina*, *Empetrum nigrum*, and of some other plants, which after having been frozen up all the winter, are exposed, on the first melting of the snow, full of juice and in high flavour. Shortly afterwards, when the callow young require food, the parents obtain abundance of grubs."

When we take into consideration the migratory habits of this bird, and the extreme high northern latitudes it affects, the fact of its occasionally occurring in Europe is not so startling as it would otherwise appear: a single glance at a globe will in fact make it plain to our readers, that when migrating from these high latitudes, a slight deviation from its regular course would carry it on to the continent of Europe, where, as we have before stated, it is occasionally seen. In the third part of his 'Manuel' M. Temminck states that it has been killed in Germany; and M. Brehm informs us that it has been seen in the neighbourhood of Vienna. In its affinities we are inclined to consider this bird as a true *Merula*, or as belonging to that section of the *Merulidæ* which includes the Common Blackbird, and we also find that its habits, manners, song, and nidification are much in accordance with those of that bird. "So much," says M. Audubon, "do certain notes of the Robin (the American name for the *Merula migratoria*) resemble those of the European Blackbird, that frequently while in England the cry of the latter, as it flew hurriedly off from a hedge-row, reminded me of that of the former when similarly surprised, and while in America the Robin of that country has in the same manner recalled to my recollection the Blackbird of England."

The sexes are alike in plumage, but the tints of the female are somewhat paler, and she is also smaller in size.

Head and sides of the face deep sooty black; round the eye a circle of white; all the upper surface fuliginous grey tinged with brown on the shoulders; wings and tail blackish brown externally edged with grey; two outer tail-feathers tipped with white; chin white spotted with brownish black; breast and under surface reddish orange, each feather delicately fringed with grey; vent and under tail-coverts mingled white and grey; bill yellow; irides hazel; feet pale brown.

We have figured an adult of the natural size.

BLACK-THROATED THRUSH.

Turdus atrogularis, *(Temm.)*

Drawn from Nature & on Stone by J. & E. Gould. Printed by C. Hullmandel.

BLACK-THROATED THRUSH.

Turdus atrogularis, *Temm.*

La Merle à gorge noire.

This fine bird is one of the ornithological rarities of the Fauna of Europe: it is, however, a species of common occurrence in the Himalaya mountains, whence we have received numerous examples, differing in no respect from those taken in Europe. Although M. Temminck states that it is a native of Hungary and Russia but rare in Austria and Silesia, we have only seen two native-killed specimens, which are in the collection at Vienna, and one of them was, we believe, killed in the neighbourhood of that city: young birds are also said to have been taken in Germany. From the circumstance of most collections from the Himalaya mountains containing examples of this bird, the fact is clearly established that the northern and higher regions of Asia constitute its native habitat. Our knowledge of this species is so limited that we are unable to state with certainty whether the black gorget is characteristic of the summer plumage, or whether when once acquired it is permanent: we suspect the latter to be the case, as we have received specimens in various stages of plumage, some of which were totally devoid of the black throat, while others had it partially developed; in all probability these last were females or immature birds.

The whole of the upper surface is brown with a slight tinge of red, the outer edges of the wing-feathers being somewhat lighter; throat and chest dull black, each feather being slightly margined with white; belly white; under tail-coverts rufous brown; bill dark at the point and yellow at the base; feet brown.

The Plate represents a male and a young male or the female of the natural size.

FIELDFARE.

Turdus pilaris, *(Linn.)*

Drawn from Nature & on Stone by J & E. Gould

Printed by C. Hullmandel

FIELDFARE.

Turdus pilaris, *Linn.*

Le Merle litorne.

The Fieldfare is only a winter visitant of the British Islands and the temperate parts of Europe, arriving in autumn and departing northwards in the spring, its native habitat being the regions adjacent to the arctic circle, such as Sweden, Lapland, the Northern parts of Russia, Norway, &c., where, according to Mr. Hewitson of Newcastle, it is very abundant. From this gentleman's interesting and very valuable work on the eggs of British Birds, we have made the following extract: "We were soon delighted by the discovery of several of their nests, and were surprised to find them, so contrary to the habits of other species of the genus with which we are acquainted, breeding in society. Their nests were at various heights from the ground, from four feet to thirty or forty feet or upwards, mixed with old nests of the preceding year. They were, for the most part, placed against the trunk of the spruce fir; some were, however, at a considerable distance from it, upon the upper surface and towards the smaller end of the thicker branches. They resemble most nearly those of the Ring Ouzel: the outside is composed of sticks, and coarse grass and weeds gathered wet, matted together with a small quantity of clay, and lined with a thick bed of fine dry grass. The eggs are five and sometimes six in number, very like those of the Blackbird and Ring Ouzel. The Fieldfare is the most abundant bird in Norway, building as above described in society, two hundred nests or more being frequently found within a very small space."

After the breeding season is past, and when the severities of winter set in over those northern regions, vast flocks congregate together and pass gradually southwards till they find a locality affording the necessary means of subsistence; hence, in our locality they spread themselves over fields and pasture lands in search of worms, grubs, and insects, retreating to thick hedges, where various berries supply them food, when the snow precludes their other means of support. Unlike the Song Thrush they are shy and wary, not allowing themselves to be approached, but taking wing and wheeling off in a body to some distant spot. This shyness of disposition, together with the harshness of their note, assimilates them strongly to the Missel Thrush, which in fact they closely resemble, except in their gregarious habits.

The sexual differences in the Fieldfare are so trifling as to be scarcely perceptible; indeed, it requires anatomical examination to ascertain the distinction.

The Fieldfare generally leaves us in March or April, and, as far as we know, there are no instances of their having bred in our island. Their powers of song are very moderate, and their common call note very like that of the Shrikes.

Their flesh is by many held in considerable esteem, and hence they are often eagerly pursued by the gunner, a circumstance which, if we mistake not, conduces much to their timid and suspicious habits.

In size, the Fieldfare is next to the Missel Thrush, but possesses a style of colouring peculiar to itself. The head, lower part of the neck and rump cinereous grey; the top of the back and wing-coverts chestnut brown; space between the beak and the eye black; a greyish white streak passes above the eyes; the throat and breast light rufous brown, with lanceolate black spots; the feathers of the flanks are blotched with black and bordered with white; the abdomen pure white; tail black, the outer feathers being inclined to grey; bill bright orange, with a black tip; tarsi black.

The Plate represents an adult bird of the natural size.

MISSEL THRUSH.

Turdus viscivorus. *(Linn).*

Drawn from Life and on Stone by J & E. Gould.

Printed by C. Hullmandel.

Genus TURDUS. *Auct.*

Gen. Char. *Beak* moderate, emarginated, compressed and arched at its point. *Upper mandible* slightly notched. *Gape* furnished with a few bristles. *Nostrils* basal, lateral, oval, partly closed by a naked membrane. *External toe* joined at its base to the middle one, which is shorter than the tarsus. First *quill-feather* very short; third or fourth the longest.

MISSEL THRUSH.

Turdus viscivorus, *Linn.*

La Draine.

This bird has derived its appellation from the alleged circumstance of its feeding upon the berries of the Misseltoe; and we place it at the head of the genus as being the largest, and exhibiting the generic characters in the greatest perfection. Of all the Thrushes, the present species is the most extensively spread over the older continent, being not only found in Europe, but also in that altitude of the Himalaya Mountains and the high lands of Asia which afford a temperature similar to our own. Thinly dispersed over the British Isles, the Missel Thrush is a solitary and unsocial bird, differing considerably in its habits from the common favourite, which delights to dwell within the cultivated precincts of our shrubberies and gardens. Affecting remote situations, it retires from the haunts of human society to pasture lands, wide commons or meadows skirted by orchards or groves, feeding, like its generic companions, on snails, worms and the larvæ of insects, during the months of spring and summer, but resorts to berries, especially those of the mountain ash, the haw, and, according to authors, the misseltoe, when autumn and winter deprive it of more esteemed fare. It is one of our earliest breeders, the commencement of March being the season of incubation; the place of nidification being sometimes orchard trees, at others those of more lofty growth, such as the elm or oak; and the nest, with a view to its concealment, is artfully placed either close against the stem or in a fork of one of the larger branches, being composed on the outside of coarse lichen, gray moss, or such dried vegetables as are found on the spot and accord with the colour of the tree: the materials are carelessly interwoven. Within this outside covering is a layer of mud neatly lined with fine grasses; the female laying five eggs of a pale blueish white spotted with dull red.

As this bird is one of our earliest breeders, so also may the note of the male be heard the earliest in the spring, while, perched at the top of some tall tree, he serenades his mate with loud discordant sounds, which consist of monotonous unpleasant notes, repeated by the hour together. The Missel Thrush is very pugnacious during the breeding season, attacking all birds indiscriminately should they intrude within a certain distance of his nest. After this period we have seen the Missel Thrush collected in small companies, but never in such congregated numbers as the migratory species of the genus.

The male and female differ from each other so little in size and colour, that one description will serve for both: the young, however, have their feathers edged with a darker colour, which soon disappears, when the plumage assumes the adult colouring.

The weight of the Missel Thrush is near five ounces; its total length eleven inches.

The whole of the upper surface is of an ashy brown; between the beak and eye the feathers are grayish white. The under-surface is white, more or less tinged with yellowish red varied with barb-shaped brown spots, which become more oval on the lower parts; wing-coverts edged and tipped with white; the three outer tail-feathers ending in a lighter colour; beak and legs yellowish white, the former the darkest.

The annexed Plate represents an adult male in its spring plumage.

SONG THRUSH.
1. Turdus musicus; *(Linn.)*

REDWING.
2. Turdus Iliacus; *(Linn.)*

Drawn from Life & on Stone by J. & E. Gould.

Printed by C. Hullmandel.

SONG THRUSH.

Turdus musicus, *Linn.*

La Merle grive.

This universal favourite appears to inhabit every country in Europe, which may be considered its true habitat. It may be taken as a typical example of the true Thrushes, which, as a tribe, are numerously dispersed over a great portion of the globe: the temperate countries of America afford us several examples of this particular form; both Asia and Africa as well as the Indian Islands having also their melodious Thrush. It is very generally dispersed over the wooded districts of the British Islands, and is particularly partial to shrubberies and thick hedge-rows. It is by no means fearful or suspicious of man, confidently venturing within the precincts ot gardens and orchards, where its bold, varied and energetic song secures it good-will and protection. It builds its nest early in the spring a few feet from the ground, in any tree or shrub within the immediate vicinity of its haunt. Its nest is outwardly constructed of coarse moss intermingled with dead leaves and grasses; the inside is neatly covered with a composition of cow-dung, light vegetable mould, and clay; and without any other lining the female deposits her eggs, which are usually four or five in number, of a beautiful blue colour spotted with black.

The habits of the Song Thrush differ materially from those of the Redwing and Fieldfare. It is not gregarious; and although numbers annually arrive here from more northern countries, they scatter themselves singly over the fields and thickets in search of that food which the season affords. As the severities of winter approach, numbers appear to continue their migrations still further south; but it rarely occurs that our island is left entirely destitute.

The young at an early age assume the markings and general appearance of the parent birds: the difference in plumage which the sexes present is so trifling as to be scarcely distinguishable.

Their food consists of worms, insects, snails (more especially the *Helix nemoralis*), and fruits.

The head and upper parts are of a brownish olive; sides of the neck and breast of a pale yellow, the latter varied with arrow-shaped spots of rich brown; centre of the belly white; under wing-coverts pale reddish-orange, but neither so deep or decided in colour as in the Redwing; base of the bill and legs light brown; tip of the bill inclining to black.

Our Plate represents an adult bird in mature plumage.

REDWING.

Turdus Iliacus, *Linn.*

La Merle mauvis.

This bird is strictly a migratory species; for although it frequently remains with us for the greater part of the year, viz. from October till May or June, it invariably retires to the pine forests of Norway and Lapland for the purpose of breeding. It is rather inferior in size to the Thrush, and unlike that bird is gregarious, visiting us in the autumn in flocks of considerable numbers, and when the weather is mild frequenting pasture lands, feeding on insects, worms, &c. On the approach of frost, it subsists on the berries of the white thorn, mountain-ash, and ivy, the last of which it is very partial to, particularly in the spring. We have seen this plant in the month of May, entirely stript of its fruit by the Redwing. We have often known this bird perish from starvation when the winter has set in early and severe, in which case they subsist as long as there are any berries remaining on the trees before named; but when this resource fails, they have not strength to proceed further south, and inevitably perish.

The habits of the Redwing are much more shy and suspicious than those of the Thrush. It evinces a great partiality for tall trees and woods, and never lives in low hedges and bushes like the Thrush. Its song is similar to that of the Thrush, with the exception of being less powerful; it often sings sweetly before it leaves us for its more congenial and favourite breeding places.

Head and whole of the upper surface olive brown; the space between the bill and the eye dark brown, intermingled with yellow; a streak of yellowish white extends over the eye; sides of the neck and flanks white with obscure blotches of brown; belly pure white; under wing-coverts reddish orange; legs light brown; the sexes present no external differences.

We have figured an adult male of its natural size.

NAUMAN'S THRUSH.
Turdus Naumannii, *(Temm.)*

Drawn from Nature & on Stone by J & E. Gould. *Printed by C. Hullmandel.*

NAUMANN'S THRUSH.

Turdus Naumannii, *Temm.*

Le Merle Naumann.

So directly intermediate is this bird between the Fieldfare and Redwing, that had we not seen numerous examples we should have been inclined to consider it either an accidental variety or a hybrid produced between these two well-known birds. Although it does not occur in Europe so frequently as is mentioned by M. Temminck, still we have seen it in Continental collections; one in particular we recollect to have observed in the Museum at Munich, which was killed near that city: it appeared to be fully adult, and was in beautiful preservation. From the little information we could obtain respecting it, we can state with certainty that it is a species of the greatest rarity, visiting only the eastern portions of the Continent alone, and at very indefinite periods. It is strange that we have never observed a single specimen among the vast collections which have been sent home from India during the last few years, although that country directly intervenes between the eastern portions of Europe and Japan, where it is abundant. Two fine examples collected in the latter country have been transmitted to us by our valued friend M. Temminck; these specimens offer little or no difference from the bird in the Munich Museum, of which our Plate is an accurate representation.

Of its habits and nidification nothing is known; but in its general economy, food, &c., it doubtless resembles the other members of the group.

The sexes are scarcely distinguishable by their plumage; the female, like those of the Fieldfare and Redwing, being only somewhat smaller in size, and rather less bright and decided in her markings.

Top of the head, ear-coverts, and upper surface deep brown, each feather being edged with reddish, which red tint becomes more conspicuous on the rump, scapularies, and secondaries; a stripe of pale buff passes from the base of the bill over the eye; throat and upper part of the breast very pale buff; the sides of the neck ornamented with fine arrow-shaped markings of brown; chest and flanks dark brown, each feather broadly bordered with greyish white; belly white; under tail-coverts buff; tail and primaries dark brown, the outer edges of the tail-feathers tinged with reddish; feet light brown; bill light brown at the base, passing into dark brown at the tip.

The Plate represents a male of the natural size.

PALLID THRUSH.

Turdus pallidus; *(Pall.)*

Drawn from Nature & on Stone by J. & E. Gould.

Printed by C. Hullmandel.

PALLID THRUSH.

Turdus pallidus, *Pall.*

Le Merle blafard.

The claim of this bird to a place in the Fauna of Europe, says M. Temminck, (from whom we received the specimen from which our figure was taken,) is based on the capture of three individuals, one of which was taken in September 1823, near Herzberg in Saxony. It is one of the many discoveries made by Pallas, whose merits as a naturalist are too well known to require our praise.

Like the *Turdus Sibericus* this bird is extremely common in Japan, whence, through the kindness of M. Temminck, we have received several examples: it is also spread over the whole of Siberia, and occasionally passes the boundary line and visits the centre of Europe.

Of its habits, manners, and nidification nothing is known, but in these respects it doubtless closely resembles the other members of the genus.

The whole of the upper surface is of an olive brown; ear-coverts brown with a faint line of white down the centre of each feather; tips of the wing-coverts yellowish, forming a band across the wing; sides of the throat pale reddish brown blotched with white; chest and flanks pale reddish brown, the former ornamented with numerous spots of a darker tint; throat, centre of the abdomen, and under tail-coverts white; two outer tail-feathers largely tipped with whitish on their inner web; bill and feet light brown.

We possess other specimens in which the spots on the breast are wanting, and the white of the throat and abdomen is less pure; but whether this difference is occasioned by sex or age we are unable to determine.

We have figured the bird in both states of the natural size.

WHITE'S THRUSH.

Turdus Whitei; *(Eyton.)*

Drawn from Nature & on Stone by J. & E. Gould.

Printed by C. Hullmandel.

WHITE'S THRUSH.

Turdus Whitei, *Eyton.*

Le Merle de White.

THREE specimens of this rare bird having been killed in Europe, two on the banks of the Elbe near Hamburgh, and a third which was shot by Lord Malmesbury at Heron Court in Hampshire, in January, 1828, we have deemed it necessary to include a figure of it in the present work.

While at Hamburgh we were fortunate enough to obtain one of the specimens taken there, from the person who had it, in a fresh state; this specimen now forms a part of the collection of T. B. L. Baker, Esq., of Hardwicke Court, Gloucester, who doubtless values it as one of the greatest rarities of his collection. Although we have placed this bird in the genus which comprehends the true Thrushes (the type of which is the *Turdus musicus*, Linn.), still we doubt not that this bird, with the *Turdus varius* of Dr. Horsfield and another from New South Wales, will be found to constitute a well-marked and distinct group among the *Merulidæ*. From the greater length of their wings, we are led to believe that these birds possess very considerable powers of flight, and that in all probability they are strictly migratory in their habits. When compared with the true Thrushes a considerable difference may be observed in the form and length of the tail, and also in the tarsi and toes.

In its general size the *Turdus Whitei* exceeds by almost a fifth the *Turdus varius*, while the bill is much smaller: the length of the wing in Mr. Baker's specimen of *Turdus Whitei* is six inches and three quarters, while that of the *Turdus varius* and of the species from New South Wales scarcely exceeds five inches and a half. The great difference in size and the smaller bill will be sufficient to establish the specific value of this fine bird, which has with much propriety been dedicated to the celebrated and kind-hearted White, whose work on the Natural History of Selborne is not only fraught with instruction, but has given a decided impetus to the study of this branch of knowledge in this country. This species is very common in Japan, and is in all probability dispersed over a great part of Southern Siberia.

Crown of the head, back of the neck, back, rump, and upper tail-coverts light yellowish brown, each feather tipped with a crescent-shaped mark of blackish brown; wing-coverts and tertiaries dark brown tipped with buffy brown; spurious wing dark blackish brown, crossed in the middle by a band of buff; primaries dark brown on their inner webs, and buffy brown on the outer; four central and two outer tail-feathers pale brown, the latter tipped with whitish; the remainder blackish brown; throat, centre of the abdomen, and under tail-coverts white; the remainder of the under surface pale buff, each feather passing into deep buff near the tip and terminating in a crescent of blackish brown; bill and feet light brown.

Our figure is of the natural size.

SIBERIAN THRUSH.
Turdus Sibericus; *(Pall.)*

Drawn from Nature & on Stone by J. & E. Gould.

Printed by C. Hullmandel.

SIBERIAN THRUSH.

Turdus Sibericus, *Pall.*

Le Merle à sourcils blancs.

This fine and rare Thrush was first described by Pallas as an inhabitant of Siberia, but on referring to the works of this author, his account of this species is so meagre that no information whatever is given relative to its habits and manners, except that, like the other members of the genus, it possesses considerable powers of song.

Were we more intimately acquainted with its economy, we doubt not that it would be found to differ in some slight degree from that of the common species, as from the peculiar silkiness and dark colouring of its plumage, together with its shorter tail, we cannot fail to observe a considerable difference in their form; and in all probability the *Turdus Sibericus* will prove to be the intermediate form connecting the members of the genus *Petrocincla* with the true Thrushes, especially if, on investigation, it should be ascertained that the bird evinces a partiality for rocky situations.

From Siberia, where it inhabits the wooded mountains, it has been known occasionally to stray into Russia and the Crimea, and hence it becomes necessary to add it to the list of European birds. Besides these localities we can also state that it is a very common bird in Japan.

The female may be readily distinguished from the male by having a lighter-coloured head and neck, which, with the whole of the under-surface, is spotted much after the manner of the typical members of the genus *Turdus*.

The specimens from which our figures are taken were kindly forwarded to us by M. Temminck, and may be thus described:

Male: forehead, crown of the head, and nape deep black; over each eye a broad stripe of white; all the remainder of the plumage brownish black, each feather edged with slaty grey; bill and feet brown.

Female: forehead, crown of the head, and nape deep brown; chin and throat pale buff; sides of the face and neck and all the under surface slaty grey irregularly spotted with buffy white; tips of the feathers on the centre of the abdomen, vent, and under tail-coverts tipped with dull white; primaries and secondaries brown; the remainder of the plumage as in the male but lighter.

We have figured an adult male and female of the natural size.

WATER OUZEL OR DIPPER.

Cinclus aquaticus; *(Bechst)*.

Genus CINCLUS.

Gen. Char. *Beak* slender, slightly bent upwards, compressed, cutting edges bending inwards; upper mandible notched at the tip. *Nostrils* at the sides of the base, naked, cleft lengthwise, partly covered by a membrane. *Legs* short; toes three before, one behind, the outer toe joined at its base to the middle one. *Wings* short; the first quill not half the length of the second, which is also shorter than the third and fourth.

WATER OUZEL.

Cinclus aquaticus, *Bechst.*

Le Cincle plongeur.

The genus *Cinclus*, as far as our knowledge at present extends, is very limited, including only three species, of which one is a native of the Himalayan mountains, another of Mexico, and the third (the subject now under consideration,) is peculiar, we believe, to Europe, where it is dispersed among the mountainous districts of the Continent and also of our own island.

The lonely, secluded, and indeed local situations in which this bird resides, have prevented our becoming familiar with its habits and manners, which, from their extraordinary and novel character, merit a more strict attention than they have hitherto received. We here allude to the power which the Water Ouzel possesses of diving and remaining submersed beneath the water while in search of food,—habits which, though they have generally attracted the notice of naturalists, have not received that close and philosophical scrutiny which the subject deserves; and we would recommend, to those who have the opportunity, a close study of this bird in a state of nature; for, however unqualified it may appear to be for such habits, it is undoubtedly capable of descending to the bottom of streams and rivers, for the purpose of prosecuting its search after insects and larvæ which are the inhabitants of the stony bottoms of mountain streams.

As far as the fact of its submersion goes, we have ourselves many times witnessed it; but have never been able to mark unobserved the actions of the bird under water, so as to say whether it is by a powerful effort that it keeps itself submersed, or whether it is completely at its ease as some have asserted.

The Water Ouzel is a spirited and restless little bird, full of life and activity, flitting from stone to stone along the borders of streams; and it is especially fond of perching upon any rock that happens to be elevated in the centre of the current, where, conspicuous by its white breast, it may be observed dipping its head and jerking its tail in a manner not unlike that of the Wren, at one moment pouring forth a lively twittering song, (and that even in the depth of winter, when the earth is covered with snow,) and at the next diving down, and rising again at a considerable distance. When so disposed, its flight is straight, low and rapid,—in fact, much like the Kingfisher, and it is equally solitary in its habits. It is, however, seldom found in the same situations as the Kingfisher, the latter being a frequenter of streams which flow through a fertile country, while the Water Ouzel is peculiar to the rapid and limpid streams which descend the mountain sides, and run through glens at their base.

This interesting little bird builds its nest in the fissures of the rough stones and ridges which are common in such localities, and among the large loose stones of the margin. The nest is ably constructed of the various mosses and grasses nearest to hand, and covered with a dome, like that of the Wren: the eggs are from five to seven in number, of a pure and delicate white. The birds having arrived at maturity, neither undergo any peculiar changes in the plumage, nor exhibit external sexual differences. The young, however, are more brown on the upper surface, and the white extends over the whole of the abdomen, interrupted by little markings of brown, which become darker as they proceed.

In our islands we must look for the Water Ouzel in Wales (where we had the pleasure of obtaining the individuals from which the figures in the accompanying Plate were taken), Derbyshire, Yorkshire, and all the northern hilly counties. On the Continent it is extensively spread among the alpine and mountainous districts from Russia to Italy.

The upper surface is of a strong blackish brown, each feather having its outer edge black; the throat and chest are pure white; the abdomen rufous; the beak black; irides hazel.

Our Plate represents an adult, and a young bird of the year, of the natural size.

BLACK-BELLIED WATER OUZEL.
Cinclus melanogaster; *(Temm.)*

Drawn from Nature & on Stone by J & E Gould.　　Printed by C. Hullmandel.

BLACK-BELLIED WATER OUZEL.

Cinclus melanogaster, *Brehm.*

La Cincle à ventre noir.

A SPECIMEN of M. Brehm's *Cinclus melanogaster* having been transmitted to us by M. Temminck, we have ventured to give a figure of it, although we agree with the latter gentleman in questioning its specific value; it therefore remains for a future knowledge of its habits and manners to decide whether it may be considered as distinct, or only a variety dependent upon difference of climate and locality. In its general size and relative admeasurements it is rather less than the common species (*Cinclus aquaticus*), and is of a deeper colour both on the upper and under surfaces. According to M. Brehm it inhabits the north-eastern parts of the Continent, visiting in very severe winters the coasts of the Baltic, and is neither shy in its habits nor distrustful of the presence of man.

Its food consists of insects and their larvæ.

The head, back of the neck, and all the under surface deep chocolate black; the feathers of the back dark grey in the centre, with broad black edgings; wings and tail black; throat and chest white; feet dark brown; bill blackish brown.

We have figured an adult of the natural size.

PALLAS'S WATER OUZEL.

Cinclus Pallasii; *(Temm.)*

Drawn from Nature & on Stone by J. & E. Gould. *Printed by C. Hullmandel.*

PALLAS'S WATER OUZEL.

Cinclus Pallasii, *Temm.*

La Cincle de Pallas.

In the third part of his 'Manuel d'Ornithologie,' we find M. Temminck has included this rare species of Water Ouzel as an occasional visitant to the eastern confines of Europe, more particularly the Crimea and those portions of European Russia contiguous to the Asiatic continent. In our 'Century of Birds from the Himalaya Mountains' will be found a figure of this species as an inhabitant of the glenny streams of that fine country. Since the publication of that work we have received specimens of the young, as well as additional examples of the adult, and our present Plate is consequently rendered more complete and of greater interest by containing a representation of the bird in its young state, which on comparison will be found to possess a plumage very similar to that of the common species (*Cinclus aquaticus*). In the specimen from which our figure was taken, and which had nearly attained its full size, there was not the slightest trace of the chocolate colouring which characterizes the adult; in all probability therefore the change is effected by a total loss of the feathers early in the following spring, or at the second moult.

M. Temminck has favoured us with specimens of the Japan Water Ouzel, which differ so slightly from those killed in India as not, in our opinion, to admit of their being separated; it may be observed, however, that the Japan specimens are rather darker in colour, and that this difference is even perceptible in the young of the two species.

Whether the *Cinclus Pallasii* offers any material difference in its habits and manners from those of the British Water Ouzel we are unable to state, but in all probability they are very similar.

The adult is of an uniform chocolate brown, with the feet and bill black.

The young has the whole of the plumage of a fuliginous grey with numerous crescent-shaped markings of pale greyish white, which are most numerous on the throat, giving it a whitish appearance; feet and bill black.

We have figured an adult and a young bird of the natural size.

ROCK THRUSH.

Petrocincla saxatilis, *(Vig)*.

Drawn from Life and on stone by J & E Gould.

Printed by C. Hullmandel.

Genus PETROCINCLA.

Gen. Char. *Beak* stout, straight, the ridge arched at the point. *Nostrils* basal, round, partly covered with hairs. *Wings* of middle length; the first *quill-feather* very short, or almost spurious, the third the longest, the second a little shorter. *Feet* moderate, somewhat strong; the *acrotarsi* and *paratarsi* perfect. *Tail* equal.

Type of the genus, *Turdus saxatilis,* Linn.

ROCK THRUSH.

Petrocincla saxatilis, *Vigors.*

Le Merle de Roche.

The Rock Thrushes, of which the present may be taken as a good example, differ so much from the more typical birds of the family, in form as well as in habits, manners and the localities they frequent, as to justify their being raised to the rank of a genus. This was hinted at by M. Temminck, who formed them into a section, which section has been subsequently established as a genus by Mr. Vigors, under the name of *Petrocincla.*

These birds, instead of dwelling in groves and woods,—a circumstance which so peculiarly characterizes the Thrushes in general,—affect the rugged and inaccessible declivities of rocks and mountains, for which their form is adapted; the shortness of their tails and the length of tarsi indicating them to be among those birds which live more exclusively on the open ground. In many respects they manifest a relationship to the *Saxicolæ,* between which and the rest of the *Merulidæ* they seem to constitute a link of union, forming also a close alliance with various groups of ground Thrushes from other portions of the globe.

The present species is an inhabitant of the central and eastern portions of Europe, confining itself almost exclusively to the rocky and mountainous districts, especially the Alps, the Apennines, the Pyrenees, and some of the higher mountains in France. In such situations it incubates, constructing a nest of moss and herbage in clefts of the rock, among masses of loose stones, or in old ruins, laying four eggs of a pure greenish blue. M. Cuvier, in the short notice he gives of this bird in his *Règne Animal,* states that the male is distinguished by its beautiful song; but into any minute details respecting its manners and peculiarities, we do not profess ourselves able to enter, and it is a matter of regret, that those who have had so many opportunities of becoming acquainted with its habits, &c., in its native haunts, should have given us such meagre accounts respecting them.

The examples which have come into our hands prove it to be a species that undergoes several remarkable changes of plumage, which we cannot better explain than by availing ourselves, in our text, of the description given by M. Temminck, who appears to have paid a close attention to the subject.

"In the adult male, the head and neck are of an ashy blue, darkening on the upper part of the back into black clouded with blue; below this a large white space extends as low as the upper tail-coverts, which are also black. The shoulders black; the quill-feathers dark brown; the tail ferruginous red, except the two middle feathers, which are rufous brown; the chest and whole of the under surface bright ferruginous, each feather, especially those of the inferior tail-coverts, being obscurely tipped with white; beak blackish; legs brown. Length seven inches and a half."

"The females have the whole of the upper parts brown; on the back are several large whitish markings bordered with brown; the throat and sides of the neck of a pure white;" but it often happens that the feathers of that part are edged with ashy brown; the rest of the inferior surface is of a reddish white, with fine transverse bars at the tip of each feather; tail of a light red, with the two middle feathers of an ashy brown.

The young of the year are, again, altogether different. The whole of the upper parts are ashy brown mottled with whitish grey; the end of each feather marked with a white spot; the quill-feathers and coverts are darker, the feathers forming the coverts having a grey border and whitish ends; tail red, lighter in colour at the extremity; the under parts very similar in colour to those of the adult female, but with more of white varied with red, and a multitude of irregular markings of brown.

Its food consists of Scarabæi and other insects, as well as wild berries.

Our Plate represents a male and female in full plumage.

BLUE THRUSH.
Petrocincla cyanus, *(Vig:)*

Drawn from Life & on Stone by J. & E. Gould.

Printed by C. Hullmandel.

BLUE THRUSH.

Petrocincla cyanus, *Vig.*

Le Merle bleu.

LIKE its congener, the Rock Thrush (*Petrocincla saxatilis*), the present species is a native of the rocky and mountainous districts of Europe, particularly towards the south, being very abundant in Piedmont and the Apennines, and also of common occurrence throughout Spain, Sardinia, Italy and the Levant. It is met with also in the South of France, but is rare in Switzerland. India and China produce a bird in every respect identical, with the exception of size, those which are received from Asia being considerably smaller than their European representative. Although the congenial habitat of the Blue Thrush is the rocky scenery of mountain chains, among which it breeds and remains throughout the year, still in many of its characters it seems to constitute a link between the more typical form of the genus *Petrocincla* and that of the true Thrushes, which latter it approaches in the proportions of the tarsi and tail. In the typical *Petrocinclæ*, (*P. saxatilis*, for example,) the tarsi are strong and very elongated; but the tail is short, a conformation in harmony with strictly terrestrial habits. In the present bird, the tarsi are more moderate and the tail more developed; still, however, as its habits, style of plumage and general outline declare, it is in every sense a member of the genus in which it is now placed.

The Blue Thrush is shy and solitary, dwelling with its mate in the still and sequestered recesses of the rocks, in the clefts of which it builds its nest, though this is not always the case, as it often chooses the crumbling walls of lonely towers or buildings, and sometimes the holes of trees, in which to rear its young. The eggs are dull greenish white. Its food, like that of its congener, consists of grasshoppers, large insects in general, and wild berries.

The male and female exhibit considerable difference in their plumage, the young males of the year resembling the latter. In the adult male, the whole of the upper surface is of a deep greyish blue, many of the feathers being margined with grey; the wings and tail are black; the under surface is of a lighter blue than the upper, with obscure narrow bars of brown edged with white on the chest and abdomen; the beak and tarsi are black.

The female has the whole of the upper surface brown, obscurely barred with ash colour; the wings and tail blackish brown, each feather having a blueish margin; the throat light brown, the feathers tipped and edged with black; the chest and under surface varied with light brown, grey and black, in pointed scales and transverse bars.

The young males may be seen in various stages between this style of colouring and the rich blue of the bird in its maturity.

The Plate represents a male and female in their adult plumage, of the natural size.

BLACK WHEATEAR.

Saxicola cachinnans; *(Temm.)*

Drawn from Nature & on Stone by J. & E. Gould.

Printed by C. Hullmandel.

BLACK WHEATEAR.

Saxicola cachinnans, *Temm.*

Le Traquet rieur.

Although the most proper situation for this fine species is doubtless among the true Wheatears (*Saxicolæ*), yet its greater size, robust bill, and more short and rounded wings, indicate a departure from the typical form, and an approximation to some other group, which at present we have not been able distinctly to make out; we suspect, however, that the group to which it will ultimately be found to lead is one of the terrestrial division of the family of *Merulidæ*. Though one of the birds of Europe, we cannot include it in our native fauna: it, is indeed, confined to the southern portions of the Continent, and is common at Gibraltar, where it is known annually to breed; it is also found in all the rocky and arid districts of Spain, Sicily, and the islands of the Mediterranean, as well as on the opposite coast of Africa. Judging from its form alone, we should be led to consider that the present species is not migratory, a supposition which is confirmed by its being a resident in countries where its food is ever abundant, and by its never having been known to visit the more northern districts of Europe, to which the Long-winged Wheatears are periodical visiters. We have said, that the present species is an inhabitant of the northern coast of Africa, which country also produces another closely allied to it, differing only in having a more lengthened wing, and the top of the head of a pure snow white. These two birds are by many ornithologists considered as one species, and that the white-crowned one is the adult of the present bird. We have ourselves carefully examined both these birds, and have no hesitation in declaring that they are truly distinct, differing not only in colour, but in relative admeasurements, the white-crowned species having a body of the same size, but a wing nearly an inch longer, being in all respects a typical Wheatear. Whether this African species is also a native of any part of Europe, we have as yet had no opportunity of ascertaining. The confusion between the two species has evidently arisen from the circumstance, that the young of the white-headed or African species does not possess the white on the top of the head, and in this state cannot, except by a narrow scrutiny, be distinguished from the young or the female of the true *Saxicola cachinnans*.

In its manners, the Black Wheatear is shy and timid, avoiding the presence of man, and confining itself to arid rocky places, where it is rarely disturbed by his presence. Its food consists of insects, beetles, &c.

The general plumage of the male is black; the rump white, as are the tail-feathers, except at their tips, which with the whole of the two middle feathers are black; beak and tarsi black; irides dark brown.

The female resembles the male in the distribution of her colours, but the black is much less pure and strongly inclines to brown.

The young resemble the female in their plumage.

We have figured an adult male and female of the natural size.

PIED WHEATEAR.

Saxicola leucomela; *(Temm.)*

Drawn from Nature & on Stone by J & E Gould.

Printed by C Hullmandel.

PIED WHEATEAR.

Saxicola leucomela, *Temm.*

Le Traquet leucomèle.

While the downs, commons, and barren heaths of our island are enlivened by the presence of the Common Wheatear (*Saxicola Œnanthe*, Bechst.), and the plains and deserts of Spain, Italy, and the southern districts of Europe in general are equally so with *Saxicola aurita* and *stapazina*, the northern portions, which include Russia and Lapland, are to be enumerated among the places in which the present fine species habitually takes up its station; and although so little is known respecting the natural habits of this bird as to leave us in a state of uncertainty regarding its migration, as well as the localities it chooses in which to incubate and rear its young, yet we may reasonably suppose that its general economy is in strict unison with that of its congeners. The *Manuel* of M. Temminck informs us, that so exclusively boreal is this species that it is never seen in temperate climates; which circumstance will lead us to infer that Siberia, Upper Tartary, and the most northern portions of Asia will hereafter prove to be countries of which nature has destined this bird to be a native, and the limited numbers which occasionally visit Europe to be individuals traversing the outer limits of their appointed range.

The female we have never yet seen; her colouring, however, will be readily understood when we state that where the male is black, the same parts in the opposite sex are of a dull brown; and the parts which are white in the former are of an obscure light brown in the latter.

It is said to construct a nest in rocks and old buildings, sometimes on the borders of rivers: of the number and colours of its eggs nothing is at present known.

The adult male has the top of the head, back of the neck, rump, base of the tail, breast, and under parts, with the exception of the vent, which is light rusty brown, of a pure white; the rest of the plumage being of an equally fine black, thus forming a strong contrast; legs and bill black.

The Plate represents an adult male of the natural size.

WHEATEAR.

Saxicola œnanthe. *(Bechst.)*

Drawn from Life & on Stone by J. & E. Gould.

Printed by C. Hullmandel.

WHEATEAR.

Saxicola Œnanthe.

Le Traquet moteux.

The Wheatear is one of the summer migratory birds which annually visit the British Islands, arriving in March, when it disperses itself over wild heaths, moors, fallow grounds, and rabbit warrens, making a nest of moss and vegetable fibres, lined with hair or wool, generally in a hole on the ground or among loose stones, but not unfrequently in the cleft of a rock or at some distance in a deserted rabbit burrow, and laying five or six eggs of a uniform pale blue. The extreme delicacy of the Wheatear has caused it to be much sought after as a luxury for the table, for which purpose incredible numbers are annually taken. Mr. Pennant, in his British Zoology, states that at Eastbourne, in Sussex, the number annually taken amounts to 1840 dozen. They are principally caught by shepherds, who from the nature of their occupation have every opportunity of studying the habits of this bird, so as to contrive the most successful mode of securing them,which is generally effected by nooses of horsehair.

In September, previous to their departure to the Continent, they make the downs of our southern counties a place for their general assemblage, where they wait for a favourable wind to carry them over the intervening channel.

In its habitat, the Wheatear is especially confined to open and bare grounds, seeking neither the covert of the furze nor the hedgerow, as is the case with the other British species of *Saxicola*,—a circumstance which, perhaps, in connexion with a trifling modification of form, has induced some authors, viz. Brisson, Stephens, &c., to separate it from *S. rubetra*, the Stonechat, and the Whinchat, *S. rubicola*, and advance it to the rank of a genus, to which they have given the name of *Vittaflora*, the propriety of which we leave to others to determine, as it is not so much our object to enter into the minutiæ of generic divisions, as to give a faithful portraiture and history of each species, with a view to their natural arrangement on comprehensive principles.

The Wheatear is a pleasing and elegant bird in its plumage; and its manners, though retired, are lively and active. Hopping and springing from clod to clod, and occasionally breaking out into short flights in pursuit of insects, it becomes conspicuous from the snow-white mark across the base of the tail.

Beside the softer insects which it captures on the wing, Coleoptera and their larvæ form its diet, to which worms, &c., are also added.

Although not generally classed among our song birds, nevertheless the Wheatear is not without its vocal powers, warbling a soft and sweet strain, not unfrequently while quivering on the wing a few yards from the earth: occasionally its notes rise to a bolder and more elevated pitch; and when kept in confinement, a matter of no great difficulty, it charms us with its simple song, continued through the depths of winter.

In the adult male, the bill is black; the irides dark hazel; from the base of the bill a white line extends over each eye, and beneath it a broad black band passes which includes the orbits and ear-coverts; upper part of the head and back cinereous gray; rump and tail-coverts white; the two middle tail-feathers black; the rest black two thirds of their length from the base; wings blackish brown; each feather edged with a lighter rust-coloured border; throat and neck beautiful buff, becoming lighter as it proceeds downwards; tarsi black.

In the female, the under parts are brown; the forehead inclining more to gray; the black parts in the male, including the mark across the eye, are here exchanged for deep brown; the edges of the wing-feathers are more or less ferruginous; the white at the base of the tail is less extensive, and the neck and chest are reddish, becoming lighter as it approaches the under surface.

The young of the year of both sexes somewhat resemble the adult female; but a tinge of red pervades the whole of the plumage, and especially the edges of the quill- and tail-feathers. Total length about six inches.

We have figured a male and female in their spring plumage.

RUSSET WHEATEAR.

Saxicola Stapazina; *(Temm.)*

Drawn on Stone from Life by J & E. Gould.

Printed by C. Hullmandel.

RUSSET WHEATEAR.

Saxicola stapazina, *Temm.*

Le Traquet stapazin.

In making the *Saxicola stapazina* a different species from the *S. aurita*, we rely not so much on our own observation as on the opinion of M. Temminck, who assured us personally that he had every reason for considering them as distinct species; and in his "Manuel d'Ornithologie" he remarks, that the European habitat of the *S. stapazina* is more limited than that of *S. aurita*, being restricted solely to the rocky borders of the Mediterranean, the South of Italy, Dalmatia and the Grecian Archipelago; that it is rarely seen in the North of Italy, and never in the central districts of Europe. It is to be regretted, that in consequence of the peculiar localities in which alone this bird is found, our opportunities for studying it during its various changes are very limited: we have, however, exerted ourselves to obtain as many specimens as possible, and we now possess a series of examples, killed at different seasons of the year, upon which we rely for our description. Unlike the Common Wheatear, which exhibits so marked a difference in the plumage of the sexes, the Russet Wheatear, in the adult stage, differs rather in the purity than in the decided contrast of colours which distinguishes the male and female; but, like the *Saxicolæ* in general, each sex, after the autumn moult, loses, by the gradual action of the air and light, as the spring approaches, the rich rufous tone of colouring by which the plumage is at first characterized, the tints becoming gradually paler and the black of the wings deeper, the brown tips of the feathers being worn off.

In habits and manners, the *Saxicola stapazina* is a true example of its genus, preferring, like the Wheatear, wide elevated downs, where it obtains its food, seldom perching upon trees, and never retiring to the woodlands for shelter. Of its nidification nothing positive is known.

In the adult plumage of spring, the male is thus distinguished. From the beak to the eye, and from thence over the ear-coverts, extends a band of black, of which colour are the throat, scapulars and quills; the top of the head, the rump and under parts are pure white; the back of the neck and back are light rufous; the tail white for three parts of its length and black at the tip, with the exception of the outer feathers, which are almost wholly black, and the two middle ones, which are quite so.

Immediately after the autumn moult, the top of the head and back of the neck have a shade of ash colour; the breast is reddish, gradually passing into white, and the black scapulars and quills are edged with rufous.

The young males of the year resemble the female, in which the tints are altogether of a redder hue; the dark feathers of the throat and region of the eyes being brownish black, the quills and coverts edged with reddish, and the breast reddish white.

The Plate represents an adult of the natural size just after the autumn moult, and a bird of the first year, killed at the same season, differing only in the rufous edging of the wing-feathers.

BLACK-EARED WHEAT EAR.

Saxicola aurita, *(Temm.)*

Drawn from Life and on Stone by J. & E. Gould. *Printed by C. Hullmandel.*

BLACK-EARED WHEATEAR.

Saxicola aurita, *Temm.*

Le Traquet oreillard.

In a remark subjoined to a description of the *Saxicola aurita*, in M. Temminck's "Manuel d'Ornithologie," that learned naturalist observes, that had he not a thorough conviction of the distinction between this and the bird previously described by him, (*S. stapazina*,) he should be ready to admit their specific identity, differing as they do in one point only, viz. the black throat of the latter being exchanged in the former for white or whitish rufous, the black band from the beak over the ear-coverts being alone retained: in the rest of the plumage the agreement is precise. For our own part, we confess, that were it not that M. Temminck expresses himself positively on this subject, and asserts that " the *stapazina* in its different stages has the throat and a part of the neck *always* of a deep black or blackish," we should have hesitated, the difference being less than is known to occur in many birds under the varying circumstances of age, sex, or season. We, however, follow the opinion of so distinguished an ornithologist, supported, as we doubt not it is, by positive proofs,—and therefore describe the species as truly distinct.

In habits and manners the Black-eared Wheatear agrees with its allied congener, inhabiting the hilly districts of the South of Europe: it is, however, more common in the North of Italy than the Russet Wheatear. Though in the centre of Europe it is never seen, the borders of the Mediterranean, the Apennines, Sardinia and the Neapolitan States are abundantly supplied with this species. Of its nidification we have no accounts upon which we can rely.

Adult male in spring: From the beak to the eye and thence over the ear-coverts extends a band of black; head and rump pure white; back of the neck and back light reddish brown; throat and under parts white; tail white for three parts of its length and tipped with black, excepting the outer feathers, which are nearly all black, and the two middle feathers, which are entirely so; wings black.

The adult female nearly resembles the male in all her markings, having the head and upper surface reddish brown; throat whitish; breast reddish, becoming lighter below; rump white; and the wings blackish brown, each feather being finely edged with reddish.

After the autumn moult, the plumage exhibits the deep tints and rufous edgings to the feathers which characterize the preceding species at the same time. The young of the year differ little from the adult female, and exhibit only obscure traces of the ear-mark; but their plumage is more equally tinted with rufous.

The Plate represents an adult male in the spring and autumn plumage.

WHINCHAT.

Saxicola rubetra, *(Bechst)*.

Drawn from Life and on Stone by J. & E. Gould.

Printed by C. Hullmandel.

WHINCHAT.

Saxicola rubetra, *Bechst.*

Le Grand Traquet.

Among the smaller migratory birds which visit us on the return of spring, the Whinchat is one of the most pleasing and elegant; it seldom, however, favours us with its presence before the middle of April, frequenting, in pairs, the pasture lands and commons of every part of England, but is more scarce in Devonshire and Cornwall, especially the western portions of those counties. Though not a distinguished songster, its simple and hurried notes are by no means unpleasing, and well accord with its active and sprightly manners. In some of its habits it is not unlike the Fly-catchers, perching on a stem of grass or dock, darting at the insects as they pass by, and returning again to its station. But its length of tarsi indicates the bare and open ground of meadows and commons to be its peculiar province; hence it is not found to frequent woods or thick coppices, as is the case with our songsters in general: shy and timid, it seldom allows itself to be approached, but with a quick and lively action flits forward to the next bush or elevation of earth, incessantly watching the intruder; and, if again disturbed, repeating the same short flight; still, however, keeping within the neighbourhood of its residence for the season, and where, on her sheltered nest, the female is carrying on the process of incubation. During this period the male bird displays great restlessness and apprehension if the nest be approached, flitting from spray to spray, jerking its tail repeatedly, and uttering its querulous note, which may be represented by the two syllables *u—tick*, the latter of which is frequently reiterated, and the whole note is sounded so distinct and clear as to be heard at a considerable distance. It builds its nest on or near the ground, and forms it principally of coarse grasses lined with finer fibres: the eggs are in general five or six in number, of a greenish blue, minutely speckled with light reddish brown at the large end.

The Whinchat seems to be universally spread over the northern portion of the European continent, its favourite localities being the same as in England, viz. mountainous heaths and extensive pasture lands.

Although a general similarity exists between the plumage of the sexes, the males may always be distinguished by their brighter and more strongly contrasted colours, and by the conspicuous white stripe over the eye and on the wing.

As the autumn advances and insects become scarce, the Whinchat dissappears, passing over to the more southern countries, and not improbably to the Levant, Syria, and the northern coast of Africa, where its supplies of insect food are still abundant.

In the male, the bill is black, furnished at its base with a few bristles; a broad black streak beginning at the bill passes through the eyes and covers the ear-feathers, above which extends another line of white; crown of the head, back, and wing-coverts of a dark brown, the edges of each feather being of a light ferruginous colour; chin white; throat and breast orange-brown; belly, vent and thighs pale buff; tail short, the bases of the outer feathers white, the rest black.

In the female, the streak over the eye is much less conspicuous; the cheeks instead of being black are of the same colour as the rest of the head; the general plumage is duller, the marks less distinct, and the white mark on the wing totally wanting; legs and toes black. Total length about five inches.

Our Plate represents a male and female.

STONE-CHAT.

Saxicola rubicola. *(Bechst)*.

Drawn from Life & on Stone by J. & E. Gould. *Printed by C. Hullmandel.*

STONECHAT.

Saxicola rubicola, *Bechst.*

Le Traquet pâtre.

M. Temminck, whose knowledge of European birds cannot be questioned, states in his "Manuel d'Ornithologie," that the *Saxicola rubicola* is a bird of passage in Europe, but stationary in Africa: however this may be, it is certainly stationary in England, and may be observed at all seasons on commons, moorlands, and shrubby heaths, from one extremity of the British Isles to the other. It is a species possessing a wide range of habitat, as examples from India and Africa present no specific differences. Its habits and manners are somewhat in unison with its allied congener the Whinchat, but it is even more restless and noisy, flitting from bush to bush, or rock to rock, and not unfrequently perching on the tops of the flower of the thistle or highest twig of the whinbush, at the same time uttering its singular monotonous notes, which may be compared to the clicking of two stones struck together at repeated intervals.

The present bird and the Whinchat (*Saxicola rubetra*, Bechst.) present many points of difference, both in form and habits, from the rest of the genus: instead of being confined almost exclusively to the ground, as is the case with *Saxicola œnanthe*, *stapazina* and *aurita*, they give the preference to low bushes and shrubs, as above noticed, on which they habitually perch, constituting in this respect an intermediate grade between the genuine *Saxicolæ* and the true woodland *Sylviadæ*, or rather, perhaps, the *Muscicapidæ* (Flycatchers), which they resemble in the abruptness of their actions and in their manner of darting from their perch at insects on the wing, in pursuit of which they appear incessantly occupied. These, indeed, with larvæ and worms, constitute their food.

The Stonechat builds its nest at the bottom of bushes, or among the crevices of rocks: the eggs are pale green with a few blotches of light red.

The male and female offer a decided contrast in their colouring.

In the male, the head, throat and tail are of a deep black; the sides of the neck, the scapulars and rump of a pure white; the back deep black, each feather having a light reddish margin; wings blackish; breast deep rufous, becoming paler on the under surface.

The female has the upper surface of a brownish black, each feather having a yellowish red border, as have also those of the wings and tail, which are brown; throat black slightly dotted with white and reddish; the white space on the side of the neck and scapulars is less extensive, and the rufous of the chest less bright.

The young male closely resembles the adult female.

The Plate represents a male and female in perfect plumage, of the natural size.

REDSTART.

Phœnicura ruticilla; *(Swains)*

Drawn from Life & on Stone by J. & E. Gould.

Printed by C. Hullmandel.

Genus PHŒNICURA.

Gen. Char. *Bill* rather slender, somewhat widened at the base; compressed towards the tip, which is deflected and emarginated. *Tomia* of the mandibles, before the nostrils, bending inwards. *Gape* slightly bearded. *Nostrils* basal, oval, lateral, pierced in a membrane, and partly concealed by the feathers of the forehead. *Wings* rather long, with the first quill very short; the second inferior to the third; the fourth the longest of all. *Tail* of mean length, slightly rounded or square; coloured more or less with reddish brown. *Legs* having the *tarsi* longer than the middle toe. *Toes* slender, the outer toe joined at its base to the middle one; the former and the inner toe short, nearly equal in length and each reaching only to the second joint of the middle one. *Claws* not much hooked; that of the hind toe the longest.

REDSTART.

Phœnicura ruticilla, *Swains.*

Le Bec-fin de Murailles.

The genus *Phœnicura*, as instituted by Mr. Swainson, though the term itself is somewhat exceptionable, forms a well-defined and natural group, of which the present species may be considered a typical example. All the individuals of the genus appear to be confined to the Old World, several of the species being restricted to Asia, and those which may be considered as European being all migratory, and retiring on the approach of winter to a warmer climate.

The *Phœnicura ruticilla* is distinguished by the beautiful contrasts and richness of its colouring, as also by the sprightliness and animation evinced in the vigilant pursuit of its prey, while every action is accompanied by a peculiar vibratory movement of the tail, repeated for a considerable time on its alighting.

Familiar with man, this interesting visiter frequents gardens and orchards, fearlessly building in situations as if expressly to court observation,—for example, between the branch of a fruit tree and the wall against which it is nailed, the gardener's tool-house, or the holes of an old building or out-house, or, indeed, in any convenient aperture. The eggs are five or six in number, of a beautiful greenish blue colour. During the time of incubation, the male, conspicuous by the band of white on his forehead and the deep red of the tail, may be observed assiduously engaged in the capture of the softer winged insects, which he seizes while flying, darting after them from one resting-place to another with great celerity; he does not, however, return after each sally to the same perch again, like the Flycatchers, but continues a system of irregular pursuit.

As a songster, the Redstart holds no inferior place, though its song is hurried, and the notes neither rich nor powerful; still it never fails to excite feelings of pleasure from its simple sweetness.

The male in the adult plumage has the head and upper part of the back fine blueish ash colour, a broad white band extending from eye to eye across the forehead; the throat black; the breast, rump and lateral tail-feathers of a brilliant rufous; the under surface whitish; the under tail-coverts light rufous; and the two middle tail-feathers brown.

The female is distinguished by the general greenish brown of her plumage; the upper surface being more tinged with reddish, the throat cinereous merging into reddish brown on the under parts, and the tail dull rufous.

The young birds, like those of the Redbreast, are brown mottled with white: by degrees, however, they lose this plumage, and before leaving us in autumn, the males begin to acquire traces of the distinguishing style of colouring, the head having a tint of grey and the throat showing indications of black, while the upper parts acquire an obscure grey colour, each feather having a reddish margin. At all stages the beak and tarsi are black.

The Plate represents a male and female of the natural size.

BLACK REDSTART.

Phœnicura Tithys; *(Vard & Selby)*.

Drawn from Nature & on Stone by J. & E. Gould.

Printed by C. Hullmandel.

BLACK REDSTART.

Phœnicura tithys, *Jard. and Selby.*

Le Bec-fin à rouge queue.

Since the discovery of this species as an occasional visitant, as recorded by us in the "Zoological Journal," vol. 5. No. 17. p. 102., we have ascertained the fact of several other examples having been killed in different counties: and here we may mention that a fine specimen was shot at Brighton, another at Bristol, and a male bird was killed on Teignmouth sands in Devonshire, on the 7th of January 1833, by L. Sulivan, Esq., who placed the specimen in our hands. Still the occurrence of the *Phœnicura tithys* in our island must be considered as a circumstance of extreme rarity, though in some parts of the Continent it is as abundant as its nearly allied congener the Redstart with us. According to M. Temminck, it is found in the northern provinces of Europe, especially in rocky situations, a fact borne out by Mr. Sulivan's observations on the specimen he killed, which was flitting about the rocks on the Devonshire coast. In Holland and the flat lands of the Continent it appears to be nearly as rare as in England. In our late journey through Prussia, we observed it all along the road between Frankfort and Berlin. Its nest, M. Temminck says, is placed in the clefts of rocks, or in the fissures of towers and other old buildings; the eggs being six in number, of a pinky white.

The male and female offer considerable difference in the colouring of their plumage, the former of which may be thus described:—

The space between the beak and the eye, the cheeks, the throat and breast, deep black, which fades off into blueish ash on the belly and flanks; the upper parts are more inclined to dark grey; the forehead inclining to white; the rump and tail bright red, the middle feathers of the latter being brown; the greater coverts of the wings bordered with white; beak, irides and tarsi blackish brown.

The female has the upper parts of a dull brownish grey; the lower parts light grey; the coverts and quill-feathers bordered with grey; the rump and the tail-feathers dull red; beak, irides and tarsi as in the male.

We have figured an adult male and female of the natural size.

BLUE THROATED WARBLER.

Phœnicura Suecica.

Drawn from Life and on Stone by J. & E. Gould.

Printed by C. Hullmandel.

BLUE-THROATED WARBLER.

Phœnicura Suecica, *Jardine & Selby.*

Le Bec-fin gorge bleue.

THE scientific authors of "Illustrations in Ornithology" were induced to separate several species of Warblers from the very extensive genus *Sylvia* of Latham, on account of their general resemblance to our well-known Redstart. The term *Phœnicura* was applied to them as a generic distinction, and several reasons have induced us to adopt the genus, and consider the Blue-throated Warbler as belonging to this new subdivision. The species of this small group appear to be intermediate in their nature between those belonging to the genera *Saxicola* and *Curruca.* Like the Wheatear, the Blue-throated Warbler is considered a delicate article of food, and in the vicinity of Alsace numbers are captured for the use of the table; in its habits, in the situation often chosen for its nest, and the colour of its eggs, it exhibits a general resemblance to the Redstart.

The Blue-throated Warbler, somewhat resembling our well-known Robin in its form, is found thinly scattered over the countries of Europe, from Sweden to the Mediterranean, but is most plentiful in the central parts of the Continent. Throughout Germany and the Northern territories it is a migratory bird, like many others of the Warblers, appearing in April, and departing in September. On their arrival they frequent thick hedges, small woods and the borders of forests, building their nests in holes of trees, sometimes in cavities between stones near water, or on the banks of rivulets among roots which the action of the stream has laid bare. The nest is formed of dried bents and moss, with a few dead leaves, and lined with various sorts of hair. The female lays five or six eggs of a delicate pale greenish blue; the male is remarkable for his attention to his mate, and has an agreeable song, which is sometimes heard in the night.

The length of an adult bird is nearly six inches; the top of the head, all the upper parts of the body and wings are uniform clove brown; the beak black; over the eye a pale streak; throat and fore-part of the neck ultramarine blue, with a well-defined spot of pure white in the centre; beneath the blue colour is a black bar, then a narrow line of white, and still lower a broad band of bright chestnut; belly dirty white, flanks and under tail-coverts light reddish brown; the two middle tail-feathers clove brown, throughout their whole length, all the others on both sides have the basal half bright chestnut, the other half black; legs and claws brown. The female resembles the male in the uniform colour of the upper parts; the white patch on the throat, descending from the beak, occupies a much larger space; the blue colour on the sides of the neck is mixed with black; the successive bars of blue, black, white and chestnut towards the bottom of the neck in front are much less perfectly defined; and the belly and flanks more inclined to brown.

The Blue-throated Warbler is rather a rare bird both in the western parts of France and in Holland, still more rare in this country, only one instance of its occurrence being on record. This specimen was shot in May 1826, by Mr. T. Embleton, on the boundary hedge of the Newcastle Town Moor, and by him presented to the Museum of the Literary and Philosophical Society. The first notice of the capture of this interesting addition to our Fauna appeared in a Synopsis of the Newcastle Museum, by G. T. Fox, Esq., of Durham, (pages 298 and 300). This bird was considered a young male, and was probably obtained soon after its arrival.

We have figured a male and female of the natural size.

ROBIN.

Erythaca rubecula; *(Swains.)*

Drawn from Nature & on stone by J. & E. Gould.

Printed by C. Hullmandel.

Genus ERYTHACA, *Swains.*

Gen. Char. *Bill* broad, rather depressed at the base, gently narrowing towards the tip, where it is slightly compressed; of mean strength, with the upper mandible deflected at the tip, and emarginated. *Nostrils* basal, lateral, oval, pierced in a large membrane, and nearly concealed by the projecting feathers of the antiæ; gape beset with thick bristly hairs. *Wings* having the first quill very short, the second double the length of the first, the third shorter than the fourth and fifth, which are nearly equal, and the longest in the wing. *Legs* with the tarsi longer than the middle toe; the outer toe joined at its base to the middle one; the outer and inner toes short, nearly equal in length, and each only reaching to the second joint of the middle one. *Claws* not much hooked; that of the hind toe the longest. Form short and compact.

ROBIN.

Erythaca rubecula, *Swains.*

Le Bec-fin rouge-gorge.

We may consider this lively and familiar bird as strictly indigenous to Europe, since among the numerous and extensive collections received from Northern Africa, India, and China we have never observed a single example, neither is it mentioned (as far as we are aware,) by any writer as an inhabitant of those countries; it appears, however, to extend eastward as far as the border line of Asia Minor, one example, and one alone, having been received in a collection from the shores of the Black Sea. In Europe the middle and northern regions are those in which it appears to be most abundant and over which it is universally spread; there are none, indeed, who are not well acquainted with its habits and manners, and with whom it is not a favourite. One of those species whose fearless confidence in man lead it to frequent his gardens and the precincts of his house, its sprightly manners and its animated song, which is poured forth morning and evening even throughout the autumn and colder part of the year when other songsters are silent, render it a most welcome visitor to his habitation. Attractive as the Robin is in all its habits and manners, still it is of a quarrelsome and pugnacious disposition: two males seldom agree to live within the range of the same garden or within a given distance, the stronger always driving away its weaker antagonist; in this respect the Robin differs from many other birds that flock together in winter, and even from its nearly allied species, such as Redstarts and Wheatears, which migrate at the close of summer to a warmer climate, whilst it braves our coldest winters with the utmost impunity. During the greater part of the year its food consists of worms, grubs, the softer caterpillars, and other small insects, together with berries and fruits when in season; but in the depth of winter, when its natural food cannot be procured, it subsists upon crumbs and other refuse.

The sexes, which are alike in plumage, appear to continue mated throughout the whole of the year, and commence the task of incubation at an early period before many of our summer visitors have even arrived, in consequence of which they have generally two broods in the course of the year. The places chosen for the site of the nest is entirely according to circumstances, being sometimes a bank at the root of a tree, at others in the side of a house, in a hole in a wall, in the tool-house of the gardener, &c. The nest is constructed of moss, leaves, grass, the stalks of plants, or any material near at hand, generally lined with hair; the eggs are five or six in number, of a whitish grey with reddish spots. The young during the first three months of their existence would hardly be recognised as the progeny of the Robin, so much do they differ from the adult birds in the colouring of their plumage; a change of feather takes place in the winter, when their new dress resembles that of the adults and being then fully competent to shift for themselves, they are driven by their parents to a distance and compelled to locate elsewhere.

The top of the head and the whole of the upper surface is of a soft olive brown, the wings and tail being darkest; face, throat, and breast fine ferruginous red; the rest of the under surface dull white; bill, irides, and tarsi blackish brown.

The young when fully fledged have the upper surface of a deep brown, thickly speckled with dots of yellow; the chest is slightly tinged with ferruginous, each feather having a dark brown margin; under surface greyish white.

The Plate represents an adult male and female, and a fully fledged young bird, of the natural size.

ALPINE ACCENTOR.

Accentor Alpinus: *(Bechst)*

Drawn from Nature & on Stone by J & E. Gould.

Printed by C. Hullmandel.

Genus ACCENTOR, *Bechst.*

GEN. CHAR. *Bill* strong, straight, of mean length, and drawn to a fine point; the tomia of both mandibles bending slightly inwards, and the upper mandible emarginated. *Nostrils* basal, naked, and pierced in a large membrane. *Legs* strong. *Toes* three before and one behind; the outer one joined at its base to the middle toe. *Wings* having the first quill very short, and the second a little shorter than the third, which is the longest.

ALPINE ACCENTOR.

Accentor alpinus, *Bechst.*

Accenteur pegot ou des Alpes.

THE genus *Accentor* is extremely limited in the number of its species, and, with the exception of an undescribed bird from the Himalayan mountains, the members are confined to Europe.

The native habitat of the Alpine Accentor are the bleak and mountainous parts of the Continent, and, as its name implies, it gives preference to the Alps, where it dwells in districts of the most abrupt and rocky nature. It is extremely common in Switzerland and the Tyrol, ascending in summer to their most elevated portions, and seeking shelter, as winter advances, in their valleys and central regions. Several specimens have been taken in England, and if we recollect right the Rev. Dr. Thackeray, Provost of King's College, Cambridge, informed us that he observed two examples of this singular and rare bird in the garden of King's College; from the great interest he has always taken in the study of our native birds, two more welcome visitors could scarcely have come under his notice: one of them, we believe the male, was obtained, and now enriches the extensive and valuable collection of this worthy gentleman.

Its food consists of insects and their larvæ, worms, grubs, &c. It is also said to destroy grasshoppers and small locusts, which abound in alpine regions.

It breeds in the holes and fissures of the rocks, laying four or five greenish blue eggs, which, though a little larger, are not otherwise unlike those of the common Hedge Accentor of England.

In this well-defined and very natural group, we find the sexes of all the known species to be so strictly similar in the colouring of their plumage as to present no outward difference in their markings; neither do the young offer any material deviation, possessing as they do at an early age the general style of dress, but wanting that brilliancy and decisiveness of marking which characterize the adults.

Our Plate represents a male and female, corresponding in every particular with the bird taken at Cambridge, which, as Mr. Selby informs us, formed the subject of the Plate in his admirable work on the birds of the British Islands.

The crown and upper surface grey brown, the feathers on the back having their centres darker; scapularies and tertials deep brown edged with chestnut brown; greater and lesser wing-coverts black with a white space on the tip; quills blackish brown, with lighter edges; flanks chestnut brown, each feather being edged with greyish white; tail dark brown, each feather having a yellowish white spot on the inner web at the tip; legs light flesh brown; upper mandible dark horn-colour, the under one much lighter.

The figures are of the natural size.

HEDGE ACCENTOR.
Accentor modularis; *(Cuv.)*

Drawn from Nature & on stone by J. & E. Gould *Printed by C. Hullmandel*

HEDGE ACCENTOR.

Accentor modularis, *Cuv.*

L'Accenteur mouchet.

In every garden and in every hedgerow may this familiar but obscurely coloured bird be seen, not only throughout the whole of Great Britain, but nearly the whole of Central Europe. Though strictly belonging to the *Sylviadæ*, it is one of the few that make our island a permanent place of residence; it is also one of the hardiest of our small birds, and appears to brave the severest winters with indifference. It may be observed when the ground is frozen, and even covered with snow, as lively and alert as at other times in search of its food, which lies concealed on the surface of the earth, or among the dead leaves on banks, bottoms of hedgerows, &c.; often, indeed, will it mingle with the common Sparrow and the Robin, entering the farmyard, and approaching within the precincts of human habitation, and displaying great confidence and familiarity. Its actions and manners are strictly terrestrial, which is to be accounted for principally from the circumstance of its food, being mostly obtained on the ground: it progresses by a succession of short hops, inquisitively prying among the grass and leaves in search of insects, small worms, the seeds of plants, &c. During the spring the male pours forth its song, which although not characterized by any great compass of scale, is nevertheless agreeable, and is not entirely suspended during the winter months; this fact is confirmed by Cuvier, who informs us that it cheers that season with its pleasant song: we also learn from this celebrated naturalist, that although a winter visitant in France, it retires northwards in spring to breed, which is certainly not the case in our own island, as its nest and beautiful blue eggs are well known to every schoolboy. It is an early breeder, frequently beginning to build in the month of March. The nest is usually placed in the thickest part of the hedgerow, and very frequently among furze and evergreens; it is generally composed of moss and wool, intermingled with fine roots and slender bits of twig, with a lining of hair: the eggs are four or five in number, of a beautiful azure blue.

We have never seen this bird in any collection made out of Europe; and, in fact, so exclusively are the species of the genus *Accentor* confined to this portion of the world, that we have never observed more than a single example of any species in foreign collections; this was a new one from the Himalaya mountains.

The plumage of the sexes is so strictly similar that it is almost impossible, without actual dissection, to distinguish the male from the female; the former, however, has the breast of a more decided grey tint, which feature is also more conspicuous during the spring.

The general colour of the whole upper surface is deep dusky brown, with blotches of a darker colour disposed over the back; earcoverts brown minutely dotted with white; under surface greyish white with a few dashes of brown on the flanks; legs light brown; bill blackish brown

The Plate represents a male and female of the natural size.

MOUNTAIN ACCENTOR.

Accentor montanellus, *(Cuv.)*

Drawn from Nature & on Stone by J. & E. Gould. *Printed by C. Hullmandel.*

MOUNTAIN ACCENTOR.

Accentor montanellus, *Temm.*

L'Accenteur montagnard.

THE most singular circumstance connected with the history of this little bird is, that the specimen from which our figure was taken is the only example we have ever seen, after having visited nearly all the continental collections.

On the authority of M. Temminck we give as its habitat the eastern portions of the middle of Europe, and the same latitudes in Asia. He adds that it was found by Pallas in eastern Siberia and in the Crimea, and that it is somewhat common in the Neapolitan States, in Dalmatia, and in the middle of Hungary.

The specimen above alluded to forms a part of the fine collection of the Imperial Cabinet of Natural History at Vienna, and was killed near the river Krems in Austria, by the late Rev. M. Kratki, curate of Mausling, in the year 1790. Before we had an opportunity of examining this "*rara avis*," we were inclined to believe it might be a variety of the common species (*Accentor modularis*), but we are now fully satisfied that our suspicion was groundless, and we can safely add our testimony to its specific value: it is moreover a typical example of the genus, and in affinity closely resembles the common Hedge Sparrow, from which it may be readily distinguished by the conspicuous stripe of buff over the eye, and by the general tawny hue of its under surface.

We cannot close our account of this interesting bird, without expressing our warmest thanks to the Directors of the Imperial Cabinet of Vienna, who, solely for the promotion of science, have encountered the risk of forwarding this valuable specimen from Vienna to London, for the purpose of enabling us to include a figure of it in "The Birds of Europe."

Crown of the head, and a broad stripe commencing at the base of the bill and running towards the back of the head deep brownish black; over each eye a broad and conspicuous stripe of buff; back and scapularies reddish ash with large longitudinal dashes of reddish brown; wings brownish ash bordered with reddish, the tips of the greater and lesser coverts yellowish, forming a double band across the wing; tail brown; all the under surface dull buff, varied on the breast by small dashes of brown, and on the flanks with longitudinal spots of reddish ash; bill yellow at the base and brown at the tip; feet brownish yellow.

Our figure is of the natural size.

REED LOCUSTELLE.
Locustella fluviatilis.

Genus LOCUSTELLA.

GEN. CHAR. Those of SALICARIA, excepting that the beak is deeper at the base and runs more to a point; rictus bristles rudimentary; the hind claw longer, remarkably slender and clean made; and the tail broader and more decidedly graduated.

REED LOCUSTELLE.

Locustella fluviatilis.

Le Bec-fin riverain.

SINCE we learn from M. Natterer that this bird closely assimilates to the Grasshopper Warbler, *Locustella sibilans*, in habits, manners, note, particularly in a shrill inward tone producing the effect of ventriloquism, place of resort, food and nidification, and as, conjoined to these affinities, we find that they agree still more closely in structural form, which differs considerably from that of the birds forming the genus *Salicaria* of Selby, particularly in the lengthened hind claw and in the shape of the beak, we feel no hesitation in assigning to these nearly allied species a distinct generic situation, under the restored title of *Locustella*. The specific term of *fluviatilis* as applied to the present bird conveys an erroneous impression, for M. Natterer informs us that although it resorts to low situations, it nevertheless does not confine itself to reed-beds, but rather prefers swampy coppices and thickets. It is extremely scarce in the western portion of the European continent; but is plentiful in Austria and Hungary, and is very common in the island-gardens in the Danube near Vienna.

The sexes exhibit little or no difference of plumage, nor does it appear to undergo any periodical changes.

The upper parts are olive clouded with brown; under surface lighter; the throat and breast whitish spotted with brown; under tail-coverts greyish white; bill and tarsi brown.

We have figured an adult male of the natural size.

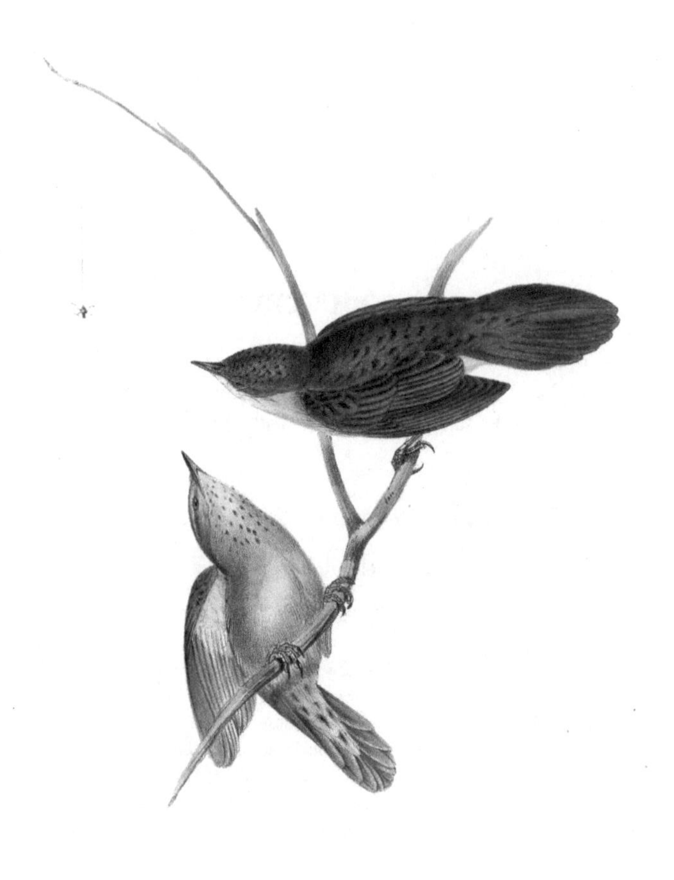

BRAKE LOCUSTELLE.
Locustella avicula; *(Ray)*.

Drawn from Nature & on stone by J & E. Gould.

Printed by C. Hullmandel.

BRAKE LOCUSTELLE.

Locustella avicula, *Ray*.

La Bec-fin locustelle.

This species, to which the modern name of Brake Locustelle has been applied, as more appropriate than those of Grasshopper Lark and Grasshopper Warbler, by which it is known to most of our readers, is one of the migratory birds of our island, where it arrives in the month of April, and although not an uncommon bird, its secluded and shy disposition renders a sight of it extremely difficult, and were it not for its peculiar and sibilant ringing cry, repeated for many minutes and producing a kind of ventriloquism, its presence would seldom be detected. By the term Brake Locustelle our readers are at once informed of the favourite localities to which this species gives preference, in contradistinction to its near ally the Reed Locustelle, whose habits lead it to frequent wet and swampy situations, reed-beds, &c.

It is pretty generally distributed over our island, but becomes more scarce as we proceed northward. Mr. Selby states that he has known it for some years past as a visitant of several low and damp situations in Northumberland, which would appear to be its limit in that direction: according to Montagu it is also an inhabitant of Ireland.

The nest of the Brake Locustelle is constructed among the densest bramble or furze bushes, and is so secretly placed as to be very seldom found; it is formed of moss and the dried stems of ladies' bedstraw, and greatly resembles that of the Whitethroat, but is thicker and more compact in texture: the eggs, which are four or five in number, are pinkish grey with numerous specks of a deeper tint.

Upper surface brown, tinged with olive; the centre of each of the feathers, except those on the rump, dusky brown; throat white, bounded by a circle of small oval brown spots; breast and flanks pale brown tinged with olive, fading into greenish white on the middle of the belly; under tail-coverts greyish white with black shafts; quills and tail dusky, margined with pale brown tinged with olive; bill brown; legs and feet pale yellowish brown.

The sexes are not distinguishable by their plumage, further than that the male has the spots on the throat more conspicuous.

We have figured a male and female of the natural size.

WILLOW LOCUSTELLE.
Locustella luscinoïdes.
Sylvia luscinoïdes; *(Savi.)*

Drawn from Nature & on Stone by J & E. Gould.

Printed by C. Hullmandel.

WILLOW LOCUSTELLE.

Locustelle luscinoïdes.

Sylvia luscinoïdes, *Savi.*

Le Bec-fin des saules.

This elegant little bird, the *Sylvia luscinoïdes* of Professor Savi, constitutes the third example, as far as is hitherto known, of the restricted genus *Locustella*, of which our Grasshopper Warbler is the British representative: it is confined to the southern regions of Europe, where it appears to be very limited in the range of its habitat. According to Professor Savi it arrives in Tuscany about the middle of April, taking refuge among the willows, reeds, and other luxuriant plants of marshy districts, which afford it a secure retreat. It readily admits of approach, may be seen among the lower branches near the ground, and also runs nimbly on the earth among the reeds.

Its food consists of insects and their larvæ.

The *Locustella luscinoïdes* is one of the birds lately added to the Fauna of Europe, and we have yet to learn all that respects its nidification and winter retreat, which we may reasonably conclude is Africa.

Head, all the upper surface, and tail reddish brown, the latter almost imperceptibly barred with lines of a darker tint; throat whitish; stripe over the eye, sides of the neck, and all the under surface pale buff; bill dark brown; feet pale brown.

Our figure is taken from a specimen kindly lent to us for the purpose, by the Directors of the Imperial Cabinet at Vienna, and represents the bird of the natural size.

CREEPING LOCUSTELLE.
Locustella certhiola.

Drawn from Nature & on Stone by J & E Gould. Printed by C. Hullmandel.

CREEPING LOCUSTELLE.

Locustella certhiola.

Sylvia certhiola, *Pall.*

Le Bec-fin trapu.

We beg leave to express our sincere thanks to Professor Lichtenstein, one of the Directors of the Royal Museum of Berlin, for the obliging manner in which he has entrusted to our care the original specimen of this rare bird, from which Pallas took his description, and which we are given to understand was mounted by that celebrated naturalist himself. In form and general colour, and doubtless in its habits, this species strictly belongs to the genus *Locustella*, from the two other species of which it may be distinguished by its larger size, and by the greyish white termination of all its tail-feathers. Of its habits and manners no account has been recorded, and we only know that it is found in Southern Russia, where it appears to be very scarce. We may here remark, however, that the secluded habits of this race prevent their being observed unless they are very closely watched, and that by the eye of one accustomed to the investigation of the manners of birds in a state of nature.

Feathers on the crown of the head, back of the neck, back and wing-coverts olive brown tinged with red, each feather being dark brown in the centre; primaries and tail reddish brown, the latter tipped with greyish white; throat white; stripe over the eye, sides of the head, and all the under surface buff, which becomes very pale on the breast and centre of the abdomen; bill and feet light brown.

The figure is of the natural size.

GREAT SEDGE-WARBLER.
Salicaria turdoïdes; *(Selby)*.

Drawn from Life and on Stone by J. & E. Gould.

Printed by C. Hullmandel.

Genus SALICARIA, *Selby.*

Gen. Char. *Bill* straight, subulate, expanded at the base, with a distinct *culmen,* compressed towards the tip, which is slightly deflected and emarginated. *Tomia* straight, those of the under mandible being gently inflected. *Nostrils* basal, lateral, oval and exposed. *Forehead* narrow and depressed. *Wings* rather short; the first quill nearly abortive; the second just shorter than the third, which is the longest of all. *Tail* rather long and rounded. *Legs* having the tarsi longer than the middle toe. *Feet* rather large and stout; the hind toe large and strong. *Claws* moderately curved, long and very sharp, that of the hind toe being double in size and strength to any of the others.

GREAT SEDGE WARBLER.

Salicaria turdoïdes, *Selby.*

Le Bec-fin rousserolle.

Under the generic title *Sylvia,* Dr. Latham has included all the soft-billed birds,—an immense multitude, differing in characters and manners. This arrangement, to a certain extent, has been adopted by M. Temminck, with an advance, indeed, towards those subdivisions so imperatively demanded by the laws of nature. The first section is that termed "*Riverains,*" and comprehends a tribe whose habits lead them to frequent the borders of lakes, marshes and rivers, where the reeds and flags afford them an asylum, their food consisting of such insects as abound in these situations. The birds of this section now form the genus *Salicaria* of Mr. Selby, which we concur in the propriety of thus instituting, as it is at once natural and necessary. Of this genus the present bird may be considered a typical example; for though larger than the other species, yet in form, habits and manners it strictly assimilates with them, frequenting the morasses of Holland in great abundance, as also the low flat lands of France, even in the neighbourhood of Calais, though, strange to say, it does not appear to cross the Channel to England.

The *Salicaria turdoïdes* is a delightful warbler, whence it has obtained its specific term: its notes are hurried and chattering, like those of the Sedge Warbler, but louder, in accordance with its superior size, and of a richer tone.

Like the rest of the genus, its food consists of gnats, the smaller *Libellulæ,* and other aquatic insects. Its nest is situated among the stalks of the growing reeds, like that of our well-known Sedge Warbler. The eggs are five in number, obtuse, greenish white spotted with black and ash colour.

There is no sexual difference of plumage,—a circumstance which characterizes nearly the whole of the species belonging to this genus.

The upper surface, wings and tail, with the exception of a white stripe over the eye, of a uniform light brown; the under surface white delicately tinged with the same colour; beak brown, darkest along the culmen and at the tip; tarsi light brown.

The Plate represents an adult of the natural size.

OLIVE-TREE SALICARIA.
Salicaria Olivetorum; *(Strickl.)*

Drawn from Nature & on Stone by J & E. Gould. *Printed by C. Hullmandel.*

OLIVE-TREE SALICARIA.

Salicaria Olivetorum, *Strickl.*

For the knowledge of this new and elegant species of Warbler we are indebted to H. E. Strickland, Esq., by whom it was discovered during the spring of 1836, in Zante, one of the Ionian Islands; and we feel assured that this addition to the European Fauna will be viewed with considerable interest by all lovers of Ornithology, but by no one more than ourselves, who have for a long time entertained a belief that more new species will yet be found to inhabit the smaller islands of the Mediterranean, particularly those in the Grecian Archipelago.

From its being nearly allied to the Great Sedge Warbler, Mr. Strickland has provisionally placed this bird in the genus *Salicaria*; but it will, if we mistake not, together with two or three others, be found sufficiently distinct from the smaller members of that genus to warrant their separation under a distinct generic title. We have no recollection of having seen this bird in any collection either from Africa or India; which leads us to believe that the southern parts of Europe constitute its true and native habitat. Mr. Strickland having obliged us with some short notes on this species, we have considered it best to transcribe them here in his own words.

"This bird belongs to that division of the *Salicariæ* in which the tail is but slightly rounded, and the colours sombre and uniform; including the *Sylvia arundinacea*, Lath.; *S. palustris*, Bechst.; *S. Turdoides*, Meyer, and other foreign species.

"I first noticed this bird, in May, 1836, at Zante, where it is by no means rare; but from its shy and restless habits I was only able to procure two specimens, both of which were males. One of these I gave to M. L. Coulon of Neufchatel, and the other is in my collection. It frequents the olive-groves, and is less aquatic in its habits than some of its congeners. Its note is a rambling warble, closely resembling that of *S. arundinacea.*

"From the proximity of Zante to the Morea, it is probable that this bird exists there also, but it seems hitherto to have escaped the notice of ornithologists.

"The male has the whole upper plumage greyish brown, with a tinge of olive; the space between the bill and the eye lighter; primaries and secondaries dark brown, the latter edged with whitish; tail slightly rounded, and of a dark brown, with the outer feather on each side margined all round, and the two next slightly tipped with white; under surface greyish white becoming darker on the flanks; chest and under tail-coverts tinged with yellow; legs and feet lead colour; beak, orange yellow at the base, darker towards the tip; irides hazel.

Total length 6 inches; wing, $3\frac{1}{2}$; tail, 3; tarsus, $\frac{7}{8}$; bill, from gape to tip, $\frac{5}{8}$."

Our figure is of the natural size.

REED WARBLER.
Salicaria arundinacea; *(Selb.)*

Drawn from Nature & on stone by J. & E. Gould. *Printed by C. Hullmandel.*

REED WREN.

Salicaria arundinacea, *Selby*.

Le Bec-fin des Roseaux, ou Efarvatte.

THIS species, which is by no means uncommon in the British Islands, is, notwithstanding, much more local in its habits than its near ally the Sedge Warbler (*Salicaria phragmitis*, Selby), from which it may at all times be distinguished by its larger size, and by the uniform tints which pervade the upper surface. It bears a striking resemblance in most of its habits and manners to the species alluded to above, arriving in the British Islands at the same period, which is generally in the third week in April, when it retires to thick reed-beds, plantations of osiers, and the swampy borders of rivers. Its note, which is varied and pleasing, is not so harsh as that of the Sedge Warbler, but is delivered in the same kind of hurried and rapid manner. It also offers a little difference in its nidification, constructing a deep upright nest of the seed-tops of reeds and long grass, lined with the finer parts of the former, and which is almost invariably attached to the stems of several reeds, which are so intertwined as to form a firm support. The eggs are four or five in number, of a greenish white, spotted and blotched with brown and dull green.

On the Continent it appears to be universally spread in all the temperate latitudes wherever extensive lowlands covered with aquatic herbage afford it a shelter.

Its food consists of aquatic flies and their larvæ.

This species is very abundant in Holland, and is also found in some parts of France and Germany, but is still more rare in the South of Europe.

The whole of the upper surface is of a dull green with a tinge of brown, the edges of the quills being paler; throat, breast, and belly yellowish white, of a deeper tint upon the breast and flanks; between the mouth and the eyes a pale streak; eyelids pale yellowish white; legs dusky brown; bill pale brown.

We have figured a male and female of the natural size.

MARSH WARBLER.

Salicaria palustris.

MARSH WARBLER.

Salicaria palustris.

Le Bec-fin verderolle.

In its general contour this little bird so closely assimilates to the *Salicaria hirundinacea*, that it would be impossible from the simple examination of preserved specimens to discover that they are specifically distinct: in their habits and manners and in the places to which they resort they are also very similar; but the yellow lining of the mouth, the enlarged size of the bill, and the greener tint of the plumage, are points by which the Marsh Warbler may at all times be distinguished from its near ally. "Another characteristic of this species," says M. Temminck, "is its song, which is singularly varied; it has also considerable powers of mimicry, and readily imitates the song of other birds most completely, particularly that of *Sylvia hippolais*, as well as the notes of *Charadrius minor*, and the piercing cry of the *Hæmatopus ostralegus*."

It generally inhabits humid and marshy situations in the neighbourhood of water, bordered with willows and reed-beds, but is also frequently observed perched upon the high stems of hemp and bushes. It is common in all the middle parts of Europe, and is abundant on the banks of the Po and the Danube, and also in some parts of Switzerland, Germany, and Holland.

The nest is constructed with much art, is of a spherical form, and is placed on the ground among the roots of willows, reeds, and bushes; the eggs are four or five in number, of a clear ash, covered with spots of a bluish ash.

Its food consists of insects and small berries.

Crown of the head, all the upper surface, wing-coverts, secondaries, and tail greenish olive; primaries blackish brown; stripe over the eye, throat, and all the under surface yellowish olive; bill yellow at the base, black at the tip; feet lead colour.

We have figured an adult of the natural size.

SEDGE WARBLER.
Salicaria Phragmitis. *(Selby)*.

Drawn from Nature & on Stone by J & E. Gould.

Printed by C. Hullmandel.

SEDGE WARBLER.

Salicaria Phragmitis, *Selby*.

Le Bec-fin Phragmite.

THE Sedge Warbler may be distinguished from its near ally the *Salicaria arundinacea* by the conspicuous stripe which passes over the eye, by its smaller size, and by the less uniform style of colouring which pervades the back and upper surface.

The habits and manners of the two species are so strictly similar as regards the situations they inhabit, and every other particular, that the inexperienced naturalist would be in doubt as to which of them was before him. The bird here figured is by far the most numerous and widely distributed: arriving in the British Islands early in spring, it retires to marshes, banks of rivers, ponds, and lakes, where the luxuriant foliage abounding in such situations affords it a retreat at once secure and in direct unison with its habits. It soon commences the task of nidification by constructing a nest of coarse grasses, intermingled with moss on the outer side, while a lining of finer grasses completes the inner: the eggs are five or six in number, of a pale grey blotched all over with pale brown. The situation of the nest varies considerably, being often placed among the reeds which border the water, while at other times it is situated on the overhanging branches of the willow; and it not unfrequently happens that wet ditches, concealed by thick brambles, afford the Sedge Warbler a secure asylum, in which case the nest is placed in the centre of the thick herbage. In its disposition this bird is restless and noisy: its song is a confused strain, which is poured forth both night and day. It displays great powers of mimickry; among its notes may be distinguished those of the lark, the nightingale, sparrow, and linnet, jumbled together in a hurried babble: its varied song is uttered with greater vehemence when disturbed or irritated. Being somewhat secluded in its habits, it is not so frequently seen as heard. During its residence with us it seldom flies further than from bush to bush, or from one reed-bed to another; yet the power of extended flight has not been withheld from this little warbler, for as soon as its insect food diminishes, and the herbage which has afforded it shelter during the summer undergoes the least decay, the sedge bird is directed by the impulse of nature to seek its subsistence in countries of a warmer latitude, where it may still find an abundant supply.

The Sedge Warbler is found throughout the British Islands, as well as in nearly every portion of the Continent, being especially abundant in France, Germany, and Holland.

The sexes offer no differences in the plumage, and the young assume the adult colouring from the nest.

The top of the head is deep brown; above the eye is situated a distinct yellowish white stripe; back and wings olive brown, the centres of the feathers being darker; rump and upper tail-coverts yellowish brown; throat white; whole of the under surface yellowish white, becoming stronger on the flanks; primaries and tail-feathers brown; bill and legs brown.

The Plate represents a male, of the natural size.

1. MOUSTACHED WARBLER.
Salicaria Melanopogon.

2. AQUATIC WARBLER.
Salicaria aquatica.

Drawn from Nature & on stone by J & E. Gould.

Printed by C. Hullmandel.

MOUSTACHED WARBLER.

Salicaria melanopogon.

La Bec-fin à moustaches noires.

THIS little warbler, now so common in European collections, appears to have been unknown to M. Temminck when he published the second edition of his 'Manuel' in 1820; subsequently, however, he gave a figure of the adult male in the 'Planches Coloriées', 245. fig. 2. and has given a description in the third part of his 'Manuel', which appeared in April 1835.

From the little information acquired respecting this bird, it would appear to be principally an inhabitant of the Roman States, where M. Cantraine informs us it is abundant in the marshes among the *Arundo speciosa*: "I have killed it in November near Rumbla, in the circle of Ragusa, at Ostia, and near the lake of Castiglione, where it is very common; it is always in the marshes, and in the thickets that border them, clinging to the rushes and uttering a very loud cry; it descends to the surface of the water and walks upon the aquatic plants." From this account of M. Cantraine we gather sufficient information to satisfy us that in its general habits and economy the Moustached Warbler strictly resembles those members of the genus *Salicaria* that inhabit our island.

Of its nidification nothing is known.

Its food consists of small coleopterous insects.

The sexes offer no perceptible difference in the markings of their plumage.

The top of the head and all the upper surface is of a dark brown, with a slight tinge of reddish brown, particularly on the margins of the wing-feathers, and a black mark down the centre of each of the feathers on the back; a greyish white stripe passes from the bill over the eye and extends to the posterior part of the head, below this is a conspicuous band of dark brown, which passes through the eye and over the ear-coverts; throat and under-surface greyish-white, becoming brown on the flanks and sides of the chest; bill black at the tip and yellow at the base; legs and feet brown; irides yellow.

The upper figure in our Plate represents an adult male of the natural size.

AQUATIC WARBLER.

Salicaria aquatica.

La Bec-fin aquatique.

ITALY and the eastern portions of the Continent appear to be the true habitat of this species, which, although differing considerably in its markings and colour from the Moustached Warbler, resembles it in many particulars; like that it is also a marsh bird and dwells among the thick reed-beds that border the sides of rivers. It is very abundant in Italy and Piedmont; it sometimes extends its visits to France and Germany, and, but very rarely, to Holland.

The nest is artfully constructed among the stems of aquatic plants, and the eggs are four or five in number, of a yellowish ash colour marked with very fine spots of greyish olive.

The sexes are alike in plumage.

A band of yellowish white passes over each eye and a stripe of the same colour down the centre of the head; the intermediate spaces dark brown; sides of the neck, scapularies, and all the upper surface yellowish brown with large, longitudinal dark brown spots; wings and tail dark brown, each feather in the former strongly edged with yellowish brown; ear-coverts brown; throat and all the under surface of a light fawn colour, which is somewhat darker on the flanks; feet and legs pale brown; bill dark brown at the tip, yellowish at the base.

The lower figure represents this bird of the natural size.

RUFOUS SEDGE WARBLER.
Salicaria galactotes; *(Mihi)*

Drawn from Nature & on Stone by J & E Gould.

Printed by C. Hullmandel.

RUFOUS SEDGE WARBLER.

Salicaria galactotes, *Mihi.*

According to M. Temminck, the introduction of this lovely species to the fauna of Europe is due to M. Natterer of Vienna, whose researches have conferred so much honour on himself and benefit to science at large.

Although we possess several fine specimens of this rare species, still, from the want of an intimate knowledge of it, we are yet in doubt as to the true situation it ought to occupy in a scientific arrangement. From a careful examination of the specimens in our possession, in comparison with the birds of M. Temminck's section denominated *Bec-fins riverains*, which have been formed by Mr. Selby into the genus *Salicaria*, we are led to assign it to that group; at the same time, we suspect that it may hereafter be found to form an intermediate link between the genus *Salicaria* and an allied group, the species of which, instead of inhabiting reed-beds and swampy situations, frequent the tall grasses of dry and sandy places, and of which the *Sylvia cisticola* is an example. On referring to the work of our valued friend M. Temminck, we find him expressing the same doubts respecting this bird which we ourselves entertain; his words are, "I know not whether this species inhabits reed-beds and the borders of waters: I have therefore arranged it provisionally in this section; for the knowledge alone of its manners and its habits can determine truly the place to which it should be assigned, whether in the section of *Riverains* or, on the contrary, of *Sylvains*."

The Rufous Sedge Warbler is a native of the southern provinces of Spain, and probably also of the opposite shores of Africa. M. Natterer discovered it at Gibraltar, and killed two pairs at Algesiras.

The general plumage of the upper surface is lively rufous; the tail-feathers being tipped with white, above which is a larger spot or bar of deep black; the quill-feathers are light brown; a brown band goes from the beak to the eye, and a white superciliary line passes over the eye; the under surface is dull yellowish white, becoming reddish on the flanks; tarsi yellowish; beak brown; irides hazel.

We have figured an adult bird of the natural size.

FANTAIL WARBLER.
Salicaria Cisticola.

Drawn from Nature & on Stone by J. & E. Gould.

Printed by C. Hullmandel.

FANTAIL WARBLER.

Salicaria cisticola, *Mihi.*

Le Bec-fin cisticole.

Not having had personal opportunity of inspecting the habits and manners of this interesting little bird in a state of nature, we are unable to say whether it should form the type of a new genus, or whether it really belongs to that of *Salicaria,* in which we have provisionally placed it; at all events it cannot be far removed from that genus. It is a species which makes the southern and eastern portions of Europe, together with the adjoining parts of Asia and Africa, its habitat. It is generally distributed along the shores of the Mediterranean from Gibraltar to Constantinople; is common in the Greek islands and the adjacent mainlands, and is also found in Italy and Sicily. It frequents low and swampy places covered with tall grasses, and, like the Reed Wren, constructs a nest preeminently curious and beautiful, excelled by none of a similar character. Although incapable, from its small size, of entwining the larger reeds, it avails itself of the tall blades and stalks of grass, among which it places its nest; these it does not draw together in the manner of the Reed Wren, but by piercing each blade, and drawing the whole together by means of cottony threads, secured at each perforation with a knot so ingeniously executed as to appear the work of reason. Between the grasses thus secured it places the body of the nest, which is composed of vegetable fibres lined with a kind of flocculent down, collected from various plants. The eggs are four or five in number, and are said to be of a bluish flesh colour. Dr. Latham is our authority for asserting that it is found in the neighbourhood of Gibraltar, everywhere darting about with vast alacrity among the bushes. When disturbed it takes long flights, chirping all the way with a remarkably loud and shrill note, and when in motion it erects the tail and spreads it in a circle, which appears very beautiful; hence the very appropriate name of Fantail.

The male and female are so nearly alike in colour as to require no separate description; the tail of the male, however, is somewhat the more elongated of the two.

The whole of the upper surface is brown, each feather having a dark centre so disposed as to produce a multitude of longitudinal dashes; the whole of the under surface is brownish white; tail graduated, all the outer feathers having a black spot near the extremity; the tip being white; beak and tarsi light brown.

The Plate represents an adult bird in its perfect plumage, together with a nest, of the natural size.

CETTI WARBLER.

Sylvia Cetti; *(Marm.)*

Drawn from Nature & on stone by J. & E. Gould. *Printed by C. Hullmandel.*

CETTI WARBLER.

Salicaria? Cetti.

La Bec-fin bouscarle, ou Cetti.

We have not been able satisfactorily to determine the true situation of this curious little Warbler: in its general contour and also in its actions, it strongly resembles the species of the genus *Troglodytes*, or true Wrens, while at the same time it possesses many characters that ally it to the Reedlings, *Salicariæ*, among which we have provisionally placed it. M. Temminck states in his Manuel that it has been killed in England; but on this point we fear that this eminent naturalist must have been misinformed, as we ourselves have never been able to ascertain the existence of any authenticated British-killed specimen. We are inclined to consider this bird as strictly a native of the southern and eastern portions of the European continent and the northern regions of Africa. M. Cantraine informs us that he has found it in the marshes of Ostia, on the borders of the Lake Castiglione, and in the neighbourhood of Rome generally. It is abundant in Sicily, and Professor Savi states that it is a common species in Tuscany, where it breeds, building in large thickets near the ground: the nest is composed of the leaves and stems of dried grasses, and the eggs are of a reddish brown without spots.

The sexes appear to be perfectly similar in the colouring of their plumage, which may be briefly described. All the upper surface of a deep rich brown, passing into blackish brown on the quills and tail-feathers; an obscure line of greyish white over each eye; throat and under surface greyish white, with a tinge of brown on the flanks; beak and feet brown.

The Plate represents two adults of the natural size.

SILKY WARBLER.
Salicaria? sericea.
Sylvia sericea, *(Natt.)*

Drawn from Nature & on Stone by J & E. Gould

Printed by C. Hullmandel.

SILKY WARBLER.

Salicaria? sericea.

Sylvia sericea, *Natt.*

Le Bec-fin soyeux.

We have been favoured with the original specimen of this rare bird by our respected friend M. Natterer, who procured it himself on the 17th of April, 1817, " near Santa Anna, upon the river Brenta, two miles from Chioggia, where it inhabits the low bushes bordering the ditches between the vineyards near the Brenta; and has a loud and tolerably fine song."

This bird is very closely allied to the Cetti Warbler, and with that species will, we conceive, constitute a minor division among the "*Riverains*" of M. Temminck; but as we have already figured the Cetti Warbler under the generic appellation of *Salicaria*, we refrain from entering into any further details on the subject. The most striking differences which these birds exhibit when compared with the true *Salicariæ*, consist in the total absence of the stiff hairs at the base of the bill; in the more rounded form of the head; in the thicker and more silky plumage, and in the more rounded wing, the primaries of which are soft and yielding. The general contour of the Cetti Warbler, together with its slender bill, immediately reminds us of the true Wrens (*Troglodytes*); not that we have any reason to believe these groups are joined by affinity, although, particularly in the position in which we have figured it, the Cetti Warbler is remarkably Wren-like, but as we have not seen this bird in a state of nature, we are unable to say anything positive on the subject.

Of its habits, manners, nidification, and eggs nothing is known.

All the upper surface, wings, and tail greyish brown; stripe over the eye, throat, breast, and centre of the abdomen pure white; sides of the face, flanks, and under tail-coverts brownish ash; upper mandible dark brown; under mandible and legs flesh colour; irides dark brown.

We have figured an adult male of the natural size.

NIGHTINGALE.

Philomela luscinia; *(Swains.)*

Drawn from Nature & on Stone by J. & E. Gould. *Printed by C. Hullmandel.*

Genus PHILOMELA, *Swains.*

GEN. CHAR. *Bill* of mean length, straight; culmen rounded; tip of the upper mandible slightly deflected and emarginated; lower mandible as strong as the upper; gape smooth. *Nostrils* basal, lateral, round, pierced in a large membrane. *Wings* of mean length; first quill very short; second of the same length as the fifth; third and fourth nearly equal and the longest. *Tail* slightly rounded. *Tarsi* long. *Feet* adapted for perching, and also for hopping upon the ground. *Claws* moderately curved and very sharp.

NIGHTINGALE.

Philomela luscinia, *Swains.*

Le Rossignol.

IN our notice of this exquisite songster, which has been the theme of poets in every age, we shall confine ourselves rather to details connected with its habits and the localities it prefers in our own island, its migration, &c., than to the merits of its vocal powers or to indulging in strains of useless admiration.

So much attention has been lately paid by Mr. Blyth to the migration and localities frequented by this species, that we consider it only our duty to refer our readers to that gentleman's paper on the subject published in the 15th and 16th Nos. of the Analyst, which will require but little addition of our own to render the matter clear to every one.

In our island it appears to be confined to particular districts; it is plentiful in the southern and eastern counties, while Devonshire appears to be its limit westward, and Doncaster in Yorkshire in a northern direction, few if any authenticated instances being on record of its occurrence beyond that town, which is the more singular as Nightingales are common in Sweden and other countries situated further north than England.

Our own observation respecting the migrations of the Nightingale is, that after leaving our island it proceeds to the opposite shores of the Continent, and gradually makes its way southwards until it arrives in Africa, which is its ultimate resting-place during our winter mouths. We have ourselves received specimens killed in the northern districts of Africa, but have never obtained any from the central or southern parts of that portion of the globe; it would appear therefore that its distribution over that vast continent is comparatively limited. In no part of Europe is it more abundant than in Spain and Italy; from whence however, equally as from our own, it regularly migrates on the approach of winter.

The Nightingale is exceedingly shy in its habits; and inhabiting low and swampy coppices, close thickets, hedges, and similar situations, it is seldom seen, its retreat being only discovered by means of its peculiar call-note and its song, which for richness and power is unrivalled until he is mated, but as soon as this takes place his notes are only poured forth at intervals and entirely cease previous to migration, when the sexes separate, and the males precede the females by ten days or a fortnight.

The nest is placed on the ground or on a low stump, and is constructed of withered leaves, sometimes lined with dry grass: the eggs, which are of a plain yellowish brown, are from four to six in number. The young appear to be principally fed with small green caterpillars, "in all probability the larvæ of some moth, or perhaps of a *Tenthredo,* peculiar to some localities."

The food of the adult consists of insects and their larvæ, berries, and fruit.

The sexes are alike in plumage, and may be thus described:

All the upper surface deep rich brown; rump and tail reddish brown; throat and middle of the belly greyish white; sides of the neck, breast, and flanks grey; bill and legs light brown.

We have figured an adult male of the natural size.

THRUSH NIGHTINGALE.

Philomela turdoïdes; *(Blyth.)*

Drawn from Nature & on stone by J. & E. Gould.

Printed by C. Hullmandel.

THRUSH NIGHTINGALE.

Philomela Turdoides, *Blyth.*

La Bec-fin Philomèle.

In adopting Mr. Blyth's specific term for this bird, we have been influenced by the desire of paying a just compliment to a young and ardent naturalist, who has taken much pains in elucidating the habits of this, as well as of many other groups in ornithology. "This bird," says he "may be said to connect our Common Nightingale with the Tawny Thrush, or more immediately, perhaps, the Tiny Thrush (*M. parva*). It is described by Bechstein to have 'the whole plumage generally, and in all parts, deeper and darker than the common species. The head is larger and the *beak thicker*; the throat white bordered with black; the breast brown, with *darker spots*;' and it is said, also, to be considerably larger, and longer by about an inch and a half. It is a very loud songster, and sings chiefly by night, but its voice is by no means so melodious as that of the Common Nightingale. 'It has,' continues Bechstein, 'a much stronger, louder, and deeper voice, but it sings more slowly and more unconnectedly; it has not that astonishing variety, those charming protractions and harmonious conclusions of the Common Nightingale; it mutilates all its strains, and on this account its song has been compared to that of the Missel Thrush, to which however it is superior both in softness and purity. The Common Nightingale is superior in delicacy and variety, but inferior in force and strength, while the voice of the larger species is so loud, that it is almost impossible to bear with it in a room.' Its call-note, &c., as described by the same author, is also very different. 'In cages,' observes Bechstein, 'they are fed like Nightingales, but are less delicate, and generally live much longer;' which, in fact, is another approximation to the Thrushes." And this approximation, we may add, is still further indicated by the spots on the breast.

The habitat of the Thrush Nightingale extends over Silesia, Bohemia, Pomerania, Franconia, and other parts of Germany; it is said to be more abundant than the common species in Hungary, Austria, and Poland; more rare in France, and never seen in Holland. It is generally found in woods situated on the tops of hills, and also in plains, particularly those in the neighbourhood of running streams.

Like the common species it feeds upon worms, flies, moths, currants, elderberries, &c.; the nest is also built in small thickets, but more frequently placed in low and damp situations; the eggs, which are larger than those of the Nightingale, are brownish olive stained with deep brown.

The sexes do not differ in the colouring of their plumage, which may be thus described:

All the upper surface brown; tail rich brownish red; throat whitish; breast clear greyish brown with dashes of dark grey; under surface whitish ash; bill brown; legs brownish flesh-colour.

Our figure is of the natural size.

GORGET WARBLER.

Calliope Lathamii.

Drawn from Nature & on stone by J. & E. Gould. *Printed by C. Hullmandel.*

Genus CALLIOPE.

GEN. CHAR. *Bill* shorter than the head, straight, compressed laterally and pointed, with a slight indication of a notch near the tip of the upper mandible: base of the bill garnished with a few fine and short bristles. *Nostrils* basal and oval. *Wings* rather short and rounded, the first quill very short, the third and fourth the longest. *Tail* short and rounded. *Tarsi* long and slender, the hinder toe furnished with a large strong claw.

GORGET WARBLER.

Calliope Lathamii.

La Calliope.

AMONG the subdivisions into which the *Sylviadæ* are now distributed, we do not find one to which we can strictly refer the present beautiful bird; we have therefore ventured to form a new genus for its reception, taking the specific appellation of Pallas for its generic designation. We cannot perceive its immediate relationship to the Accentors, to which M. Temminck has referred it: both Gmelin and Latham have considered it to be a Thrush (*Turdus*); but although not ranging with any established genus of that family, its form is very similar to some of the smaller species of that group. In the silky character of its plumage, in the presence of the gorget, and in the great difference between the sexes, it evinces a close affinity to the Blue-throated Warbler (*Phœnicura Suecica*), but in its general form and contour it approximates to the Nightingale (*Philomela Luscinia*), to which we consider it to be most nearly allied.

In naming this species after the venerable Dr. Latham, we are influenced by a desire to render a tribute of respect to one who has laboured much in the science of ornithology, and who at an extremely advanced age is now cheerfully passing the remainder of his days in the enjoyment of every domestic felicity, universally honoured by all his contemporaries.

Although the Gorget Warbler has in a few instances been taken within the precincts of Europe, its true habitat is the north-eastern portions of Asia, being a native of Siberia, Kamtschatka, and the island of Japan. Of its nidification and general economy we know but little: it is said to have an agreeable song, which it utters while perched on the topmost branches of trees.

The male has the whole of the upper surface of a uniform olive brown; over the eye a clear stripe of white; a black space between the beak and the eye; a white moustache beneath the eye; from the base of the under mandible a gorget of fine scarlet spreads over the throat; chest greyish brown; flanks brown; abdomen whitish; bill blackish brown; feet light fleshy brown.

The female differs from the male in wanting the black space between the bill and the eye, in the absence of the white moustache, and in being destitute of the beautiful gorget.

The Plate represents a male and female of the natural size.

ORPHEUS WARBLER.

Curruca Orphea; *(Mihi)*.

Drawn from Nature & on Stone by J. & E. Gould. *Printed by C. Hullmandel.*

ORPHEUS WARBLER.

Curruca Orphea, *Mihi.*

Bec-fin Orphée.

Although the present species differs in a trifling degree from the more typical examples of the genus *Curruca*, especially in having a stouter form of beak, which is more deep than wide, we do not feel ourselves at liberty to separate it on such slender grounds, as its general habits and form overbalance the minutiæ alluded to.

The Orpheus Warbler is an inhabitant of the southern provinces of Europe, and we have more than once received it in collections from India. According to M. Temminck, it is very abundant in Italy, particularly in Piedmont and Lombardy, and the southern departments of France. It is accidentally met with in Switzerland and the adjacent districts, but never occurs in more northern latitudes. On referring to the valuable little work of Professor Savi on the Ornithology of Tuscany, we learn that it is there a migratory bird, and much resembles in habits and manners the Common Whitethroat (*Curruca cinerea*, Bechst.). Its food consists of insects and berries, and it builds in bushes often in company with others of the same species. M. Temminck states, that in addition to bushes it also selects holes in ruins, old walls, or under the eaves of isolated buildings, as a site for incubation. The eggs are four or five in number, nearly white, irregularly marked with yellowish blotches and small brown dots.

The male has the top of the head and ear-coverts brownish black ; the whole of the upper surface is of a cinereous brown, with a tinge of olive, the quills and tail being rather darker ; the outer feathers on each side of the latter are white, tinged with reddish brown, which prevails more decidedly on the flanks and under tail-coverts.

The female resembles the male, except that the head is of the same colour as the rest of the plumage.

Our Plate represents a male and female of the natural size.

BLACKCAP.

Curruca atricapilla, *(Briss.)*

Drawn from life and on stone by J. & E. Gould.

Printed by C. Hullmandel.

BLACK-CAP.

Curruca atricapilla, *Bechst.*

Le Bec-fin à tête noir.

Of the numerous migratory songsters that add a charm to our spring by their melody, the Black-cap is inferior to none, being equally distinguished for the power and variety of his own notes, as for his aptness in imitating those of other species, whether discordant or otherwise,—a habit in which he frequently indulges; and so exactly similar are the sounds he produces, that we are often deceived by the skill of the imitation: his own song is particularly liquid and melodious, and poured out with great energy. The Black-cap is common in our Island during summer, and also generally spread throughout the northern and central portions of Europe; it frequents our gardens and shrubberies, but is nevertheless a bird of shy and recluse habits, and remains concealed from observation in the thickest part of the foliage. It generally appears among us in the month of April, and departs again in September.

It is worthy of notice that the males and females of this species perform their migrations separately; and we believe this habit is almost universal among the Warblers, the males preceding the females in their arrival by about a week, commencing their song immediately, and selecting a favourable locality for the purpose of nidification, which takes place as soon as the foliage is sufficiently dense for the necessary concealment, among which they construct their nest of vegetable fibres intermingled with a little moss and grasses, in a low bush, shrub or hedgerow, the female laying four or five eggs of a dingy white, clouded with light yellow brown, spotted and occasionally streaked with darker brown. Its food consists of insects, berries, and fruits.

The male has the top of the head and occiput of a deep black; the throat and under parts of a lightish grey; the upper parts, wings and tail, of a dull olive grey; feet and beak blueish ash. The female has the top of the head reddish brown; the other portions of the plumage like that of the male, but rather more obscure, the under parts having a slight tinge of red pervading the grey.

We have figured a male and female of their natural size.

GARDEN WARBLER.

Curruca hortensis. *(Bechst.)*

Drawn from Nature & on Stone by J & E Gould. *Printed by C. Hullmandel.*

GARDEN WARBLER.

Curruca hortensis, *Bechst.*

La Bec-fin fauvette.

THIS unassuming and plain-coloured bird is one of the migratory species resorting to our island, where it arrives in the month of April, enlivening our gardens, coppices, and shrubberries with its cheerful notes; and with so much melody does it pour forth its strains that it has often been put in competition with the Nightingale and Blackcap.

In its habits it is shy and secluded, seldom showing itself, and its presence is often unsuspected until its song is heard.

Its range is very general over our island as well as over the whole of the temperate and southern portions of Europe. Soon after its appearance in this country the business of nidification is commenced, the nest being constructed among nettles, or any other rank herbage, and formed of roots, grasses, various other plants and moss interwoven together; the eggs, which are four in number, are yellowish grey, blotched with wood brown, principally at the larger end.

The adults of both sexes do not differ in the tints of their plumage; the young, on the other hand, have the region of the eyes lighter, and the general colour of their plumage more olive.

The adults may be thus briefly described:

Upper surface ash grey, with a slight tinge of olive; sides of the neck ash grey; throat and under surface greyish white; flanks and breast slightly tinged with brown; bill brown; legs greyish brown.

We have figured a male of the natural size.

RÜPPELLS WARBLER.
Curruca Rüppellii

Drawn from Nature & on Stone by J & E Gould.

Printed by C. Hullmandel.

RÜPPELL'S WARBLER.

Curruca Ruppellii.

Le Bec-fin de Rüppell.

THIS elegant bird having been admitted to the Fauna of Europe as a rare visitant to its eastern countries, it becomes necessary for us to illustrate it in the present work.

The figures in the accompanying plate are drawings of individuals obtained from the collection of Dr. Rüppell, to whom the species has been dedicated by M. Temminck ; and we ourselves feel considerable pleasure in being able to assist in perpetuating the name of so distinguished a naturalist, whose exertions in behalf of science cannot be too highly appreciated, and whose enthusiastic researches in the field of nature have enabled him to add so largely to our zoological knowledge.

The habitat of the *Curruca Rüppellii* would appear to be the northern and eastern portions of Africa, where it occasionally passes over the boundary line to the adjacent confines of Europe.

M. Temminck informs us that it gives preference to thickly wooded districts ; and from the general form and contour of the body, and particularly its subdued and sober tone of colouring, we may reasonably expect that its general œconomy is in unison with the birds of our own island to which the restricted term of *Curruca* has been applied, and with which we have ventured to associate it, acknowledging at the same time that the more slender and pointed bill of the present bird offers a character somewhat at variance with the genus alluded to.

The little knowledge we possess respecting the habits and nidification of this interesting bird prevents our adding to this description that information which future discovery can alone impart.

The sexes, as in many other species of the genus *Curruca*, offer a marked difference in their colouring, the male being distinguished by its black head and throat, and by the conspicuous stripe of white which passes from the base of the lower mandible to the end of the ear-coverts, which are grey: in other respects they are alike in plumage. The female has only a tendency to the black crown, and is entirely devoid of the white stripe on the cheeks.

The whole of the upper surface is deep blue grey; wings blackish grey, the outer edges of the feathers margined with brown ; the eight central feathers of the tail black, the next on each side black on their outer webs with a large white spot on the inner ones, the outer feather on each side white for three fourths of its length ; flanks grey ; under surface greyish white tinged with vinous ; bill black at the tip and brownish white at the base ; legs brown.

The Plate represents an adult male and female, of the natural size.

SARDINIAN WARBLER.

Curruca melanocephala. *(Bechst.)*

Drawn from Life & on stone by J. & E. Gould. Printed by C. Hullmandel.

SARDINIAN WARBLER.

Curruca melanocephala, *Lath.*

Bec-fin mélanocéphale.

This bird, which is closely allied in habits, manners and plumage to our Black-cap, (*Curruca atricapilla,*) has hitherto been little known, and is seldom to be met with in ornithological collections. We are able to add but little to M. Temminck's account, who informs us that its localities are very limited, as it appears to be confined to the central parts of Spain, Sardinia, and the Neapolitan States; a circumstance which seems the more probable, as the author has never yet seen it among any of the numerous and extensive collections from different parts of the globe which he has had the opportunity of inspecting.

The specimens from which the accompanying figures are taken were brought from Spain, in 1831, by Captain S. E. Cook, who observed the species to be not uncommon in the neighbourhood of Madrid, and in the interior of the country. By that gentleman they were presented to the Zoological Society of London; and the Author here takes the opportunity of expressing his obligations to the Council and Members, for the permission so kindly allowed of availing himself, on this and other occasions, of the treasures of their Museum. Its food, like that of the Black-cap, consists of flies, the larvæ of insects, and soft berries; but we are unable to say whether or not it is as sweet and charming a songster as our own British species. It builds in low bushes near the ground, the female laying five eggs of a yellowish-white, spotted with a darker colour.

The male and female differ not only in the colour of the head, but also in the general tinge which pervades the plumage. In the male, the forehead, top of the head, and ears are of a deep black; the rest of the upper surface of a dark leaden grey; the quill-feathers tinged with brown; the tail inclining to black, the outer feather on each side having its tip and outer edge dull white; the throat white; the sides grey, becoming lighter on the under surface; the legs light brown; irides brown; beak black, base of the under mandible whitish.

In the female, the top of the head is of a dark leaden grey, like the back of the male; the whole of the upper surface is tinged with olive-brown; the throat white; the sides light russet brown, becoming lighter below; the beak and legs as in the male.

The length is about five inches.

The Plate presents a male and female of the natural size.

SUB-ALPINE WARBLER.

Curruca leucopogon; *(Mihi)*.

Drawn from Nature & on Stone by J. & E. Gould.

Printed by C. Hullmandel.

SUBALPINE WARBLER.

Curruca leucopogon, *Mihi.*

Le Bec-fin subalpin.

We are indebted to the collection of the Zoological Society of London for the first examples of this species which we have had the opportunity of examining, and which were presented to the Museum by the celebrated Professor Savi, of Pisa, who, in his work on the Birds of Tuscany, states, that the *Sylvia passerina* and *S. subalpina* of M. Temminck's *Manuel* should be considered as synonymous with the *Sylvia leucopogon* of Meyer. We have also more recently been favoured by M. Temminck with fine examples of this species, and with a communication in which he acknowledges that recent observation has established the justice of Professor Savi's opinion, in considering his *Sylvia subalpina* as the old male in the livery of spring of the *Sylvia passerina*, and which, as we have before stated, Professor Savi considers synonymous with *S. leucopogon*, being the same bird in different stages of plumage. It appears, indeed, to be a bird whose changes of plumage are not yet rightly understood, even by those who have had the best opportunities of observing it in a state of nature. Professor Savi has forwarded specimens of this interesting bird in different states, one of which certainly answers to the description of the *S. passerina* of M. Temminck.

The figures which we have given to illustrate this species are those of adult birds of both sexes, killed in summer. The female will be seen to offer a contrasted difference to the male in the colour of her plumage, which offers no very great dissimilarity to that of the young, except that the blueish grey of the upper surface is of a reddish brown tinge, the flanks also in the young being of a more sandy yellow. We trust it will not be considered that we have added new difficulties by increasing the synonyms which have already been applied to this species, in adopting the generic station and title of *Curruca*, which we do provisionally, considering that it is closely allied to the typical form of that genus.

The natural habitat of the Subalpine Warbler is limited to the South of Europe, especially Italy and Sardinia: it is also known to exist in considerable abundance on the banks of the Nile, as far as Abyssinia. It frequents bushes and underwood, living upon insects, small caterpillars, &c. Of its eggs and nidification nothing is at present known.

The male, in full plumage, has the head and the whole of the upper plumage of a fine blueish grey; the wings and tail somewhat darker, the feathers of the wings having lighter edges, and the two outer tail-feathers being pure white; the throat, breast, and flanks are reddish chestnut; a white moustache passes from the angle of the beak to the side of the neck; the middle of the belly dirty white.

The female differs from the male in having the breast only tinged with pale rufous, the white moustache being much more obscure.

The young, as we have already stated, differs from the female in having the upper plumage of a more uniform greyish brown, and in being much more pale beneath.

The Plate represents the male and female of the natural size.

1. COMMON WHITETHROAT.
Curruca cinerea. *(Bechst.)*

2. LESSER WHITETHROAT.
Curruca garrula. *(Bechst.)*

Drawn from Life & on Stone by J & E Gould.

Printed by C. Hullmandel.

COMMON WHITE-THROAT.

Curruca cinerea, *Bechst.*

Le Bec-fin grisette.

Of all the migratory birds which pay their annual summer visit to this country for the purpose of making it their home during the breeding season, the Common White-throat is by far the most abundant and the most extensively distributed, every hedge-row and coppice being enlivened by its presence. It is no less abundantly found throughout the mild and southern districts of continental Europe. Like many other birds of the genus in which it is ranked, it has the power, to a considerable degree, of imitating the notes of other birds independent of its own hurried, confused, and babbling strain. During the time the female is engaged in the task of incubation, the male may be seen mounting in the air to a considerable height above the tops of the hedges with a singular jerk of the tail and raised crest, uttering its song, if a series of rapid and intricate modulations may be thought worthy of the name. Among the entangled branches and thick foliage of hedge-rows it displays the utmost quickness and address,—concealing itself from observation with great wariness: to hedges, indeed, particularly those which border lawns or broad ditches overgrown with nettles and other wild plants, the Common White-throat always appears to be partial; it generally constructs its nest in such situations, and uses for its materials dry grass and bents loosely interwoven, and though by no means remarkable for beauty or elaborate workmanship, its little building is frequently so well concealed as to require a pretty close scrutiny to discover it; the eggs are generally four or five in number, of a pale blueish white, thickly speckled with ash-grey. The young at a very early age nearly resemble their parents, between whom no difference exists either in the colour or the disposition of the markings of their feathers.

The general plumage of the whole of the upper parts is of a rufous grey, the edging of the secondaries being of a brighter tint; the quills and tail-feathers of a darker hue with less of rufous; the throat is white, the under parts white, tinged especially on the breast with a slight wash of rufous; irides hazel; beak and tarsi wood brown.

LESSER WHITE-THROAT.

Curruca garrula, *Bechst.*

Le Bec-fin babillard.

The *Bec-fin babillard*, *Sylvia Curruca*, Temm., *Babbling Warbler* of Latham, and the *Lesser* White-throat, are all synonymes of the *Curruca garrula* of Brisson. In size, as its English name implies, it is considerably less than the Common White-throat, which, however, it greatly resembles in general habits and manners, but appears to be more partial to orchards, gardens, and coppices, where it frequents the tallest trees, pouring out its babbling notes with the utmost energy, but never, we believe, rising in the air and singing at the same time, as is so remarkably the case with the preceding species; its notes are also more powerful and melodious, and its disposition more restless. It is neither so abundant nor are its local resorts so extended in the British Isles as the Common White-throat; it inhabits in preference the warmer portions of continental Europe, and in England confines its visits to the southern counties of our Island, becoming more and more scarce as we proceed northward, where beyond a certain limit it is unknown. The situations in which it builds are similar to those chosen by the Common White-throat; such as bushes, brambles, nettles, &c., where it forms a nest of bents and other grasses, lined with finer fibres partially mixed with hair. The eggs are four or five in number, of a white colour dotted and blotched with ash-grey and brown, except at the small end, which is plain. This species, as well as the preceding, leave this country in the months of August and September.

The top of the head is pure ash colour, becoming deeper on the ear-coverts; the general plumage above is greyish brown inclined to olive; the throat pure white; the under surface white very slightly tinged with brown; irides hazel; beak and *tarsi* lead colour. The male and females are alike in colour.

Our Plate represents both the Common and Lesser White-throat in their spring plumage, of their natural size.

SPECTACLE WARBLER.

Curruca conspicillata.

Drawn from Nature & on Stone by J & E. Gould.　　　　Printed by C. Hullmandel.

SPECTACLE WARBLER.

Curruca conspicillata, *Mihi.*

Le Bec-fin à lunettes.

This elegant little bird, which in form and general style of colouring is so nearly allied to the common Whitethroat, *Curruca cinerea*, is an inhabitant of the southern provinces of Europe. We have seen several specimens, killed in Spain, where, according to the observation of Capt. S. E. Cook, R.N., (see "Sketches in Spain," vol. 2. p. 264.) it is far from being uncommon. "This beautiful little bird," he observes, "is stationary in Andalusia, frequenting low and moist situations, and, I suspect, not much extended in their habitats. I found them in the marshes near Seville, where they live with the *C. cinerea*, and may easily be mistaken for them."

In addition to the above authority we have other grounds for stating the close resemblance of its habits and manners to those of our Whitethroat; and though we have no information as to its song or nidification, it may be reasonably presumed that in these respects also it exhibits a decided alliance. According to MM. Savi and Temminck, it is found in Sardinia, where it may be observed in places covered with bushes or wood: in the North of Italy and France it has never been seen.

The male has the top of the head and cheeks fine ash colour; the space between the eye and the beak black, whence a circle of the same colour surrounds the white of the eyes; the upper surface vinous ash colour; the wings blackish, edged with rufous; throat white; under surface white, tinged with vinous, which passes into reddish on the flanks; tail somewhat graduated, and brownish black, with the exception of the outer feathers, which are nearly white, the second and third being also tipped with the same colour.

The female is only to be distinguished from the male by the paler colouring of the plumage, which still preserves the markings of the male, except that the circle of black round the eyes is scarcely if at all apparent.

Our Plate represents an adult male and female of the natural size.

MARMORA'S WARBLER.
Curruca sarda.

Drawn from Nature & on stone by J & E Gould. Printed by C. Hullmandel.

MARMORA'S WARBLER.

Curruca Sarda.

Le Bec-fin Sarde.

In his " Manuel d'Ornithologie" M. Temminck informs us that we owe the knowledge of this species to the Chevalier Marmora, and that it was described " in the Annals of the Academy of Turin, on the 28th of August 1819. In its plumage and by the naked circle round the eyes, it is nearly allied to the *Sylvia melanocephala*, from which it is distinguished by its beak being more feeble and slender, like that of the *Pittchou*. It may also be distinguished by the tail, of which only the exterior feathers are edged with white, whilst in the *Bec-fin melanocephala* all the exterior barbs and the ends of the two first feathers are white. The colour of the throat also prevents these nearly allied species from being confounded."

In addition to the features pointed out by M. Temminck to distinguish this bird from *Curruca melanocephala*, we may observe that in the latter the black colouring of the head is more decided and of a deeper tint than in *Curruca Sarda*, which has the whole of the upper plumage of a uniform blackish grey.

Although this bird is somewhat rare, it may be found in most of the European collections; and from the circumstance of our never having observed it in any of the numerous and extensive collections from India and Africa which we have had the opportunity of examining, we feel convinced that it is a very local species, apparently only found in Sardinia and the most eastern portions of the Continent.

It is said to dwell in small woods, and to feed on very small flies and other insects, which attach themselves to the leaves.

Of its nidification nothing is known.

In the colouring of their plumage the sexes present but a slight difference, the male being only of a somewhat darker tint, particularly on the throat and under surface.

The head, throat, and all the upper surface is of a deep blackish grey, which is darkest on the forehead and round the eyes, the sides of the neck and flanks being much lighter, and having a tinge of vinous; middle of the belly greyish white tinged with vinous; wings and tail black, the exterior feathers of the latter edged with white; the orbits of the eyes naked and of a beautiful vermilion; base of the under mandible yellow, the remainder blue; feet yellowish brown.

We have figured a male and female of the natural size.

BARRED WARBLER.
Curruca nisoria.

Drawn from Nature & on stone by J & E Gould. *Printed by C. Hullmandel.*

BARRED WARBLER.

Curruca nisoria.

Le Bec-fin rayée.

The principal feature peculiar to this rare species consists in its comparatively large size to that of the Warblers in general, in its lengthened and elegant form, and in its grey tone of colouring, relieved by numerous transverse bars of black and white. In assigning it to the genus *Curruca*, which we do with some hesitation, we are influenced by the fact that of all the groups of the *Sylviadæ*, it is to this that it makes the nearest approach. It is not a native of the British Isles, nor, as far as we are aware, has it been found in France or Holland; it is, however, tolerably common in many parts of Germany. Never having, ourselves, seen this bird in a state of nature, we prefer quoting the observations of M. Temminck, who informs us that it inhabits bushes and thickets, is abundantly spread throughout the North, occurring in Sweden and in the provinces of the North of Germany and Hungary: it is of more rare occurrence in Austria, and it is also found in Lombardy.

Its food consists of insects, caterpillars, worms, and berries.

It builds its nest in tufted hawthorn bushes, lays four or five eggs, of a whitish colour blotched with purplish ash or pure ash colour.

M. Temminck describes the male and female as offering considerable difference in the marking of the plumage. In the specimens that have fallen under our notice this difference has appeared but very trifling. It may be best, however, to give the colouring from M. Temminck. That of the adult male is as follows:

Head, cheek, back of the neck, and back, of a deep grey, as are also the scapulars and rump, but all the feathers of these parts are terminated by a small bar of brown and another of white; the wings are of a lighter ash colour, the outer feather having a large blotch of white at its extremity; this blotch of white is not so apparent on the second, and still less so on the third and fourth feathers: all the under surface is whitish transversely barred with ashy grey; under tail-coverts are grey with large white edges; beak brown; irides brilliant yellow.

The female has the upper surface of an ash colour, clouded with brown but without transverse bars; the flanks slightly clouded with reddish; the white at the extremity of the tail is more circumscribed and less pure.

The young before their first moult have the whole of the body marked with minute transverse rays of ashy brown; irides brown.

The Plate represents a male of the natural size.

DARTFORD WARBLER.

Melizophilus provincialis; *(Leach)*.

Drawn from Nature & on Stone by J & E. Gould.   Printed by C. Hullmandel.

Genus MELIZOPHILUS, *Leach.*

GEN. CHAR. *Head* large; *bill* short, greatly arched from the base, compressed, with the tip finely emarginated; *tomia* of both mandibles inflected towards the middle; *gape* slightly bearded. *Nostrils* basal, lateral, longitudinally cleft. *Wings* short, rounded, the first feather very small, the second shorter than the third, fourth, and fifth, which are equal and longest. *Tail* long and soft. *Legs* having the *tarsi* strong, and longer than the middle toe, which is nearly equal in length to the hind one.

DARTFORD WARBLER.

Melizophilus provincialis, *Leach.*

Le Pitte-chou de Provence.

WE adopt the present genus as constituted by Dr. Leach, in the formation of which we consider him borne out by the striking difference this bird exhibits in its characters to all the other European *Sylviadæ*. Its form closely allies it to the Superb Warblers (*Malurus*, Vieill.) of New Holland, while its relationship to the Common Whitethroat, *Curruca cinerea*, is strikingly apparent: its rounded wing and very graduated tail, however, form just grounds of distinction. To the British Ornithologist, the Dartford Warbler is a bird of peculiar interest. It is a permanent resident in this island, a fact which is proved by our having received it in a recent state at all seasons of the year. It is nevertheless far from being universally distributed, being principally, if not exclusively, confined to the southern and south-western districts, where it resorts to commons, heaths, and moorlands, clothed with thick furze and heather, living in a state of complete seclusion, being habitually addicted to threading the thickest portion of the brushwood, whence it is not easily driven. In the spring, at the season of pairing and nidification, it is more lively and more frequently visible, rising on quivering wing above the tops of the furze, and uttering a hurried babbling song, much after the manner of the Whitethroat; at these times it erects the feathers of its head into a crest, and distends the throat, exhibiting various attitudes and gesticulations. The nest is composed of dry stalks and grass, intermingled with wool and vegetable fibres: it is in general placed in the thickest part of a furzebush, at a short distance from the ground. The eggs, according to Montagu, are very similar to those of the Whitethroat, being speckled with brown and cinereous spots on a greenish white ground.

The Dartford Warbler is found tolerably abundant on all the heathy commons in the immediate vicinity of London, as well as those of Bagshot, Chobham, &c., but it is more particularly abundant in Devonshire and Cornwall. In no place, however, is it to be found more plentiful than in the neighbourhood of Oakingham in Berkshire, whence specimens have been sent us by John Rogers Wheeler, Esq., whose fine and choice collection contains the most beautiful examples. On the Continent it is more abundant throughout Spain, Italy, and the South of France, than in Germany and Holland.

Its food consists of various species of insects and their larvæ, to which are added, as the season affords, berries and fruits of various kinds; at least they feed on such substances in confinement where they become tame and reconciled.

The head, the back of the neck, and the upper plumage deep grey; the under plumage deep reddish brown, with a ferruginous tint; throat mottled with white; wings and tail brown, with the exception of the outer feathers of the latter, which have white tips and exterior edges; bill yellow at the base, black at the tip; legs brown.

The female resembles the male, except that her plumage is duller, the back being dusky brownish, and the throat merely exhibiting traces of the white edging to the feathers so conspicuous in the male.

We have figured a male and female of the natural size.

WREN.

Troglodytes Europæus; *(Cuv.)*

Drawn from Life & on Stone by J. & E. Gould.

Printed by C. Hullmandel.

Genus TROGLODYTES, *Cuv.*

GEN. CHAR. *Bill* slender, slightly compressed, emarginated, curved slightly. *Nostrils* basal, oval, half covered by an arched and naked membrane. *Wings* short, rounded; first quill very short; second longer; fourth and fifth equal and longest. *Tail* short, rather rounded, and carried erect. *Legs* strong. *Toes* three before and one behind; the outer toe joined at its base to the middle one.

WREN.

Troglodytes Europæus, *Cuv.*

Le Troglodyte ordinaire.

ALTHOUGH the group to which this familiar little bird belongs is filled up by numerous species in the continent of America;—Europe, and even the older continents of Asia and Africa, present us with only one example; a species, however, which in Europe is universally diffused, inhabiting the countries which border the arctic circle as well as those of the South. In England it abounds in our hedgerows and thickets, hovering about the dwellings of man, with whose presence it seems perfectly reconciled, and near whom it is allowed to dwell unmolested. No one indeed can observe its habits and manners without becoming interested in its welfare, enlivening as it does the bleak season of winter with its tremulous, shrill and lively strains; nor is it less amusing to observe it creep like a mouse through our quickset hedges and underwood, examining the moss-covered banks and stumps of trees in search of its insect food which lies concealed among the crevices.

It seldom takes long flights, but keeps to the same local situations. It remains with us during the whole of the year, braving our severest winters with impunity. It breeds early, and its familiar disposition often leads it to build in outhouses, arbours, summer-houses, and similar situations; at other times it selects the sides of walls covered with ivy, and thickly wooded shrubs. It constructs an ingenious and curiously domed nest, of moss, leaves or grass, in fact of any material that may be at hand, and lays seven or eight eggs of a pure white, prettily freckled with reddish spots. The young on leaving the nest are extremely shy, and active in concealing themselves among the herbage and the thickest parts of bushes. The sexes offer no external differences, and the young very soon assume the adult plumage.

The ground-colour of the Wren is of a reddish brown, becoming paler and more grey beneath; the whole of the plumage is prettily barred transversely with darker brown or black; a narrow white line passes above the eyes.

We have figured an adult bird of the natural size.

1. WILLOW WREN.
Sylvia trochilus, *(Gmelin)*.

2. CHIFF CHAFF.
Sylvia hippolais, *(Lath.)*.

3. WOOD WREN.
Sylvia sibilatrix, *(Bechst)*.

Drawn from Life and on Stone by J. & E. Gould. Printed by C. Hullmandel.

Genus SYLVIA.

GEN. CHAR. *Beak* straight, slender, conical, pointed, slightly notched at the tip, sides somewhat concave, base furnished with fine hairs. *Nostrils* basal, lateral, oval. *Tarsi* longer than the middle toe. *Toes,* three before, one behind; the outer toe joined at the base to the middle one. *Wings,* the first quill-feather very short, second and fourth equal, third longest.

WILLOW WREN.

Sylvia trochilus, *Gmel.* Le Pouillot.

WE here present, on a single Plate, three little birds, which are nearly allied to each other in habits, manners, and plumage, and which form the British portion of a genus to which the generic title *Sylvia* is truly applicable. They visit our budding woodlands on the return of spring, pleasing us with their sweet and delicate notes.

The sexes present no difference in plumage; the young, however, in the autumn have a brighter and more pervading tinge of yellow. All are migratory.

The Willow Wren is by far the most abundant in England: it is also dispersed throughout the greater portion of Europe, from Sweden to Italy. At the same time its localities are less strictly confined, inhabiting not only groves, woods, and willow plantations, but gardens, hedge-rows, and commons covered with bushes. It arrives in April; but like the rest of our summer visitors, its appearance seems regulated by the temperature of the season, and the consequent abundance of its food, which consists of the soft-winged insects and their larvæ. Being more familiar than the two others, we have a better opportunity of becoming acquainted with its habits. Its song is simple, consisting of a few prolonged and softly modulated notes, frequently singing while in active search for Aphides and other insects on which it subsists.

The whole of the upper surface is of a greenish-olive: over the eye extends a faint yellow stripe; throat and breast slightly tinged with yellow; belly yellowish-white; under tail-coverts yellow; legs dull flesh-colour; wings covering about a third of the tail.

The female builds a covered nest of dried moss and grasses, lined with feathers, and artfully concealed on the ground, on the sloping side of a bank, or in a dry ditch, among thick herbage or tangled bushes; the eggs are white, marked with pale reddish spots, more thickly dotted towards the larger end.

CHIFF-CHAFF.

Sylvia hippolais. La Fauvette de Roseaux.

THE Chiff-chaff so nearly resembles the former species as to be frequently confounded with it: it may, however, be distinguished by its smaller size and darker legs; in addition to this, the streak over the eye is less apparent, and the general plumage not so finely tinged with yellow.

It is one of our earliest visitors, and is less common than the two other species, differing from both in its habits and localities. Partial to groves and tall trees, it frequents the topmost branches, where it may be heard to utter its short song, composed of two distinctly repeated notes, *chiff-chaff,* which have given origin to its usual name. In the construction of its nest, and in the colour and number of its eggs, it resembles the preceding species, the spots only being of a darker red, and fewer in number.

In addition to the three species here figured, Bewick adds another, under the title of the Least Willow Wren, which we, however, believe to be a Chiff-chaff.

A little variation frequently occurs in the size of each of these birds. The present is, however, shorter than the Willow Wren by about half an inch, and proportionally less in all its other measurements.

Its habitat is extended to the greater portion of Europe, from Sweden to France.

WOOD WREN.

Sylvia sibilatrix, *Bechst.* Le Bec-fin siffleur.

THE distinguishing characters of the Wood Wren consist in its superior size, and the more elongated wing, which extends over nearly two thirds of the length of the tail; in the bright yellow streak above the eye, the yellow edgings of the wing-feathers, and the silvery whiteness of the under surface. In other respects its plumage partakes of the general tone of colour which is found in the Willow Wren and Chiff-chaff, except that the olive-green of the back and upper parts is of a more lively tint. Secluded woods and groves are its general places of resort, where it may be easily identified by its peculiar tremulous warble, both louder and possessing more variety than its congeners, and accompanied with a singular quivering of the wings.

Its nest partakes of the character of those of the other species, both as to its structure and place of concealment, but instead of feathers, is lined most commonly with fine grass or hair. The eggs are six in number, of a white ground, dotted with purplish spots, more confluent towards the larger end.

It inhabits the same countries as the two preceding species.

MELODIOUS WILLOW WREN.
Sylvia Hippolais: (*Temm:*)

Drawn from Nature & on stone by J & E. Gould.

Printed by C. Hullmandel.

MELODIOUS WILLOW WREN.

Sylvia Hippolais, *Temm.*

Le Bec-fin à poitrine jaune.

THE bird which we have figured in the accompanying Plate appears to be the true *Sylvia Hippolais* of M. Temminck and other Continental writers, and is a species which has hitherto never been found in England; the bird known under the above name in our island being now unanimously acknowledged to be the *Sylvia rufa.* It is somewhat singular that this species, so familiar to every naturalist on the Continent, and which inhabits the gardens and hedgerows of those portions of the coasts of France and Holland that are immediately opposite our own, should not, like the rest of its immediate congeners, more diminutive in size, and consequently less capable of performing extensive flights, have occasionally strayed across the Channel and enlivened our glens and groves with its rich and charming song, which is far superior to that of either of the three other species of the group.

Although we cannot with propriety separate the present bird from the true Willow Wrens, still we cannot but be struck with the shorter and stouter contour of its body and its more robust bill; it also differs considerably in its habits and mode of nidification; all those species that inhabit England constructing a singular domed nest, which is always placed on the ground, while the species here illustrated invariably builds on trees, sometimes in the shrubs of the garden, at others in the trees of forests; laying five eggs, of a reddish white blotted with spots of darker red. Those who have not had an opportunity of listening to the song of this little tenant of the grove can scarcely form an idea of its power and melody, in which respects it is only equalled by those of the Black Cap and Nightingale.

The *Sylvia Hippolais* appears to be dispersed throughout the European continent from Sweden to the shores of the Mediterranean.

The sexes offer no difference in the colouring of their plumage.

Its food consists of insects, such as aphides and other small kinds, to which are added caterpillars, &c.

The whole of the upper surface is greenish ash; a small patch of yellow is situated between the bill and the eye; the throat, breast, and under surface pale yellow; wings and tail brown, the edge of each feather being lighter; feet and bill fleshy brown; irides dark brown.

The Plate represents an adult male of the natural size.

YELLOW WILLOW WREN.

Sylvia Icterina. *(Vieill.)*

Drawn from Nature & on Stone by J & E Gould. *Printed by C. Hullmandel.*

YELLOW WILLOW WREN.

Sylvia icterina, *Vieill.*

La Bec-fin ictérine.

In form and colouring this species of Willow Wren so nearly resembles the British members of this interesting group, that it requires an intimate knowledge of its habits, manners, and song to distinguish it from them with any degree of certainty. It is considered as distinct by the continental naturalists; and M. Temminck, in his 'Manuel,' informs us, that " the length of the tail, which is an inch longer than the wings, its forked form, and the comparative length of the quills and tarsi, are characters by which it may be distinguished from the two following species (*S. trochilus* and *S. rufa*); but it is always very difficult to recognise it by the examination of a single individual." To this we may add that the bill is even longer than that of *S. trochilus*, while the wing is quite as short, if not shorter, than that of *S. rufa*.

In a second remark, M. Temminck states that M. Cantraine has killed this species hopping among the reed beds in morasses of Ostia in the month of April; and that he himself has killed several in Holland, where it is less abundant than *S. trochilus*. It also inhabits France and the Roman States, frequenting the neighbourhood of water and morasses. It is probably more abundant than is generally supposed, being frequently confounded with the two species above mentioned.

Its food consists of the small insects and flies, which are attached to the branches and leaves of trees, &c.

Of its nidification and eggs nothing is known.

Crown of the head and all the upper surface pure olive; stripe over the eye bright yellow; sides of the neck, chest, and flanks clear yellow; throat and centre of the belly yellowish white; wings and tail ashy brown bordered with greenish olive; bill and feet brown.

We have figured an adult of the natural size.

NATTERER'S WARBLER.

Sylvia Nattereri; *(Temm.)*

Drawn from Nature & on stone by J & E. Gould. Printed by C. Hullmandel.

NATTERER'S WARBLER.

Sylvia Nattereri, *Temm.*

Le Bec-fin Natterer.

This delicate little Warbler appears to be a species of common occurrence in the district of Algesiras, in the South of Spain, where it was first discovered by M. John Natterer of Vienna, after whom it has been named. Since the period of M. Natterer's visit to Spain, about the year 1820, this indefatigable naturalist has passed sixteen years in the Brazils, whence he has lately returned to Vienna, with an exceedingly rich collection of the productions of that vast country. As a naturalist, and particularly as an ornithologist, we can speak of M. Natterer in terms of the highest praise; and it is by his discerning and ardent research during his short residence in Spain, that the Fauna of Europe has been enriched with several species which but for him would, even at this time, have been unknown to us.

The species here figured is extremely rare in all the collections of Europe: it is in every respect a true *Sylvia*, as the genus has been restricted by modern classifiers, and is allied in all its affinities to the Chiffchaff and Willow Wren of our own island.

Besides the locality above mentioned, M. Temminck informs us, in the third part of his Manuel, that it is common in Provence and Switzerland, that it has been killed in the Tyrol, but that it is never seen in the North of Europe.

Its food consists of flies, small spiders, and other insects.

M. Temminck also informs us that the nests found in Italy were placed on the ground among the grass, in hilly situations; were of a spherical shape formed externally of dead leaves, with a lateral opening; and that the eggs are of a globular form, four or five in number, and white minutely dotted with reddish.

The top of the head and upper surface ash brown tinged with olive; from the gape to the upper part of the eye extends a conspicuous stripe of greyish white; all the under surface silvery grey; wings and tail brownish ash fringed with olive; bill and feet brown.

The female differs only in having her tints less clear.

We have figured a male of the natural size.

RICHARD'S PIPIT.

Anthus Richardi; *(Vieill.)*

Drawn from Nature & on Stone by J & E. Gould. *Printed by C. Hullmandel.*

Genus ANTHUS.

Gen. Char. *Bill* straight, slender, rather subulated towards the point; the base of the upper mandible carinated, and the tip slightly bent downwards and emarginated; tomia of both mandibles pressed inwards about the middle. *Nostrils* basal, lateral, oval, partly concealed by a membrane. *Tarsi* generally exceeding the middle toe in length. *Toes* three before, and one behind; the outer toe adhering to the middle one as far as the first joint; hind claw more or less produced. *Wings*, the first quill very short, the second rather shorter than the third and fourth, which are of equal length, and the longest: two of the scapulars produced, and equal to the quills in length when the wing is closed.

RICHARD'S PIPIT.

Anthus Richardi, *Vieill.*

Le Pipit Richard.

We cannot but observe the near relationship which the birds of this group bear to those of the *Motacillæ*, or Wagtails, particularly that division of them to which the illustrious Cuvier gave the subgeneric title of *Budytes*; to this section they offer a very close resemblance in their general contour, in the lengthened form of their hind claw, and in their habit of frequently raising and depressing the tail, accompanying it at the same time with a lateral expansion of the feathers.

Although we have no certain proof that such is the case, we have some reason to believe that the northern and western regions of Africa constitute the true habitat of the *Anthus Richardi*, and consequently that the individuals which make their appearance within the precincts of Europe at such irregular intervals have been driven hither by some unusual cause; at all events the small number of examples which are recorded to have been captured in Europe is a sufficient proof that this portion of the globe is not its native country. M. Temminck states that it is often met with in Picardy; but the British Islands may, perhaps, be considered among the countries in which it has been most frequently taken, as seldom a year passes without examples falling a prey to the London birdcatchers while in pursuit of the more favourite Goldfinch and Linnet. Independently of the recorded instances of this nature, we know of two that were captured in this way during the spring of 1836, in the immediate neighbourhood of the metropolis, from one of which the present figure is taken.

In its actions the *Anthus Richardi* displays all the activity and alertness of the other members of the genus, among which it will rank as the largest and one of the most typical; it never perches on trees, but is always seen on the ground, where it runs with the greatest rapidity in pursuit of its food, which consists of flies, grasshoppers, and other insects.

Of its nidification nothing is known.

The sexes are alike in plumage, and there is no difference of colouring in summer or winter.

Crown of the head and whole upper surface deep brown, each feather margined with pale brown; stripe above and below the eye pale brown, inclining to buff; throat white, surrounded by a gorget of oblong dark brown spots on a lighter ground; breast and flanks pale brown; abdomen white, slightly tinged with brown; the middle tail-feathers deep brown with paler edges, outer feather on each side almost entirely white, the next having the shaft and base very dark brown and the remainder white; bill pale brown at the base and dark brown at the tip; legs and feet yellowish brown; tarsi long and stout; hind claw much produced and slightly curved.

We have figured a male of the natural size.

MEADOW PIPIT.

Anthus pratensis; *(Bechst.)*

Drawn from Nature & on Stone by J. & E. Gould.

Printed by C. Hullmandel.

MEADOW PIPIT.

Anthus pratensis, *Bechst.*

La Pipit Farlouse.

The Meadow Pipit is one of the smallest of its tribe, and one of the most common of our indigenous birds; it is a permanent resident in the British Islands, and is always to be found in such open situations as heaths, commons, and swampy tracts of country. Its food is always sought for on the ground, to which it is consequently restricted, and where it may be observed running nimbly along, no less lively in the depth of winter than in spring and summer. When startled from the ground it rises with a quick vibratory motion of the wings, uttering at the same time its shrill and well-known chirping cry. It is very generally distributed throughout the continent of Europe, particularly in Holland and France; it is also found in Northern Africa, and a great portion of Asia. Many authors have contended that two species, the Pipit and the Tit Lark of the older writers, have been confounded; recent observation, however, has fully proved that they are both identical with our Meadow Pipit, to which we object to apply the term Lark altogether. This erroneous impression appears to have arisen in consequence of a slight change in the colouring of the plumage at different seasons of the year; the autumnal plumage being characterized by a tint of fine greenish olive, which in the course of time gives place to a browner hue. We have ourselves had opportunities of ascertaining the correctness of Mr. Selby's observations on this point, which we take the liberty of inserting here. "In September and October, after their autumnal or general moult, they assemble in small flocks, resorting to the lower pastures, and not unfrequently to turnip fields. At this period the renewed plumage differs considerably from that laid aside, the green of the upper parts being of a much brighter tint, and the whole of the under parts more deeply tinged with yellow. In this state, the present species is to be recognised as the *Pipit Lark*. I have omitted no opportunity of becoming satisfied on this head, having examined specimens at all seasons of the year, and am thoroughly persuaded that the supposed species described as the *Pipit Lark* is in reality no more than the Common Pipit (*Tit Lark* of authors) in its renewed or winter plumage.

Its usual flight is by short and interrupted jerks; but in the breeding-season it differs, the bird then rising by a tremulous and rapid motion of the wings to a considerable height in the air, and commencing its song when at the greatest elevation, descending afterwards with motionless wing and expanded tail, in a sloping (sometimes almost perpendicular) direction to the earth, or to the top of some bush. It makes its nest on the ground, under the shelter of a tuft of herbage, forming it of dry grass interwoven with seed stalks of plants, and lined with finer grasses or with hair. The eggs are five or six in number, varying in colour, but the prevailing tint a pale brown, thickly covered with brownish purple red spots and specks. Like the Wagtails, it runs with celerity, and feeds upon flies, worms, and other insects." The nest of this bird frequently forms a receptacle for the egg of the Cuckoo.

The whole of the upper surface is dark olive green, the centre of each feather being brownish black; under surface yellowish white, spotted on the sides of the neck and breast with blackish brown; flanks white, with oblong streaks of a dark colour; tail blackish brown, the outer feather on each side having the exterior web white, and being largely tipped with the same colour, the second feather having a small white spot near the tip; bill and feet brown.

The Plate represents a male and female of the natural size.

TAWNY PIPIT.

Anthus rufescens; *(Temm.)*

Drawn from Nature & on Stone by J & E Gould.

Printed by C. Hullmandel.

TAWNY PIPIT.

Anthus rufescens, *Temm.*

La Pipit rousseline.

THIS elegant species of Pipit, to which M. Temminck has given the specific name of *rufescens*, is the same as that described by Bechstein under the name of *Anthus campestris*, which name M. Temminck states he has been induced to change to that of *rufescens*, as *campestris* is one of the synonyms of the *Anthus pratensis*, and is, moreover, a term inapplicable to either of the species.

The present is, with the exception of *A. Richardi*, the largest of the European Pipits, and may at once be distinguished by the yellow tinge pervading the whole of its plumage. It is almost strictly terrestrial, as its lengthened tarsus and hind claw clearly indicate, seldom perching upon trees: hence it is to be seen on commons, plains, and in open hilly country. M. Temminck says it is common in Germany and France, where it is migratory, visiting those countries for the purpose of incubation. It is rare in Holland, and we are not aware of any instance of its being discovered in England. We are informed that, like the rest of its genus, it constructs a nest on the ground, among herbage or loose clots of earth, laying four or six roundish eggs of a pale grey with violet and russet markings. The outward sexual differences in this species are not distinguishable, nor do the sexes offer any marked difference of plumage in the various seasons of the year.

The whole of the upper surface is of a tawny or yellowish grey, the middle of each feather having a slight dash of brown; the wings, the scapularies, and the lesser and greater wing-coverts, abruptly margined with yellowish white; throat white, as is also a large streak which passes over the eye; the whole of the under surface light Isabelle yellow; a few dusky spots commence at the lower mandible and are thinly dispersed over the chest; tail somewhat forked, the middle feathers brown, the outer ones nearly white; beak brown; legs and claws flesh colour.

The Plate represents a pair of these birds of the natural size.

ROCK OR SHORE PIPIT.
Anthus aquaticus; *(Bechst.)*

Drawn from Nature & on Stone by J. & E. Gould. *Printed by C. Hullmandel.*

ROCK OR SHORE PIPIT.

Anthus aquaticus, *Bechst.*

Le Pipit spioncelle.

In the British Islands this species of Pipit is strictly stationary, inhabiting the rocky and elevated portions of the coast during summer, and the lengthened and muddy shores of the sea during autumn and winter. It rarely leaves the neighbourhood of the coast unless when following the indentations of bays or the course of large rivers, particularly those influenced by the tide. Marine insects and worms appear to constitute its sole food, in the capture of which it displays the address and agility which characterizes all the members of this group. Its call-note and song are very similar to those of the Common Pipit, from which it may be always distinguished by its larger size, and by the dusky and more obscure colouring of its plumage. It is an early breeder, and mostly selects the clefts and ledges of rocks for the site of its nest, which is composed of various marine grasses, lined with hair and fine vegetable substances; the eggs are four or five in number, of a light yellowish grey, with reddish brown specks over the larger end, and sometimes distributed over the whole surface.

We have some reason to believe that there are two species of Rock Pipits nearly allied to each other, as we have never been able to find in any of the examples killed in the British Islands that uniform vinous tint we have observed to pervade the breast of the continental examples; neither have we been able to meet with any specimens in continental collections that strictly accord with the dull and indistinct markings of those of the British Islands; to this point we would therefore beg to direct the attention of those naturalists who may possess opportunities of investigating the subject.

Crown of the head, all the upper surface, wings, and tail dark greenish olive, the feathers of the two latter margined with paler olive; throat whitish; stripe between the bill and the eye, and all the under surface pale greenish olive; the sides of the neck and breast ornamented with large spots of dark brown; the outer tail-feather on each side white; upper mandible and the tip of the lower black; the base of the latter yellowish brown; legs and feet brown.

Our figure is of the natural size.

TREE PIPIT.

Anthus arboreus; *(Bechst.)*

Drawn from Nature & on stone by J. & E. Gould.

Printed by C. Hullmandel.

TREE PIPIT.

Anthüs arboreus, *Bechst.*

Le Pipit des Buissons.

To a superficial observer no two birds can appear more nearly the same than the Tree and the Meadow Pipits, but on a close examination it will be perceived that a marked and permanent difference exists in the structure of the hind claw, a circumstance which, although apparently very trifling, materially influences their respective habits and manners. The hind claw of the Meadow Pipit is long, slender, and almost straight, while that of the Tree Pipit is short, and decidedly curved; hence, while one naturally frequents the ground, the other is habitually arboreal: there is also another distinction between these birds, namely, the permanent residence of one in our island and the regular migration of the other. The Tree Pipit is only a summer visitant, arriving early in spring, and forming one of the numerous songsters that render this season of the year so peculiarly delightful. It generally sings while perched upon some tree rising above the hedgerow, and is often seen to ascend on quivering wing to a moderate elevation and to descend again to the branch from which it had risen, pouring forth its animated and pleasing song as it descends. As is the case with many of our migratory songsters, the strains of this bird are most powerful and more constantly uttered on its first arrival, before the female has reached our shores, and during the season of pairing; after the work of incubation has commenced, the song is comparatively but seldom heard.

Its nest is constructed under the shelter of tufts of herbage or small bushes, and is composed of moss, fibres of roots, and withered grasses, lined with fine dry grass and horse-hair; the eggs, which are four or five in number, are of a greyish white sprinkled all over with brownish purple specks.

Its food consists of flies, small beetles and other insects, and their larvæ.

The sexes are alike in plumage, and may be thus described:

All the upper surface olive green, the feathers of the head and upper part of the back having the centre brownish black; wing-coverts margined with yellowish white, forming a double bar across the wings; chin and throat white, passing into dull buffy yellow on the sides of the chest; breast spotted with oblong marks of brown; flanks spotted with brown; middle of the belly and under tail-coverts greyish white; two middle tail-feathers pointed and of an olive brown; the exterior feather with the whole of the outer and the greater part of the inner web white; tip of the second feather also white; legs and toes yellowish brown.

We have figured an adult male of the natural size.

RED-THROATED PIPIT.

Anthus rufogularis, *(Temm.)*

Drawn from Nature & on Stone by J. & E. Gould

Printed by C. Hullmandel

RED-THROATED PIPIT.

Anthus rufogularis, *Brehm.*

Le Pipit à gorge rouge.

This very interesting species of Pipit must be considered rather as a native of India and Africa than of Europe; it has nevertheless been killed within the borders of the latter division of the globe, Sicily being, as we are informed, not unfrequently visited by it. The specimens from which our drawings are taken were sent us as Sicilian examples by M. Temminck. In size and in the general style of the colouring of the upper surface this bird exhibits a close resemblance to the Common Pipit of our island, but may be distinguished from that species and in fact from all the other known members of the group, by the rufous brown of the throat, which colour frequently proceeds over the chest and abdomen. Of the three specimens forwarded to us, one has the whole of the under surface of a rich ferruginous brown, another has the throat only of this colour, and the third has this tint so slightly pervading the under surface as to be scarcely perceptible.

In its habits, manners, and nidification this bird doubtless closely resembles its congeners, but on these points nothing has as yet been recorded.

The adult in spring has the head, and the whole of the upper surface, wings, and tail blackish brown, each feather margined with olive brown, which becomes still lighter on the extreme edge; two outer tail-feathers nearly white; a stripe over the eye, the throat, and breast rich ferruginous brown, the remainder of the under surface ferruginous buff, the lower part of the breast and flanks ornamented with oblong spots of dark blackish brown on the centre of each feather; bill and feet brown.

The female has all the upper surface similarly marked but of a much lighter tint; the stripe over the eye and the throat only, rich ferruginous brown; under tail-coverts buffy white; the remainder of the under surface pale buffy white with a conspicuous mark of blackish brown down the centre of each feather; bill and feet brown.

The young resembles the adult male, but has the light margins of the upper surface not so extensive; is destitute of the ferruginous throat; and has the breast and flanks very numerously marked with oblong spots of dark brown; bill and feet brown.

Our Plate represents male and female of the natural size.

PIED WAGTAIL.

Motacilla alba, *(Linn)*.

Drawn from Life and on stone by J. & E. Gould

Printed by C. Hullmandel

Genus MOTACILLA, *Auct.*

Gen. Char. *Beak* slender, cylindric, straight; the upper *mandible* angulated between the nostrils, its tip laciniated. Scapulary feathers long. Hind claw shorter than the toe, nearly straight. *Tarsi* elevated. *Tail* elongated.

PIED WAGTAIL.

Motacilla alba, *Linn.*

La Bergeronette grise.

The birds supplying the subject of our present notice, though very common, are general favourites: their form is slender and elegant, their habits inoffensive, and all their varied motions so graceful and active as to insure for them our notice and regard; and the generic term applied to them, truly indicative as it is of one of their very constant actions, is most happily chosen.

The whole length of a fully grown male bird is about seven inches and a half; and in summer the upper part of the head and neck, the back, middle tail-feathers, chin, throat and breast, are black; the forehead, the space round the eyes, the cheeks, and sides of the neck white; lower part of the breast, the belly and under tail-coverts also white; the flanks grey; wing-feathers black, with broad external edges of white forming conspicuous bars in the coverts of the quill-feathers; the two outer tail-feathers on each side white; legs and beak black. After the breeding season, a second moult takes place; the chin and throat become white, leaving only a gorget of black; and the feathers on the upper parts of the neck and back are dark grey. In this state it remains during the winter, till the following spring bringing on a partial moult, invests it again with the plumage first described.

Young birds of the year have the crown of the head and all the upper parts ash-grey, inclining to blueish grey; wing-coverts broadly edged and tipped with white; on the upper part of the breast a crescent-shaped patch of dusky ash, the ends of which extend upwards to the ears in a narrow line on each side of the neck; throat, cheeks, under parts of the neck and lower part of the breast, dirty white; abdomen nearly pure white; quills blackish, with an oblong white spot on their inner web.

The old birds choose various localities in which to place their nests. It is sometimes built on the ground, or in a hole in a pollard-tree or old wall; sometimes on a grass-grown bank; and we have found it, concealed with great care, in a depression on the side of a hay-rick. The nest is formed of moss, dried grass, roots and wool, with a lining of feathers and hair: the eggs, four or five in number, are nine lines and a half long by seven lines and a half in breadth; the ground-colour greyish white, minutely specked all over with ash-brown. The birds frequent the margins of shallow streams and ponds, and are often seen to wade a short distance to secure the larvæ and flies on which they feed. When the young are able to quit the nest, the parent birds entice them by their example over newly mown meadows, and the smooth lawns and grass-plots of pleasure-grounds. Here, secure of their footing, every motion is agile and graceful, darting and running with the rapidity of thought in pursuit of their various insect food. Their flight is undulating, performed by short jerks; and their notes, which are most frequently uttered while on wing, are far from disagreeable. With the decline of the year these birds partially desert the more inland districts, and resort in considerable numbers to the marshy margins of rivers within the influence of the tides.

The Pied Wagtail is common and stationary over the whole of the southern part of the European Continent. It also remains during winter dispersed over the southern counties of England; yet we learn from Mr. Selby and Bewick, that even so far north only as Durham, it migrates southward in October, and does not again make its appearance till the following March; and Mr. Low in his Natural History of Orkney tells us, that it continues there the shortest time of any of the migratory birds that come to build, and is never to be seen after the end of May. It is also known to migrate still further north; but, as might be expected, the higher the degree of latitude attained, the shorter is the duration of the visit.

M. Temminck, in his Manual of the Birds of Europe, includes two species of Pied Wagtails, the *M. lugubris* of Pallas, and the *M. alba* of Linneus. We have been so disappointed hitherto in our attempts to obtain both species from continental collectors, as almost to induce us to suspect their existence as distinct. All the specimens we have as yet been able to procure, prove to be referable to our own single species under some one of its various appearances; and we may add, that the last edition of the *Fauna Suecica* contains only *M. alba*.

The Wagtails, as they are usually and very appropriately termed, are common to the Old World. India supplies several species, some of which so closely resemble those inhabiting Europe, as to have been considered only as varieties, if not identical. Neither of the continents of America, however, if we recollect rightly, has as yet been ascertained to possess a single species of true *Motacilla*.

Our Figures represent two birds of the natural size, in the plumage of winter and summer.

WHITE-WINGED WAGTAIL.

Motacilla lugubris, *(Pall.)*

Drawn from Nature & on Stone by J & E. Gould. *Printed by C. Hullmandel.*

WHITE-WINGED WAGTAIL.

Motacilla lugubris, *Pall.*

La Bergeronnette lugubre.

On referring to the accompanying description of the White Wagtail, our readers will find that considerable confusion exists respecting the three species of Pied Wagtails inhabiting Europe; there cannot, however, be the slightest doubt that the present bird is very distinct from the one so commonly dispersed over the British Islands, and also from the White Wagtail, so abundant in France and the temperate portions of the continent of Europe: independently of its larger size, and the white colouring of its wings, the *Motacilla lugubris* may be readily distinguished from its European congeners by the conspicuous black mark between the bill and the eye. Its true habitat is also much more eastwardly than that of the other two species, being scarcely ever known to advance westward of the central parts of Europe. M. Temminck informs us that it is common in the Crimea and in many parts of Hungary, and that it is very sparingly dispersed in Italy, Provence, and Picardy. It is also found in Egypt, and in all probability in many parts of Asia Minor and the southern parts of Siberia. It is extremely common in Japan, where, according to M. Temminck, it frequents the streams of mountain valleys.

In its food and general economy, as well as in its seasonal changes, it strictly resembles the other members of the genus.

In summer, the back part of the head and neck, a line from the bill to the eye, and from the eye to the occiput, back, rump, six middle tail-feathers, throat and chest are black; the shoulders, tips and outer edges of the primaries dark greyish brown; the remainder of the wings white, with the exception of the tertiaries, which are brown in the centre; bill and feet black.

In winter the upper part of the throat is pure white; and the back and scapularies uniform grey instead of black.

We have figured two birds, one in the plumage of summer, and the other in that of winter.

WHITE WAGTAIL.

Motacilla alba; *(Linn.)*

Drawn from Nature & on stone by J. & E. Gould. *Printed by C. Hullmandel.*

WHITE WAGTAIL.

Motacilla alba, *Linn.*

La Bergeronette grise.

In the early part of the present work we figured the Pied Wagtail, which is a permanent resident in our island, as the true *Motacilla alba* of Linnæus: subsequent experience has, however, convinced us that the true *Motacilla alba*, a species so common in France and Europe generally, never visits our island, and it would appear that our bird is almost equally unknown in the temperate portions of the Continent. The question therefore arises as to the species to which it must be referred: we ourselves are inclined to believe it quite distinct from *Motacilla lugubris*, a species inhabiting the eastern and southern portions of the Continent, and with which it has by some naturalists been considered as identical; should this ultimately prove to be the case, the English bird will require to be characterized under a new specific appellation. The Channel appears in fact to constitute the boundary of these species, as is also the case with *Motacilla neglecta* and *Motacilla flava*.

The true *Motacilla alba* differs from our species more in its colouring than in any other respect: neither in its full nuptial dress nor at any other period does it exhibit the deep black colouring on the back which is so conspicuous in the British species; at least after the examination of a great number of specimens, we have never been able to trace the slightest indication of it at any age.

The White Wagtail frequents meadows, particularly those in the neighbourhood of running streams, also villages, cities, belfries, towers, and similar situations. It is a common species in Africa and the high lands of India.

Its food consists of flies, millipedes, various other insects, and their larvæ.

The nest is placed in any convenient situation that may offer; in the clefts of rocks, under the arches of bridges, in towers, or among hollow trees; the eggs are six in number, and are of a bluish white, spotted with black.

In spring the forehead, sides of the neck, under surface, and the two outer tail-feathers, are pure white; occiput, nape, throat, chest, middle tail-feathers, and upper tail-coverts, black; back and sides pure ash colour; wing-coverts blackish brown bordered with white.

The female differs in having the white less clear, and the black mark on the occiput not so extensive.

In winter the throat and front of the neck are pure white; on the lower part of the neck a collar of deep black; all the ash of the upper parts less clear than in summer.

We have figured an adult male of the natural size.

YELLOW HEADED WAGTAIL.

Motacilla citreola: *(Pall.)*

Drawn from Life and on Stone by J. & E. Gould.

Printed by C. Hullmandel.

YELLOW-HEADED WAGTAIL.

Motacilla citreola, *Pall.*

La Bergeronette citrine.

A PAIR of this very rare and beautiful species of Wagtail has been very obligingly lent to us for this Work by Lieut. Col. W. H. Sykes, whose rich collection of birds from the Dukhun has at all times been open to our inspection, and has afforded us opportunities of ascertaining the fact, that many of our rare European birds, which more exclusively belong to the eastern portions of that continent, are in reality natives of the western countries of Asia, whence, it would appear, they occasionally migrate to the adjoining districts of Europe, so as to form a connecting link uniting the productions of Europe to those of Asia; and this appears to be the case with the bird before us, whose rare and uncertain visits, according to M. Temminck, to Oriental Russia and the Crimea have afforded little opportunity of ascertaining either its habits or the circumstances connected with its nidification; but we may reasonably expect, from its close affinity in appearance to our well-known Yellow Wagtail (*Motacilla flava*, Ray,), its habits and manners, as well as its mode of nidification, that it is somewhat similar. In the Proceedings of the Committee of Science and Correspondence of the Zoological Society of London, Lieut. Col. Sykes briefly mentions, that this bird has the habits, manners, aspect and size of *Budytes melanocephala*, and, like it, is solitary and only found in the vicinity of rivers; but Colonel Sykes did not see the two species together. Larvæ of water insects and greenish mud were found in the stomach. Colonel Sykes expresses his belief, that this species together with the *Budytes melanocephala* and *Budytes Beema*, all possessing the long hind claw, do not habitually perch, but like other birds with a similar claw, as in the genera *Anthus*, *Alauda*, *Mirafra*, and *Fringilla crucigera*, Temm., nocturnate on the ground. We regret that neither our own observations, nor the numerous works to which we have access, will enable us to add much to this short notice; we may state, however, that we coincide with the views of M. Temminck in considering that Hungary and the Archipelago may also be occasionally visited by it.

In size the *Motacilla citreola* is somewhat inferior both to *M. flava* and *neglecta*, from both of which it may readily be distinguished, in its spring plumage, by the fine citron yellow which covers the top of the head, the cheeks, and the whole of the under surface. A crescent-shaped band of black crosses the occiput, and dark ash colour slightly tinged with grey pervades the upper plumage; the middle and the greater coverts of the wings are edged with white; the tail-feathers black, with the exception of the two outer ones on each side, which are white.

The males and females in winter, says M. Temminck, have not the black occipital band, that part being then yellow like the rest of the head.

Our Plate represents a male and female, proved to be such by dissection, in their summer plumage; the lower bird in the Plate being the male.

YELLOW WAGTAIL.

Motacilla flava; *(Ray)*.

Drawn from Life & on Stone by J & E Gould.

Printed by C. Hullmandel.

YELLOW WAGTAIL.

Motacilla flava, *Ray.*

Budytes flava, *Cuv.*

The present species as well as that we have denominated *Motacilla neglecta* have been separated by M. Cuvier from the *Motacilla* of authors, and formed into a new genus under the name of *Budytes*, in consequence of its lengthened hind claw indicating an approach to the genus *Anthus*, or Pipits. Although we see the force of M. Cuvier's views, still we feel rather inclined in this instance to adhere to the old arrangement, as the general habits and manners of the Wagtails are so much alike; at the same time a direct union may be discovered between the pied species and those of more gaudy plumage, both as regards the style of colouring and structure, as in the *Motacilla boarula*, that bird having the fine yellow plumage so characteristic of Cuvier's subdivision, together with the elongated contour of body and shortened hind claw which distinguish those to which M. Cuvier would exclusively apply the term *Motacilla*. It must, however, be allowed that the birds in question form the extreme limits of one genus, and indicate where another commences; still we think grounds like these scarcely warrant the formation of new genera, otherwise they would indeed become multitudinous. We have, nevertheless, not felt ourselves at liberty to omit the notice of the genus *Budytes*, established as it is by so eminent a naturalist as Cuvier, whose clear and comprehensive views cannot be too highly appreciated; but having stated our opinion, will leave it to the intelligent reader to decide for himself, whether to retain the old term *Motacilla*, or adopt the more recent title as a generic appellation.

This delicate and showy bird visits us early in spring, frequenting open plains and meadows, particularly fields of rising wheat, pastures for cattle, and arable lands. It is a bird sufficiently familiar to allow of close approach, and may be observed full of life and vivacity, running nimbly over the turf, catching flies and other insects, which constitute its food. Hence it is often found near flocks of sheep and herds of cattle, whence its French name Bergeronnette. It continues with us during the summer, and leaves us again early in autumn, the adults invariably leading the way, and are followed by the young at a subsequent period. Its note is rather shrill, but resembles that of the other birds of the genus. It breeds on the ground, constructing a nest of loose fibres and dried grasses lined with hair. The eggs are four or five in number, of a yellowish white colour mottled with darker yellow brown. We have every reason to believe that the present species is equally common throughout the western portions of Europe, but has been overlooked or confounded with the bird we have denominated *Motacilla neglecta.*

In the adult male, the bill is black; irides hazel; the whole of the head and back of the neck pale olive green, over the eye a streak of pure yellow; back, scapulars, rump and tail-coverts rather darker than the colour of the head; quill-feathers dusky, edged with yellowish white tail dusky, the two middle feathers tinged with olive, and the two outer feathers on each side having their external webs almost entirely white; legs black, hind claw long; the whole of the under parts bright yellow.

The female differs in having the plumage less brilliant, the yellow of the under parts being much paler.

We have figured a male and female of the natural size.

GREY HEADED WAGTAIL.

Motacilla neglecta; *(Gould).*

Drawn from Life & on Stone by J. & E. Gould *Printed by C. Hullmandel.*

GREY-HEADED WAGTAIL.

Motacilla neglecta, *Gould.*

Budytes neglecta, *Cuv.*

Having received this bird in considerable abundance from the Continent at the same period of the year in which the Yellow Wagtail visits England, we were struck with the difference between the two; we therefore diligently examined the works of Buffon, Linnæus, and Temminck, and found that each of these authors accurately describes the present bird under the name of *Motacilla flava*; but upon turning to the work of our English naturalist Ray, we found that his description of the Yellow Wagtail truly agreed with our species, and that the Continental authors had in their works applied his name to the bird now in question, a species possessing totally different characters, and with which Ray does not appear to have been acquainted. Our intention, therefore, in figuring the present bird under the new name of *Motacilla neglecta,* is to secure to the Yellow Wagtail the name originally applied to it by our distinguished countryman; recognising in the bird here figured a species distinct from the Yellow Wagtail, with which it has hitherto been confounded. To prevent any suspicion that the two birds are merely varieties, we took the trouble of procuring from various parts of the Continent, examples, both male and female, of *Motacilla neglecta,* a species not found in England, which we compared with both sexes of our British *M. flava,* and had the pleasure of placing several examples of both sorts on the table before the Meeting of the Committee of Science and Correspondence of the Zoological Society of London, on the 12th of July 1832; and in the printed Report of Proceedings, our opinions were first recorded. Although not visiting England, we presume the *M. neglecta* to be extensively spread over the Continent: we have received it in a recent state, shot in the neighbourhood of Paris. On the 28th of May last, males and females were killed by N. C. Strickland, Esq., in Sweden, who kindly favoured us with the loan of his specimens. From the account of this gentleman, their manners are very different from those of our Yellow Wagtail, running about with the tail elevated and the wings hanging down and spread; and so singular was their appearance that he was induced to make a drawing of the birds in this attitude, which he has kindly submitted to our inspection. We have also received this bird from the Himalaya Mountains. As our acquaintance with this species is extremely limited, although on the Continent it is so well known, we avail ourselves of the account of this bird in the work of M. Temminck, where it will be found described under the name of *M. flava.*

This eminent naturalist informs us that it inhabits meadows and the banks of small streams, and is found generally in the northern and central parts of Europe.

The head and back of the neck are of a pure blueish ash-colour; the rest of the upper parts are of an olive green; a white line extends from the beak above the eyes, and another from the lower mandible passes below the orifice of the ears; the lower parts are of a bright yellow; the quills and middle tail-feathers blackish, edged with yellowish white, the two outer tail-feathers white; the tail slightly rounded and extending beyond the wings only an inch and nine lines; the posterior claw long and arched. Length of the bird six inches.

The female has the upper surface more clouded with ash-colour, the under surface less lively, and the throat is yellowish white.

The young, with considerable resemblance to the female, are of a dull ash-colour above, and yellowish white below; the breast is sometimes blotched or waved with brown.

Flies and aquatic insects form their principal food; their nest is built in holes in the ground in meadows, or at the roots of trees; the eggs are six in number, of a greenish olive, with light flesh-coloured blotches.

We have figured a male and female in their adult plumage, of the natural size.

GREY WAGTAIL.

Motacilla Boarula, *(Linn)*

Drawn from Life and on Stone by J. & E. Gould. *Printed by C. Hullmandel.*

GREY WAGTAIL.

Motacilla Boarula, *Lath.*

La Bergeronnette jaune.

This elegant and graceful species is indigenous in Great Britain, but performs periodical migrations within the limits of our island, visiting the southern counties during the winter, and returning northwards to its breeding places as the spring advances, when it is by no means uncommon in Scotland and the border counties.

In habits and manners the Wagtails have all a general similarity; but the present species may be especially distinguished by its finer contour, the superior sprightliness of its air, and the activity of its movements; not only running nimbly and rapidly along the ground, but perching also on trees. More solitary than its congeners, it seldom congregates, but is found dispersed in pairs along the edges of limpid streams, frequenting more rarely the dry and open pastures.

According to Mr. Selby, it arrives at its northern breeding places in April, where, on a stony bank or shelf of rock forming the margin of a stream, it constructs a nest of moss and dried grasses lined with hair; the female laying six eggs, of a yellowish grey blotched with darker shades of the same colour, and usually producing two broods in the year. The process of incubation and rearing its young being over, it retires in September, or the beginning of the following month, the parent birds and their young migrating together.

The present species is extensively spread throughout the northern portions of the European Continent, and we have received it also from the highlands of India.

The grey Wagtail, like the more typical of its family, exhibits considerable variety in its summer and winter plumage, a circumstance which we have endeavoured to elucidate in the annexed Plate, from which it will be seen that during the breeding season and summer months the plumage is not only generally brighter, but that the throat becomes black in both sexes, although in the female the hue is not so deep. It would seem that another moult takes place in autumn, after which the black of the throat entirely disappears, and a yellow tinge assumes its place, from which fact the present species has been occasionally confounded with the yellow Wagtail; it may, however, at once be distinguished by its more slender and symmetrical form of body, the grey colour of the back, and the bright sulphur hue of the rump. Above the eye there is a streak of white, and another below; this last the least conspicuous. The under parts are bright gamboge-yellow; wings dark grey, coverts tipped with white; middle tail-feathers black; outer one on each side white; the next white on its external edge: total length eight inches, of which the tail measures four.

Its food consists of insects, especially aquatic varieties, and their larvæ, in pursuit of which it may be frequently observed wading fearlessly in the shallow parts of the stream, displaying grace and agility in every action.

Our Plate gives the male in his summer and winter dress; but little differences exist in the plumage of the sexes.

1. FIRE CRESTED WREN.
Regulus ignicapillus, *(Brehm.)*

2. GOLDEN CRESTED WREN.
Regulus vulgaris; *(Cuv.)*

Drawn from Life and on Stone by J & E. Gould.

Printed by C. Hullmandel.

Genus REGULUS, *Cuv.*

Gen. Char. *Beak* very slender, short, straight, slightly compressed laterally; the *upper mandible* slightly laciniated towards the tip. *Nostrils* covered with two recumbent feathers. *Wings* rounded. *Tarsus* longer than the middle toe.

FIRE-CRESTED WREN.

Regulus ignicapillus, *Briss.*

Le Roitelet à triple bandeau.

The beautiful little birds which compose this genus are distinguished at once by their diminutive size, by the rich golden crown of their heads, and by the minute comb-like feather which covers their nostrils. Although so diminutive, they are a courageous, spirited, hardy, and active family; enduring, even in the cold countries of the North, the severities of the hardest winter. Their habits, food, nidification, the number and colour of their eggs, bring them in close connexion with the Titmice, while their more feeble but sweetly modulated song, and comparatively weak bill indicate their alliance to the *Sylviadæ*, from which combination of characters it may be a matter of doubt whether its true station is among the former or the latter. On this point we leave the reader to form his own conclusions; although, for ourselves, we are inclined to consider its alliance to the Titmice as based upon the most solid grounds.

We have, however, to notice a new claimant in one of the present family, to a place in the Fauna of Great Britain, which has been long known as a continental species under the name of *R. ignicapillus*. The authority we possess for adding this name to the list of British Birds rests not upon our own observation, but upon the testimony of an accurate and attentive observer of nature, the Rev. L. Jenyns, of Swaffham Bulbeck, Cambridgeshire, who exhibited a recent specimen (accidentally killed near his own residence,) before the Committee of Science and Correspondence of the Zoological Society of London, at the meeting of the 14th of August 1832.

If this bird has been hitherto overlooked in England, the omission has arisen from its close similarity to the common species; we have therefore figured both in one Plate, that their differences may be more clearly perceived. The true habitat of the Fire-crested Wren appears to be confined to the southern portions of Europe, being found in abundance in France, Belgium, and the eastern provinces. In its habits, manners, food, and nidification, it strictly resembles the Golden-crested Wren, from which it differs in plumage in the following points. Its crest is more fiery; the sides of the neck and top of the back more tinged with a golden lustre; and in the alternate stripes of black and white, which occupy the sides of the face, both above and below the eye; the under surface is also rather more grey; its size is the same, or as nearly so as possible.

GOLDEN-CRESTED WREN.

Regulus vulgaris, *Cuv.*

Le Roitelet ordinaire.

The Golden-crested Wren is the smallest of the European Birds, and is generally dispersed in every region from the Arctic circle to the utmost limit of the warm and sultry regions of the South: in the British Isles it is to be found throughout every district, inhabiting woods, coppices and hedgerows, but especially plantations of Fir and Oak, where they appear to be companions with several of the species of the Tits, particularly *Parus cœruleus* and *ater*, with which they may be generally observed engaged in scrutinizing the highest and outmost branches of trees, clinging with ease and dexterity to the under surface, prying inquisitively into every crevice in search of insects and their larvæ, which, with tender buds and small seeds, constitute their food. We have observed that this species, as well as the Long-tailed Titmouse, which it also sometimes accompanies, is in the habit of traversing with a certain degree of order and regularity considerable extent of district, returning nearly at a given time to the same locality, so as to perform a circuit of several miles in the course of a single day. Their common call, which is constantly repeated as if to keep the family together, is a weak but shrill cry, so closely resembling that of the Creeper (*Certhia familiaris*), as scarcely to be distinguished from it. The song, however, which is poured out at the season of nidification is plaintive, sweet and melodious. It constructs a beautiful little rounded nest of moss and lichens, warmly lined with feathers, which is artfully suspended on the under surface of a fir branch, usually near its extremity, and among the thickest foliage, laying from seven to ten very small eggs of a yellowish white colour.

The plumage of the male is of a uniform olive green on the upper surface, the wing-primaries and the tail being brown; the secondaries barred with black and white; the head is ornamented with a beautiful silky golden crest, with an outward border of black capable of elevation or depression at pleasure; the space between the eye and the base of the beak is white, while in the Fire-crested Wren the same part is black; the whole of the under surface is grey, more or less tinged with olive; the beak black; tarsus greenish yellow.

We have figured a male and female of the Golden-crested Wren, and a male only of the Fire-crested, omitting the female of the latter, as there is no distinguishing characteristic.

DALMATIAN REGULUS.

Regulus modestus.

Drawn from Nature & on Stone by J. & E. Gould.

Printed by C. Hullmandel.

DALMATIAN REGULUS.

Regulus modestus, *Mihi.*

A SINGLE specimen of this interesting little bird has been sent to us by the Baron de Feldegg of Frankfort, to whom our acknowledgements are due, not only for this instance of his liberality in consigning to our care, at the risk of loss and injury, a bird probably unique in the collections of Europe, but for many other similar instances of disinterested generosity.

The only history of this bird which we have been able to collect was that written on the label attached to it by the gentleman above mentioned, and is as follows: "I shot this bird, which on dissection proved to be a male, in Dalmatia in the year 1829." We were informed at the same time that it was not known to any of the German ornithologists, and consequently that it had not received a specific title; this we have ventured to give, and suggest the term *modestus*, in allusion to its chaste plumage, and to the absence of the crest, which forms so conspicuous a feature in the other species of the genus, with which we have carefully compared it, and have no hesitation in assigning it a place amongst them as a distinct and genuine species. Judging from its plumage, we believe that the example is fully adult.

Its most conspicuous character are the three yellow stripes which ornament the head; the brighter and most highly coloured of these marks, contrary to what obtains in any of the other *Reguli*, being that over each eye, while the coronal stripe is palest, and consists of feathers similar in length to those which cover the rest of the head.

Our Plate represents the bird in two different attitudes, to exhibit more clearly its characters and colouring.

With the exception of the stripes on the head, the whole of the upper surface is delicate olive green, becoming abruptly paler on the rump; the quills and tail are brown, edged with pale yellow, which is more conspicuous on the secondaries; two transverse bands of the same colour cross the shoulders; the whole of the under surface is pale greenish white; bill and tarsi brown.

The figures are of the natural size.

THE

BIRDS OF EUROPE.

BY

JOHN GOULD, F.L.S., &c.

IN FIVE VOLUMES.

VOL. III.

INSESSORES.

LONDON:

PRINTED BY RICHARD AND JOHN E. TAYLOR, RED LION COURT, FLEET STREET.

PUBLISHED BY THE AUTHOR, 20 BROAD STREET, GOLDEN SQUARE.

1837.

LIST OF PLATES.

VOLUME III.

Note.—As the arrangement of the Plates during the course of publication was found to be impracticable, the Numbers here given will refer to the Plates when arranged, and the work may be quoted by them.

INSESSORES.

* Named erroneously in the letter-press Black and White Lark.

LIST OF PLATES.

* Since the completion of the work Captain S. E. Cook has informed me, that I have made one or two slight mistakes in my account of the Azure-winged Magpie, *Pica cyanea;* and adds that it "is stationary in Spain, and not migratory as thought by M. Wagler. There is also a mistake respecting its inhabiting willows, which are rare in Spain, and not found at all where this bird is most frequent." It is "common in all the royal parks and chaces in New Castile, but is by far the most numerous in the Sierra Morena, in some parts of which it is so abundant, as to be very destructive in the olive grounds."

† Named erroneously on the Plate Picus tridactylus.

‡ Named erroneously Sitta rufescens.

GREAT TIT.

Parus Major; *(Linn.)*

Drawn from Nature & on Stone by J. & E. Gould.

Printed by C. Hullmandel.

Genus PARUS.

GEN. CHAR. *Bill* strong, short, somewhat conical, slightly compressed, sharp-pointed, and hard. *Nostrils* basal, round, covered with reflected bristly feathers. *Feet* with three toes before and one behind; the fore ones divided to their origin; the hind toe strong and armed with a long hooked claw. *Wings,* the first quill of moderate length or almost obsolete; the second shorter than the third; the fourth and fifth longest.

GREAT TIT.

Parus major, *Linn.*

Le Mésange charbonniere.

THE Great Tit, as its name implies, is one of the largest and most typical of the native examples of the present genus; and it is also certainly one of the most beautiful, from the contrasts of its colours, which are brilliant and decided. In its habits and manners, as well as the places it frequents, it strictly agrees with its congeners. It is distributed throughout the whole of the wooded districts of Europe, being stationary in almost every locality: in the British Islands it is certainly so. In severe winters it often leaves the hedges and fields for the warm thickets, coppices and gardens, and not unfrequently farm-yards, where it becomes bold in its endeavours to obtain a subsistence. Its summer food consists of insects and their larvæ, together with the buds of trees and fruits; to these it adds the scattered crumbs from the cottage-door, of whatever matters they may by chance consist, whether animal or vegetable, its digestive powers being apparently adapted to a great variety.

On the approach of spring it becomes noisy and restless, betaking itself to the top branches of high trees, where it utters its harsh note for the day together: the note greatly resembles the noise made by the filing of a saw, or the creaking of a gate on rusty hinges.

It builds a nest in the holes of decayed trees, in the crevices of walls, often in the deserted nest of a crow, a bed of cow's hair and feathers being the receptacle of the eggs; these vary in number from eight to fifteen, and are of a white colour spotted with reddish brown.

The sexes offer but little difference of plumage, the female having less brilliancy of gloss.

The head, throat, and lower part of the neck glossy black; occiput white; back olive green; rump grey; under parts fine yellow, with a black mesial streak; tarsi bluish grey; bill black.

Our Plate represents the male and female of the natural size.

1. SOMBRE TIT.
Parus lugubris; *(Nat.)*

2. SIBERIAN TIT.
Parus Sibericus; *(Gmel.)*.

Drawn from Nature & on Stone by J. & E. Gould.

Printed by C. Hullmandel.

SOMBRE TIT.

Parus lugubris, *Natt.*

Le Mésange lugubre.

We have figured on the accompanying Plate two species of Tits, nearly allied to each other in form, colour and native locality, neither of which approach the British Islands, nor even the more temperate parts of the European continent. The first is the *Parus lugubris*, a species that may at all times be distinguished by its greater size, exceeding, although but in a small degree, our well-known *P. major*: it is, however, clothed with plumage less gaudy, being entirely devoid of those contrasts of black, white, and yellow, which characterize the plumage of that species.

M. Temminck informs us that the *Parus lugubris* is almost restricted to the European confines of the Asiatic border, and that, although pretty common in Dalmatia, it has never been observed in Austria or any part of Germany. The manners, habits and food of this species we believe to be similar to those of its British congeners; but we have no details to offer respecting them from our own experience, nor has any author to which we have access given any particulars respecting them.

The male and female are alike in plumage, and may be thus described:—The whole of the upper surface of a brownish ash colour, becoming deeper on the top of the head; the secondaries and tail-feathers slightly margined with whitish; throat brownish black; the cheeks and the whole of the under surface white, slightly tinted with brownish grey; beak and feet lead colour.

SIBERIAN TIT.

Parus Sibericus, *Gmel.*

Le Mésange à ceinture blanche.

Although the *Parus Sibericus* has no great attraction as regards beauty of plumage, it has in its shape and general form a more elegant and graceful contour than the *P. lugubris*. In size it is considerably smaller, having at the same time a longer and a graduated tail, offering, though in a slight degree, a relationship to the Long-tailed Tit, so commonly dispersed over Europe; and we have to regret that the extreme rarity of the Siberian Tit, in Europe at least, prevents our ascertaining whether its habits and manners offer any approximation to those of the bird just referred to. M. Temminck, in his *Manuel*, informs us that it is an inhabitant of the most northern parts of Europe and Asia, migrating in winter to some of the provinces of Russia; and we received from Sweden the specimens from which our figures were taken.

The plumage of *Parus Sibericus* may be thus detailed:—The upper surface is of a deep ash colour, tinged on the back with brown; the quills, secondaries and tail-feathers edged with white; throat black; cheeks and upper part of the chest pure white; under parts greyish white, washed with rufous on the flanks; bill and tarsi lead colour.

Our Plate represents these two rare species of the natural size.

TOUPET TIT.

Parus bicolor; *(Linn.)*

Drawn from Nature & on Stone by J & E. Gould. *Printed by C. Hullmandel.*

TOUPET TIT.

Parus bicolor, *Linn.*

La Mésange bicolore.

There can be no doubt that the northern regions of America form the true habitat of this species; we have, however, seen specimens which were undoubtedly killed in Russia, and therefore no longer hesitate in classing it among the occasional visitants of the European continent; nevertheless it is there extremely rare and is confined to the regions adjacent to the arctic circle. In the works of Wilson and Audubon its manners are described as resembling those of the other members of the genus. "It moves along the branches," says the latter gentleman, "searches in the chinks, flies to the ends of twigs, and hangs to them by its feet, whilst the bill is engaged in detaching a beech- or hazel-nut, an acorn, or a chinquapin, upon all of which it feeds, removing them to a large branch, where, having secured them in a crevice, it holds them with both feet, and breaks the shell by repeated blows of its bill. . . . It resorts to the margins of brooks to drink, and when unable to do so, obtains water by stooping from the extremity of a twig overhanging the stream; it appears to prefer this latter method, and is also fond of drinking the drops of rain or dew as they hang at the extremity of the branches." The same author also informs us that its notes, which are usually loud and mellow, are rather musical than otherwise; that it is somewhat vicious in its disposition, and occasionally attacks and destroys smaller birds by repeated blows on the head until it breaks the skull.

The nest is constructed of all kinds of warm materials, and is generally placed in the holes formed by the Downy and other species of Woodpecker, but is occasionally placed in a hole dug by the bird itself for that purpose. The eggs, which are from six to eight in number, are of a pure white, with a few red spots at the larger end.

The sexes are so much alike as to be scarcely distinguishable.

Forehead black; sides of the head brownish black; all the upper surface uniform grey; under surface greyish white, tinged with yellowish brown on the flanks; bill black; irides dark brown; feet lead colour.

We have figured an adult of the natural size.

AZURE TIT.

Parus cyanus; *(Pall.)*

Drawn from Nature & on Stone by J. & E. Gould. *Printed by C. Hullmandel.*

AZURE TIT.

Parus cyanus, *Pall.*

Le Mésange azurée.

THIS beautiful little Tit is a native of Siberia, whence it frequently strays into the northern parts of Europe, such as Russia and Poland, and it has been known to penetrate so far south as Germany. Like the rest of its family it dwells in woods and forests, generally in the most retired parts; it is not so much to be wondered at, therefore, that its history is shrouded in obscurity, when we consider how little intercourse naturalists have hitherto had with the remote countries which it inhabits.

Were we allowed to judge from analogy, we might very reasonably conclude that its manners and its disposition are in strict unison with those of its near relative the Blue Tit (*Parus cœruleus*) of England.

For the specimens from which our figures were taken, and which we believe to be the only examples in England, we are indebted to the liberality of the directors of the Royal Museum of Berlin.

Like the rest of its race, the sexes of the Azure Tit offer little or no difference in the colouring of the plumage.

Nothing is at present known respecting its nest or eggs.

The forehead, throat, and breast are white; a band of deep blue extends from the eye round the back part of the head; the back and rump are fine blue grey; the tail-feathers fine deep blue with white tips, and the outer one on each side wholly white; the wings deep blue, the secondaries largely tipped with white; a band of the same colour crosses near the shoulders; primaries grey, brown on their inner webs and white on the outer; feet and bill lead colour.

The Plate represents a male and female, of the natural size.

BLUE TIT.

Parus cæruleus, *(Linn.)*

Drawn from Nature & on stone by J. & E. Gould. Printed by C. Hullmandel.

BLUE TIT.

Parus cœruleus, *Linn.*

La Mesange bleue.

Few birds can be more familiar to our readers than the Blue Tit, the habits and manners of which every one must have repeatedly noticed, since of all the species it is the most common in our gardens and around the precincts of our habitations, and it is exceeded by none in its sprightly actions and in the address and activity with which it searches the extremities and shoots of trees in quest of its insect food. The mischief it does to the tender buds of trees, in stripping off their envelopes, has rendered it very obnoxious to the gardener, although doubtless the benefit it confers by the destruction of insects more than compensates for the injury. Like the rest of the British Tits, it is a permanent resident in our island, braving the severity of our hardest winters, against which it is peculiarly defended by the full downy plumage which invests the whole of the body. On the approach of spring its simple note may be heard in our woods and gardens, which is a true sign that its pairing-season has already commenced, and that the mated birds are preparing for the task of incubation. The situation chosen for the nest varies according to circumstances; most frequently it is in the hole of a tree, the chinks of a wall, and even the interstices of old posts or palings; it is generally constructed of moss lined with feathers and hair: the eggs are white, speckled with dark red.

The young assume the colouring of the adults at an early age, and quickly follow their parents in their assiduous search after insects and their larvæ. The family group keep united until autumn at least, when they all separate, going in winter in single pairs, or passing the colder months singly or in company with other small birds.

The sexes are so closely alike in colouring as to offer no decided difference; the tints of a male are, perhaps, somewhat the brightest.

On the Continent they are widely distributed, and exhibit the same habits and manners that they are observed to do in the British Islands.

The top of the head is fine cœrulean blue; the forehead, stripe over the eye, and cheeks white; a black stripe passes from the bill, through the eye and surrounds the white of the cheeks; the upper surface is delicate olive green; the wings and tail blue, the secondaries being slightly tipped with white, and the primaries dark brown; the whole of the under surface yellowish green; tarsi and bill blueish lead colour.

The Plate represents a male and female of the natural size.

1. COLE TIT.
Parus ater; (Linn.)

2. MARSH TIT.
Parus palustris; (Linn.)

Drawn from Nature & on Stone by J & E. Gould.

Printed by C. Hullmandel.

COLE TIT.

Parus ater, *Linn.*

La Mésange petit charbonnière.

The Cole Tit appears to give a preference to woods of birch, oak, and pine, in hilly and mountain districts; nevertheless, it frequents, in tolerable abundance, hedgerows, shrubberries, and gardens, over nearly every portion of Europe.

In its habits and manners it is remarkably quick and active, searching with great assiduity among the twigs and buds of trees for insects and their larvæ, upon which it feeds, while every action is animated and sprightly. It braves with indifference not only our severest winters, but even those of the northern portion of the Continent. Its nest is placed indifferently according to circumstances, being sometimes formed in the hole of a decayed tree or old wall, and at others on the ground: it is composed of moss and wool, generally lined with hair. The eggs are from six to ten in number, of a pure white, sparingly dotted with reddish brown.

The sexes offer little or no difference in their plumage, and the young assume at an early period the colouring of maturity.

The top and sides of the head are black; a white mark occupies the occiput; throat black; sides of the face white; upper surface grey, with a slight tinge of brown; wings and tail brownish black, the former having two transverse bands of white; flanks and under-surface white slightly tinged with rusty brown; bill black; tarsi lead-colour; irides hazel.

MARSH TIT.

Parus palustris, *Linn.*

La Mésange nonnette.

The Marsh Tit is slightly superior to the Cole Tit in size, and differs from it also in the situations it frequents, giving the preference to low tracts of land, covered with thickets, in the neighbourhood of swamps and marshes; it is also found in orchards, gardens, and similar localities. It appears to be more abundant in Holland than in any other country; it is, however, very generally spread, and is found in very high northern latitudes. In England it is as equally diffused as the rest of the genus, and is as active and sprightly in its habits, prying in search of food with the same dexterity and adroitness. It constructs its nest in the holes of trees, and lays from ten to twelve eggs, of a white colour, dotted with reddish brown spots. It may be observed that the Marsh Tit and the Cole Tit often associate together during winter, and it not unfrequently happens that the Crested Wren and other small birds join their company.

The sexes offer no distinction, except that the colours of the female are more obscure.

In the male the top of the head and back of the neck are deep black; the upper surface, wings, and tail are greyish brown, the latter being somewhat the darkest; breast black; cheeks and throat white; the under surface white, clouded with dusky brown; bill black; legs lead-colour; irides dark hazel.

The upper bird in our Plate represents the Cole Tit, the lower the Marsh Tit, both of the natural size.

CRESTED TIT.

Parus cristatus; *(Linn:)*.

Drawn from Life & on Stone by J. & E. Gould.

Printed by C. Hullmandel.

CRESTED TIT.

Parus cristatus, *Linn.*

La Mesange huppée.

In our attempts to discover this interesting species of Tit in this country we have been unfortunately disappointed, notwithstanding every work which has hitherto been published on British Ornithology has enumerated it as one of our indigenous birds. In his work on British Birds, Mr. Selby states that he has been informed by Sir W. Jardine that the Crested Tit has been found in some plantations near Glasgow, where it annually breeds. Upon the testimony of this distinguished naturalist, we feel ourselves bound to agree in the propriety of its admission into the list of our native Fauna, hoping that at no distant day we shall receive further information on the subject. Rare as it is with us, it is very common in some parts of Europe, being abundant in the pine forests of all its northern regions, and especially where juniper trees are plentiful. M. Temminck, however, assures us that it is scarce in Holland: we know it to be common in Germany, France, and the Alpine regions.

In habits and manners it resembles the rest of the smaller Tits, feeding on insects, berries, and the seeds of evergreens.

M. Temminck says it builds its nest in the holes of trees or walls, or in the abandoned dreys of Squirrels and Pies. The eggs are as many as ten in number, of a white colour, marked on the larger end with spots of blood red.

The sexes offer no external difference in plumage.

The head is furnished with a beautiful crest, capable of erection, consisting of long white feathers, having their centres black; the cheeks and sides of the neck white, bounded before by a band, which passes from the throat to the sides of the neck, and behind by a similar band passing from the back of the head in the same direction; the ear-coverts are also bounded by a narrow line of black; the general plumage above is delicate brown, and below white, slightly tinged with brown.

The Plate represents a male of the natural size.

LONG-TAILED TIT.

Parus caudatus; *(Linn.)*

Drawn from Nature & on Stone by J. & E. Gould.

Printed by C. Hullmandel.

LONG-TAILED TIT.

Parus caudatus, *Linn.*

Le Mésange à longue queue.

INDEPENDENTLY of the deviation from the form of the rest of the genus which this little Tit exhibits, its habits, mode of nidification, and food, also tend to place it in an isolated situation, and it is questionable whether it might not be with justice considered as the type of a new and distinct genus.

Few persons who have been accustomed to observe the habits of our native birds can have failed to be struck with the peculiar actions of this bird, which is continually wandering from tree to tree and hedgerow to hedgerow, diligently traversing every branch in quest of insects and their larvæ, which constitute almost its sole food: these peregrinations appear to be repeated day after day over a given circuit, and it often happens that at the same hour, on several successive days, they may be found at the same place; during the breeding and summer seasons there is, however, an exception to this general rule, which leads us to suspect that these wandering parties consist of the broods of single nests which continue to associate together till the following summer causes them to separate into pairs for the purpose of breeding. Noted as birds of this genus are for their active and restless habits, the Long-tailed Tit is conspicuous among them as being the most agile and expert; clinging in every possible attitude to the branches with the utmost ease, and prying into every bud and crevice, even along the under sides of the twigs, with the strictest scrutiny.

Among the nests of our British birds, that of the Long-tailed Tit is pre-eminent for beauty and the ingenuity displayed in its structure: in form it is oval, and domed over at the top, and is generally fixed in the forked branches of a low tree or tall bush in a dense hedgerow; it is composed externally of moss, lichen, fibres, and wool, admirably interwoven together, and is lined internally with feathers: in this secure and warm receptacle, the female lays her eggs to the number of twelve or twenty, white in their colour, with obscure reddish spots at the larger end.

Though not a songster, this interesting bird utters during the season of love a few simple, twittering notes; but these cease with the summer, a chirping call being its only note during the rest of the year. Its flight from tree to tree is tolerably rapid, but cannot be maintained for any distance.

The top of the head is white; a black mark passes through each eye to the occiput, and joining there with that of the opposite side, runs in a broad streak down the back, passing off at the edges into a rose red, which is the tint of the upper surface; the quills are black; the secondaries edged with white; cheeks and throat greyish white, under parts pinky grey; tail long and graduated, the four middle feathers black, the two next tipped with white, the rest with the outer webs white also; beak and tarsi black.

The female does not differ from her mate in colouring.

We have figured a pair of the natural size.

BEARDED TIT.

Parus biarmicus; *(Linn.)*
Calamophilus———; *(Leach)*.

Drawn from Life and on Stone by J. & E. Gould. *Printed by C. Hullmandel.*

Genus CALAMOPHILUS, *Leach.*

Gen. Char. *Beak* nearly as in the genus *Parus*, but the upper *mandible* at its tip is somewhat curved. *Tail* elongated, wedge-shaped. *Legs* very slender.

BEARDED TIT OR REED BIRD.

Calamophilus biarmicus, *Leach.*

La Mesange moustache.

Dr. Leach was induced to separate this very interesting and elegant bird from the genus *Parus*, in consequence of its differing in several minor characters from the other species of that genus, particularly in the situation it affects as a place of abode and nidification; constructing a nest on or near the ground in wet and marshy places: its food is also very different, consisting of the seeds of reeds, with aquatic insects and minute shelled-snails, for the trituration of which it is furnished with a strong muscular gizzard. It inhabits England as well as most of the temperate countries of Europe, but is more particularly abundant in the low and marshy districts of Holland, France and Germany. Its disposition is timid, and its manners shy and retired, dwelling in situations both local and difficult of access; a circumstance which, until lately, has prevented naturalists from giving any minute details respecting its peculiar habits. We are indebted to Mr. Hoy, an intelligent observer of nature, for the best account of this bird yet published, from which, as given in "The Magazine of Natural History," vol. 3. p. 328, we take the liberty of making the following extract.

"The borders," says Mr. Hoy, "of the large pieces of fresh water in Norfolk called Broads, particularly Hickling and Horsey Broads, are the favourite places of resort of this bird; indeed it is to be met with in that neighbourhood wherever there are reeds in any quantity, with fenny land adjoining. During the autumn and winter they are found dispersed, generally in small parties, throughout the whole length of the Suffolk coast, wherever there are large tracts of reeds. I have found them numerous, in the breeding season, on the skirts of Whittlesea, near Huntingdonshire, and they are not uncommon in the fenny district of Lincolnshire; whether they are to be met with further north I have had no means of ascertaining, but they do not appear to have been noticed north of the Humber. It begins building in the end of April. The nest is composed on the outside of the dead leaves of the reed and sedge intermixed with a few pieces of grass, and invariably lined with the top of the reed, somewhat in the manner of the nest of the Reed Wren (*S. arundinacea*, L.), but not so compact in the interior. It is generally placed in a tuft of coarse grass or rushes near the ground, on the margin of the dikes, in the fen; sometimes fixed among the reeds that are broken down, but never suspended between the stems. The eggs vary in number from four to six, rarely seven, pure white, sprinkled all over with small purplish red spots, intermixed with a few small faint lines and markings of the same colour; size about the same as that of the Greater Tit, but much more rounded and shorter. Their food during the winter is principally the seed of the reed; and so intent are they in searching for it, that I have taken them with a birdlime twig attached to the end of a fishing-rod. When alarmed by any sudden noise, or the passing of a hawk, they utter their shrill musical notes and conceal themselves among the thick bottom of the reeds, but soon resume their station, climbing the upright stems with the greatest facility. Their manners in feeding approach near to the Long-tailed Tit, often hanging with the head downwards, and occasionally assuming the most beautiful attitudes. Their food is not entirely the reed-seed, but insects and their larvæ, and the very young shelled-snails of different kinds, which are numerous in the bottom of the reedlings. I have been enabled to watch their motions when in search of insects, having, when there has been a little wind stirring, been often within a few feet of them, quite unnoticed, among the thick reeds. Was it not for their note betraying them, they would be but seldom seen. The young, until the autumn moult, vary in plumage from the old birds; a stripe of blackish feathers extends from the hind part of the neck to the rump. It has been said that the males and females keep separate during the winter; but I have always observed them in company; they appear to keep in families until the pairing time, in the manner of the Long-tailed Tit; differing in this respect, that you will occasionally find them congregated in large flocks, more particularly during the month of October, when they are migrating from their breeding-places."

To this interesting account we may add, that they are to be met with occasionally on the banks of the Thames, from the thick reed-beds of Erith in Kent throughout the course of the river to Oxford; but their visits are by no means regular, or to be calculated on with certainty.

The total length of the male Bearded Tit is about six inches; the beak orange, the upper mandible longer and overhanging the under; irides yellow; feet black; crown of the head, nape, and cheeks delicate ash colour; between the base of the beak and the eyes is a black mark, which proceeds down the side of each cheek, and terminates in a fine and lengthened moustache; throat white; breast vinous grey; the sides of the breast, the back, and the four middle tail-feathers fine reddish orange; primaries brown externally, edged with white; secondaries the same colour as the back, with a black longitudinal stripe; vent black; tail graduated, and about three inches in length.

The female is rather less than the male, of a more uniform ferruginous colour, with a few dashes of black on the upper part of the neck and back, and has a faint yellowish white instead of a black moustache.

We have figured a male and female of the natural size.

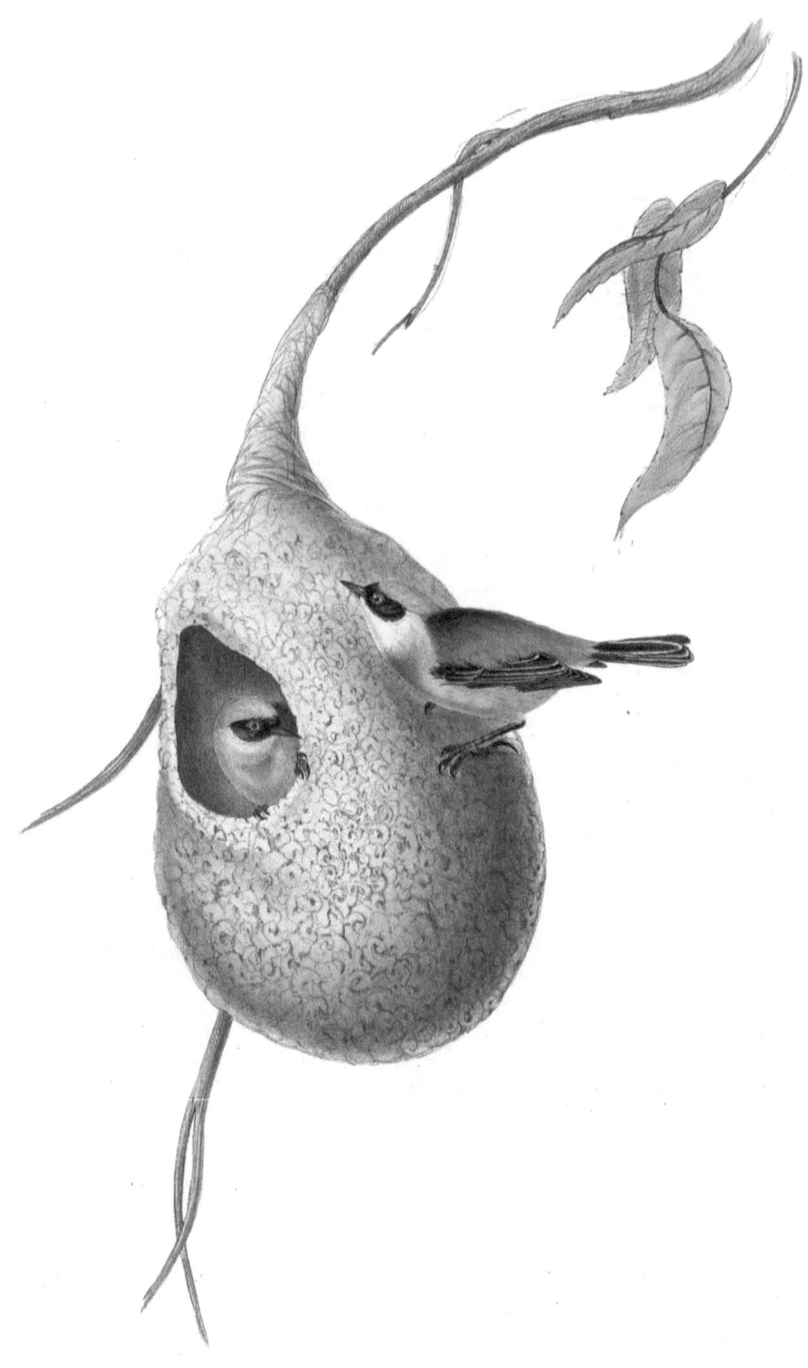

PENDULINE TIT.

Ægitalus pendulinus; *(Boje)*.

Drawn from Nature & on Stone by J. & E. Gould.

Printed by C. Hullmandel.

Genus ÆGITALUS, *Boje.*

GEN. CHAR. *Beak* moderate, very acute, the upper mandible straight, middle toe free from the base: hind claw large and strong. *Tail* truncate and moderate.

PENDULINE TIT.

Ægitalus pendulinus, *Boje.*

Le Mésange rémiz.

WE quite agree with M. Boje in the propriety of assigning this elegant little bird to a separate genus, distinguished by several minute particulars from that of *Parus*. In habits, manners, and the localities it frequents, it bears a great resemblance to the Bearded Tit (*Calamophilus biarmicus*, Leach); but in the form of its beak and tail, and in its mode of nidification, it not only differs from it, but also from every other species of the family. In this little bird, whether we regard its elegant hanging nest or its chaste plumage and sprightly form, there is much to attract attention: it is not, however, among the natives of our island, but must be sought for in the southern and eastern provinces of Europe. It is tolerably abundant in Italy and the South of France, and is also found in some parts of Russia, Poland, and Hungary, everywhere frequenting the borders of rivers and inland sheets of fresh water, where willows, reeds, and luxuriant herbage afford it shelter. Its food is said to consist, like that of the Bearded Tit, not only of seeds, but also of aquatic insects, and the animals inhabiting small freshwater shells.

Proverbial as are the Tits for the beauty and skilful structure of their nests, none are more remarkable and curious than that of the present species: it is constructed of the soft down of the willow or poplar; and this substance, which closely resembles cotton wool, is interwoven together with admirable ingenuity, so as to form a flask-shaped nest, with a lateral opening into the internal chamber. It is suspended at the extremity of a drooping branch of a willow or any similar tree overhanging the water. The eggs are six in number, of a pure white, marked with a few red blotches.

The sexes in the adult state offer but little difference in the colour of their plumage; the markings of the female, particularly the black band across the face, are however more obscure, and the young, besides being of a lighter colour, want the black mark entirely.

The plumage of the adult male is as follows: a black band extends across the forehead, encircles the eyes, and spreads over the ear-coverts; top of the head light grey; throat lighter; the upper surface chestnut brown, more intense on the middle of the back, fading off to buff; breast chestnut, becoming lighter as it spreads over the abdomen; wings and tail brownish black, each feather having a lighter margin.

Our Plate represents a pair of these birds, and their nest, of the natural size.

WAXEN CHATTERER.

Bombycivora garrula; (Temm^k)

Drawn from life and on stone by J & E Gould.

Printed by C. Hullmandel.

Genus BOMBYCIVORA, *Temm.*

GEN. CHAR. *Beak* short, straight, elevated, the upper *mandible* slightly curved towards its extremity and furnished with a very marked tooth. *Nostrils* basal, ovoid, open, concealed by short stiff hairs directed forward. *Toes,* three before and one behind, the external and middle toes united. *Wings* moderate, the first and second *quill-feathers* equal and longest.

WAXEN CHATTERER.

Bombycivora garrula, *Temm.*

Le Grand Jaseur.

THE birds composing the genus *Bombycivora,* as restricted by M. Temminck, are by no means numerous, three species only having as yet, we believe, been discovered. The present beautiful example, which is the largest, and may be considered the most typical of the genus, is the only one Europe affords us; it is also to be found in the northern regions of America, although much less common than the smaller allied species peculiar to that Continent. The rare and uncertain visits of the Waxen Chatterer to our Island afford us but little insight into its history, as it is in winter only that small flocks now and then appear, driven probably by the severity of the season in northern climes to a more southern retreat.

Its true habitat appears to be the regions of the arctic circle, whence it emigrates to the adjacent districts both of Asia and Europe. Dr. Latham informs us that it is plentiful both at St. Petersburg and Moscow in the winter, but is observed to come there from parts further north, and to depart again to the arctic circle in spring. It is never known to breed in Russia, is scarce in Siberia, has not been found beyond the Lena, and is mentioned as a Tartarian bird by Frisch, who says it breeds among the rocks; and nothing more, we believe, is known respecting its habits and nidification. The formation and general structure of its beak indicate it to be a true berry-feeder, and we accordingly find that during its visits here its food consists of the berries of the mountain ash, the haw, the privet, &c.

The general colour of the body of the male bird is of a dull vinous ash, with a bright ferruginous tinge on the forehead and cheeks; the feathers of the head prolonged into a beautiful crest; beak and tarsi black; the throat, the feathers of the nostrils, and a band which passes from the beak through the eye, black; primary quill-feathers brownish-black, each feather being marked on the inner margin of the tip with a yellow line; secondaries tipped with white and having the shaft prolonged and furnished with singular appendages resembling red sealing-wax, whence its name; upper tail-coverts ash-coloured, the under ones ferruginous; tail black tipped with a yellow band.

The male and female offer but slight external differences, both having the wax-like appendages to the secondaries; they are, however, less numerous in the female, and are altogether wanting in the young.

Our Plate represents a male in the adult plumage and of the natural size.

BLACK LARK.

Alauda Tartarica; *(Pall.)*

Drawn from Nature & on stone by J. & E. Gould.

Printed by C. Hullmandel.

Genus ALAUDA, *Linn.*

GEN. CHAR. *Bill* subconic, short, the mandibles of equal length, the upper one slightly convex. *Nostrils* basal, lateral, oval, partly concealed by small reflected feathers. *Feet,* three toes before, and one behind; the anterior ones entirely divided; the claw of the hind one long and nearly straight. *Wings,* the first quill very short, or wanting, the third the longest; tertials in most instances shorter than the quills. Coronal feathers generally produced, and capable of being erected.

BLACK AND WHITE LARK.

Alauda Tartarica, *Pall.*

L'Alouette Nègre.

THE *Alauda Tartarica* is a native of the high northern regions of the old continent, where it enjoys a most extensive habitat, being dispersed, as we have every reason to believe, over the whole of Siberia, Northern Russia, Lapland, &c., and from whence it performs periodical migrations into more temperate climes. It spreads in autumn, says M. Temminck, over the provinces of European Russia, where it dwells in small companies; hence it is necessary to include it in the fauna of Europe. Like *Plectrophanes nivalis* and *Lapponica* it is subject to very considerable and contrasted changes in the colouring of its plumage at opposite seasons: during the rigorous months of winter its clothing is remarkably thick and warm; the feathers, which are then elongated, are encircled with a band of light tawny grey, and falling closely over each other, conceal the black colouring of the base of each feather. On the approach of summer a decomposition takes place in the lighter portions of the feathers, which gradually break off, and leave the bird in the height of summer of a jet black, which style of dress continues until the autumn, when a moult takes place, and the bird again assumes its usual winter clothing. The lower figure in our Plate represents the bird in the winter dress, while the upper illustrates the nearly completed plumage of summer, when, as will be readily perceived, many of the feathers possess the remains of the winter plumage.

The *Alauda Tartarica* is a bird of great rarity in the collections of Europe; and independent of our own specimen, which we received from Paris, we know of no other example in England.

The only difference in the outward appearance of the sexes consists in the hues of the female being somewhat less deep, and in her being rather smaller in size than her mate.

We cannot examine this and the following species, *Alauda Calandra,* without being fully impressed with the propriety of separating them into a new genus, distinct from *Alauda,* which genus is typically represented by *Arvensis*; but having already figured one species which according to our ideas would range in this division under the name of *Alauda brachydactyla,* we have considered it best, in these instances also, to retain the old generic title *Alauda.*

After what has been said above we conceive any further description of the plumage will be unnecessary: the beak is yellowish buff at the base and black at the tip; the feet and legs are black.

The figures are of the natural size.

Drawn from Nature & on stone by J. & E. Gould.

CALANDRA LARK.
Alauda Calandra; *(Linn.)*

Printed by C. Hullmandel.

CALANDRA LARK.

Alauda Calandra, *Pall.*

L'Alouette Calandre.

In its general form, robust body, and powerful bill, this bird very closely resembles the *Alauda Tartarica*; the countries, however, to which these birds resort are widely different, the *Alauda Tartarica* being almost confined to the high northern regions, while the range of the *Alauda Calandra* extends nearly to the tropics: it is very abundant in Northern Africa, and is common in Spain, Turkey, Italy, and the South of France, to the north of which countries it is seldom seen. We have little to communicate respecting its habits and manners, in all of which we believe it bears a striking resemblance to the Common Lark, *Alauda arvensis.*

It constructs its nest among the herbage, and lays four or five eggs, of a clear purple marked with large grey spots.

Its food consists of insects, seeds, &c.

The sexes are only to be distinguished by the female being rather less in size, and by the black markings on the sides of the neck being less developed.

The young, as will be seen, exhibit the usual characteristics of the genus, having the tips of all the feathers margined with yellowish grey.

The upper surface is of a sandy grey, the centre of each feather being dark brown; quills dark brown edged with whitish; throat white bordered by a black lunulated stripe, beneath which the feathers are dirty white varied with black; belly white; flanks and thighs brown; the outer tail-feather on each side white on the outer web and tipped with white, the third edged with grey and tipped with white, the fourth tipped with grey, the remainder of the tail black; bill pale horn colour; legs pale grey.

The Plate represents an adult male, and a young bird of the year, both of the natural size.

SHORT-TOED LARK.
Alauda brachydactyla; *(Temm.)*

Drawn from Nature & on stone by J. & E. Gould.

Printed by C. Hullmandel.

SHORT-TOED LARK.

Alauda brachydactyla, *Temm.*

L'Alouette à doigts courts.

This species, like *Alauda Calandra* and *Alauda Tartarica*, is distinguished by the more powerful and robust form of the bill, and by the comparative shortness of the toes, circumstances which, as we have already observed, would appear to constitute the characters of a minor group of the Larks; in neither, however, are the toes so much abbreviated as in the present instance. The members of this group would appear to be widely distributed, the largest of the genus, *Alauda Tartarica*, being a native of high northern latitudes, while the delicate species here figured makes the southern regions of Europe, and the adjacent portions of Africa, its permanent habitat. It is said to abound on the hot sandy plains of the Spanish Peninsula, and that it is no less abundant in Sicily and in some portions of Italy; in fact it is found along the whole of the borders of the Mediterranean. It occurs occasionally in the South of France, but this appears to be the boundary of its range northwards.

Its nest is constructed on the ground like that of the Sky Lark, and the eggs are five in number, of an isabelle yellow, without any markings.

The sexes are not distinguishable by the colouring of their plumage; the tints of the female are, however, somewhat duller than those of the male. The young during the first autumn have the outer edges of each feather margined with buff.

The male has the top of the head and all the upper parts of a yellowish or sandy brown, with the centre of each feather darker; the quills and tail of a dusky brown, the two outer feathers of the latter having their external edges yellowish white; a whitish yellow streak over each eye; throat and belly white, the chest and flanks being tinged with yellowish brown; bill and feet light brown.

The Plate represents a male and female of the natural size.

SHORE LARK.
Alauda alpestris; *(Linn.)*

Drawn from Nature & on Stone by J & E. Gould.

Printed by C. Hullmandel.

SHORE LARK.

Alauda alpestris, *Linn.*

L'Alouette à hausse col noir.

This beautiful and singular species of Lark has lately made its appearance in Britain, and some of our museums, as well as private collections, can boast of possessing species obtained in our own country: it may, however, be considered as strictly a northern species, inhabiting the higher latitudes of Europe, Asia and America, in which latter portion of the globe it especially abounds. Wilson informs us that it is one of the summer birds of passage of that continent, arriving in the North in the fall, and usually staying the whole of the winter: it frequents sandy plains and open downs, and is numerous in the Southern States as far as Georgia. During that season they fly high, in loose scattered flocks, and at these times have a single cry, almost like the Skylark of Britain. It is, however, not improbable that this species is spread over the whole of the American continents; at least we have received it from the Straits of Magellan, where it was found by Captain King. M. Temminck states that it appears as a bird of passage in Germany, and never ventures into the southern continental provinces. The regions of the polar circle appear to be its native habitat; it also incubates and rears its young in the marshy and woody districts of the eastern portions of the fur countries of North America, according to Dr. Richardson, who quotes Mr. Hutchins as his authority for stating that its nest is placed on the ground, and that it lays four or five white eggs spotted with black. On the advance of winter, it retreats to the southwards, and is common in the United States throughout that season.

It appears to frequent wild and barren districts adjacent to the shore, situations in which it particularly delights, and more especially sandy elevations covered with scanty tufts of herbage, never perching on trees, but gaining its subsistence from the seeds of grasses and the shoots and buds of dwarf shrubs.

The male and female of this beautiful species differ in the brilliancy of the plumage. In the male, the whole of the upper surface is of a vinous ash colour, each feather having a central wash of brown; the forehead is yellow, whence a slender stripe passes over the eye; above the yellow of the forehead a broad patch of black extends across the head, terminating above each eye in a tuft of elongated black feathers, like the egrets of some of the *Strigidæ*, capable of being elevated or depressed at pleasure; from the base of the bill extends a black mark, which covers the cheeks; throat and sides of the neck yellow, succeeded by a black gorget; sides vinous ash, becoming whitish on the under surface; the two middle tail-feathers brown, the rest black, the edge of the outermost being white; bill brown; tarsi black.

In the female, the black band on the head and the egrets are not very apparent, the yellow is circumscribed and dull, and the gorget small, in which respect the young are similar.

We have figured a male and female of the natural size.

CRESTED LARK.

Alauda cristata, *(Linn)*.

Drawn from Nature & on Stone by J & E Gould.

Printed by C. Hullmandel.

CRESTED LARK.

Alauda cristata, *Linn.*

L'Alouette à hausse-col noir.

Although the Crested Lark is abundantly distributed over the temperate and warmer portions of the Continent, no instance of its having been killed in the British Islands is, as far as we are aware, on record; this is the more singular, as, from the circumstance of its extending its range to many parts of the coast which are opposite our own, and from its being particularly common in the fields and plains round Calais, it might pass and repass to Dover at will. Our personal observation of this bird while on the Continent leads us to regard it as a much more solitary bird than the Skylark, to which in its general aspect it bears a close resemblance. The Crested Lark is said to congregate in flocks occasionally; but when we observed them they were scattered over the country in pairs, very frequently in the vicinity of the main roads. They may readily be distinguished from their near ally by the crest, which in the male is generally erect; they are also said to perch on trees.

The Crested Lark frequently sings as it flies, sometimes soaring to a great height: its song is varied, although we consider it to be inferior to the well-known songster that enlivens the spring-time in our islands with its cheerful and voluminous notes.

As we have before stated, its range on the Continent is very general, and M. Temminck relates that it is sometimes found pretty far north, although in no great numbers: to these localities we may add that it is an inhabitant of the northern portions of Africa, the whole of Asia Minor, and the high lands of India.

The female is less than the male, and has not so long a crest. It is said to build early, coustructing a nest on the ground somewhat like that of the Skylark: the eggs, four or five in number, are of a pale ashy brown spotted with dark brown.

The head, and all the upper surface, wings, and tail, reddish brown with darker centres; stripe above and beneath the eye, throat, belly, under tail-coverts, and the outer tail-feathers, white; sides of the face and breast white, numerously spotted with dark brown, and tinged with reddish, the spots on the latter disposed in the form of a crescent; upper mandible dark brown; under mandible light brown at the base, and dark brown at the tip; legs pale brown.

We have figured a male and female of the natural size.

SKY LARK.
Alauda arvensis; *(Linn.)*

Drawn from Nature & on Stone by J & E. Gould — *Printed by C. Hullmandel*

SKY LARK.

Alauda arvensis, *Linn.*

L'Alouette des champs.

This well-known bird, with whose brilliant and varied song we are all so intimately acquainted, is not only common in our own island, but is equally abundant over the greater portion of Europe and the adjacent parts of Asia and Africa. In spring and summer it lives in pairs, but is gregarious during winter, associating towards the close of autumn in vast flocks, which are joined by migratory visitants from more northern districts, and if the winter be severe they all progress to the warmer latitudes of Europe and of Africa.

The Sky Lark pairs early in the spring, and during the season of incubation the song of the male, almost always uttered in the air, often at a considerable elevation, is peculiarly fine and melodious; and when, as is sometimes the case, several are heard in concert, the effect of their mingled tones is exceedingly gratifying, and the more so as they are an earnest of a reviving spring, which one might almost fancy they are welcoming with songs of rejoicing. While her mate is thus engaged the female is occupied on the ground, either preparing her nest or listening to his strains of adulation. In the early part of the season, while the pairing takes place, the males may be observed chasing each other, and exhibiting considerable pugnacity of disposition. As soon as the young are hatched the most assiduous attention is paid to them by their parents, and so deeply are they engrossed in this duty, that the song of the male is less frequently heard, and all the actions of both parents are far less sprightly and animated.

The flight of the Sky Lark is vigorous, and the bird is capable both of sweeping along with immense rapidity, or of sustaining itself in the air at almost any height for a considerable length of time; in the latter case, and while pouring forth its song, the wings have a peculiar vibratory motion.

Its food consists of grain, trefoil, and insects.

The only external difference in the sexes is, that the female is less brilliant in her markings, and has a shorter crest; she is moreover somewhat less in size, and the hind claw is shorter.

The young previously to the first moult are distinguished by all the feathers being edged with light yellowish grey.

The adult male has the bill brownish black; the general plumage yellowish brown, each feather with a darker central mark; a pale streak of yellowish white over the eye; the neck and breast pale brown spotted with brown of a darker tint; the tail brown, the outer feathers having the tip and the exterior web white, and the next with the outer web only white; the centre of the belly white, and the feet and claws light brown.

We have figured an adult male and a young bird of the natural size.

WOOD LARK.

Alauda arborea, *(Linn.)*

Drawn from Nature & on stone by J. & E. Gould. *Printed by C. Hullmandel.*

WOOD LARK.

Alauda arborea, *Linn.*

L'Alouette Lulu.

The Wood Lark is a far less abundant species than the Sky Lark, from which it differs materially in its habits and manners. Its range over the continent of Europe is very general, but in our own island it appears to be most numerous in the southern counties, being, according to Montagu, most common in Devonshire. It gives preference to open fields bordering woods, or extensive plantations where large trees are dispersed abundantly, upon which, unlike its near ally, it is fond of perching, and from whence it pours forth its melodious strains, which, although full of sweetness, are less brilliant and varied than that of the Sky Lark. Its powers of flight are very considerable, and while on the wing it often utters a short and peculiar piping cry. It is strictly migratory, departing from our shores rather late in autumn, and returning in the month of April.

The sexes are alike in plumage: the young have all the feathers bordered with yellowish white, and their general tints more tawny than those of the adults.

The nest is said to be placed on the ground, generally under the covert of some tuft or shrub; the eggs wood brown, blotched with grey and darker brown.

Bill brownish black; above the eye is a conspicuous fawn-coloured stripe; all the upper surface buff brown, each feather having a black central mark; under surface yellowish white spotted upon the neck and chest with dark brown; wing-coverts tipped with white; two middle tail-feathers brown, the remainder brown tipped with white; legs pale brown.

The Plate represents a male of the natural size.

BIFASCIATED LARK.

Certhilauda bifasciata.

Alauda bifasciata; *(Licht.)*

Drawn from Nature & on Stone by J & E Gould

Printed by C. Hullmandel.

Genus CERTHILAUDA, *Swains.*

GEN. CHAR. *Bill* moderate, slender, curved. *Nostrils* roundish. *Wings* with first quill extremely short or nearly spurious; second very long; third, fourth, and fifth nearly equal and the longest. *Tail* rather short, even. *Feet* moderate; nail of the hind toe short and straight.

BIFASCIATED LARK.

Certhilauda bifasciata.

Alauda bifasciata, *Licht*

L'Alouette bifasciée.

In the third part of his 'Manuel,' M. Temminck describes this rare species of Lark as an occasional visitant to the eastern and southern parts of Europe, it having been killed both in Sicily and Provence. It would appear to be very common on the banks of the Nile, from whence we have received specimens, and it is also plentiful in Abyssinia. It differs much in its structure from the members of the genus *Alauda*, and if we mistake not, will rank with the bird characterized by Mr. Swainson under the name of *Certhilauda*, and as such we have figured it. Taking the common Sky Lark of our island as the type of the genus *Alauda*, this will be found to exhibit many points of difference, particularly in the elongated and curved form of the bill, and in the comparative shortness of the toes and nails; and although we have not been made acquainted with its habits and manners, we feel confident that they differ considerably from those of the *Alauda arvensis* and its immediate congeners, and that Mr. Swainson's views in separating it into a distinct genus will be fully substantiated.

Of its food and nidification nothing is known.

The sexes are alike in plumage and may be thus described:

The whole of the head, back of the neck, scapularies, and upper tail-coverts pale greyish brown; wing-coverts dark brown, margined with pale brown; base and tips of the secondaries white, forming a double band across the centre of the wing, the intermediate space dark brown, with pale brown edges; primaries dark brown; tail-feathers dark brown, with the exterior web of the outer feather on each side, and the extreme edge of the next, white; all the remainder edged with pale brown; stripe before and behind the eye and one from the angle of the mouth dark brown; throat, sides of the face, under part of the wings, and all the under surface dull white; the lower part of the throat and the breast ornamented with numerous oblong spots of dark brown; bill and feet yellowish.

We have figured an adult male of the natural size.

LARK-HEELED BUNTING.

Plectrophanes lapponica; *(Selby).*

Drawn from Nature & on stone by J. & E. Gould.

Printed by C. Hullmandel.

LARK-HEELED BUNTING.

Plectrophanes Lapponica, *Selby*.

Le Bruant Montain.

So little is known of the history and changes of plumage which this scarce bird undergoes, that we are left in doubt as to whether the tricoloured livery of the upper bird in our Plate, which is that of the male in summer, is exchanged in winter, as in the case of the Snow Bunting, for a more uniform and sober dress, or whether, like some of the more typical Buntings, (*Emberiza Schœniculus*, Linn., for example,) it retains its strongly contrasted colouring throughout the year. We make this observation because there have been frequently examples killed in England, all of which resembled the lower bird of the Plate. Some of these, on dissection, proved to be males, and were most probably immature birds, the migrations of which are known to be, according to the general rule, both more widely diffused and more irregular in their course than those of mature birds.

The summer retreat of the Lark-heeled Bunting, where it incubates and rears its brood, is within the limits of the arctic circle, from whence, as winter approaches, it gradually passes southwards, in Europe as far as Switzerland, and in America visiting the northern parts of the United States in considerable abundance. Its nest, according to Dr. Richardson, who observed it in the arctic regions of the American continent, is placed upon a small hillock, among moss and shrubs, and is composed externally of the dried stems of grass interwoven to a considerable thickness, and lined very neatly and compactly with deer's hair. The eggs are usually of a pale ochre yellow, spotted with brown.

In habits and manners the Lark-heeled Bunting resembles very closely the Snow Bunting, with which it is sometimes found associated; and it is worthy of remark, that the examples killed in England have been found among the vast quantities of Larks exposed for sale in the markets of London and other large towns, a circumstance indicating its almost exclusively terrestrial habits. Its food consists of grain, the seeds of various mountain plants, and perhaps insects.

The colouring of the adult male may be thus detailed:

The top of the head, cheeks, throat, and chest jet black, interrupted by a line of white, which passes from the base of the beak over the eye, behind which it dilates and extends to the occiput, bounding the ear-coverts; a broad band of chestnut passes across the back of the neck; the whole of the upper surface is brown, each feather being edged with rufous, and having a black dash in the centre; the sides of the chest and under surface white, the flanks with a few dashes of black; bill yellow, passing into black at the tip; tarsi blackish brown; irides hazel.

The female, according to M. Temminck, resembles the young bird in her general colouring, except that a band of reddish white occupies the same place as in the male, and unites with a white streak, which passes from the angle of the beak; the throat is white, bounded laterally by a broad band of brown; the breast is marked with blotches of grey and black, and the under parts are white.

The young birds, as we may presume those to have been that were taken in this country, have the whole of the upper surface brown, each feather bearing a reddish edge, and a dark central dash; the under surface dirty white, with dashes of brown along the sides.

The Plate represents an adult male, and a young bird of the year, of the natural size.

SNOW BUNTING.

Plectrophanes nivalis; *(Meyer)*.

Drawn from Life & on Stone by J & E. Gould.

Printed by C. Hullmandel.

SNOW BUNTING.

Plectrophanes nivalis, *Meyer*.

Le Bruant de neige.

THE Snow Bunting may be strictly pronounced a migratory species throughout the temperate countries of Europe, visiting them at the commencement of winter, and enlivening the bleak hills and barren shores which at this season of the year are deserted by those birds whose instinct has directed them to seek more southerly and consequently milder regions. The summer residences of this neat and chastely plumaged bird are well ascertained to be the northern hemispheres of the continents of Europe and America, over which portions of the globe it is generally and even universally diffused. The most wild and dreary spots of the northern parts of the latter continent are, according to the accounts of persevering travellers who have visited those regions, animated with the presence of the Snow Bunting. Dr. Richardson informs us, that Southampton Island, situated in the 62nd parallel, (where this species was observed by Captain Lyon,) is the most southern locality which has been discovered as its breeding-place. As soon as the task of incubation is accomplished, they commence their migrations towards warmer regions, although, by the authority of the above-mentioned traveller, they do not hasten southwards with that immediate alacrity which distinguishes the passage of many other small birds; they linger rather in the vicinity of forts by the sea-side and other exposed places, subsisting on the seeds of grasses, and performing their journey by short stages until the approach of colder weather quickens their progress; indeed the visits of the Snow Bunting to our own island seem to depend very much on the severity of the winter in their northern retreats. The Shetland and Orkney Islands are their first resting-places, whence they proceed to the Highlands of Scotland, then the Cheviot-hills, and finally distribute themselves over the southern barren districts of the British Isles. Mr. Selby informs us, that "they arrive at the latter end of October, and generally in very large flocks, which seem chiefly to consist of the young of the year with a few adults intermixed, and afterwards, if the season should be severe, small flocks are seen, principally consisting of adult male birds in their winter dress." On the Continent they annually visit the North of Germany, France and Holland, in the latter of which countries M. Temminck states them to be very abundant, particularly by the sea-side, a situation to which they evince a partiality in our own Island, especially if a flat and sandy shore prevails. From all these places, on the approach of spring, they again flock to the northern latitudes, whence they originally came.

The Snow Bunting is subjected to a considerable variety of plumage, of which either sex, age, or season is the cause. The decided and contrasted plumage represented in the lower figure is not attained until maturity, and is then only observable in the summer season, at which time the male and female offer less distinction than is given in the accompanying Plate, the upper figure of which represents an immature bird in the plumage characteristic of the greater portion of those individuals who visit England. In this state it has been called the Tawny Bunting, and regarded by many authors as a distinct species. As the lengthened hind claw would lead us to conclude, the habits of this bird induce it to frequent rocks and arid districts, where they run with great celerity and are never known to perch on trees; and from the beak being destitute of the palatine knob, it has been separated with great propriety from the other Buntings.

The situations chosen for the nests of this species are niches in the rocks of mountainous places, and sometimes upon flat shores among large stones. The nest is formed of dried grass neatly lined with hair or feathers. The eggs are six or seven in number, of a pale flesh colour, speckled with minute dots, and blotched at the larger end with reddish brown. Their food consists of the seeds of alpine plants, and the larvæ of various insects.

The adult male in summer has the head, neck, under parts, outer tail-feathers and centre of the wings pure white; the remainder of the plumage, the feet, and bill black; irides dark brown. The female at this season differs only in having the back of the head, side of the chest, and a portion of the neck and breast tinged with rufous, and the other parts of the plumage of a less pure black. The male of the first year, the female, and adult birds in winter offer but little difference in the colour of the plumage. The upper figure in the Plate represents a bird in this stage, and the colouring may be thus characterized. The top of the head, sides of the breast, margin of the scapulars, back, and tail-feathers reddish brown; the throat, breast, four outer tail-feathers, under parts, and centre of the wings white; each of the back feathers has the centre brown; the quills and middle tail-feathers are also of the same colour beak reddish brown; irides dark brown; legs black.

COMMON BUNTING.

Emberiza miliaria; *(Linn.)*

Drawn from Nature & on Stone by J. & E. Gould. *Printed by C. Hullmandel.*

Genus EMBERIZA.

Gen. Char. *Bill* conical, strong, hard, and sharp-pointed; tomia of both mandibles bending inwards, and compressed towards the point; the upper mandible narrower and smaller than the under one, and its roof furnished with a hard bony knob; base of the mandibles, or gape, forming an angle, and rather open. *Nostrils* basal and round, partly hidden by the small feathers at the base of the bill. *Feet* having three toes before and one behind; the anterior ones entirely divided. *Claws* rather long and curved. *Wings* with the first quills rather shorter than the second and third.

COMMON BUNTING.

Emberiza miliaria, *Linn.*

Le Bruant Proyer.

This well-known species of Bunting may be considered a permanent resident in the British Islands, over the whole of which it is dispersed; it is also equally abundant on the Continent, where it ranges from the regions of the arctic circle to the most southern boundary of Europe: of this fact we have received satisfactory evidence, by the inspection of examples from Trebizond in Asia Minor, which country would, however, appear to be nearly its southern and eastern limit, as we have never observed any specimens in collections either from India or Africa.

It is the largest and one of the most typical of its genus, having the palatine knob more fully developed than in any other species. Its food consists, in a great measure, of grains and seeds, which induces it to give a preference to those cultivated districts which afford the most abundant supply. Besides grain, many of the Buntings eat insects and their larvæ with avidity, and we have observed the present bird in particular feeding on the body of the large species of Chafer *Melolontha vulgaris.* During autumn and winter the Common Bunting congregates in flocks, often in company with the Lark, to which, in the flavour of its flesh, it is very similar; on the return of spring it is dispersed over the face of the country, being partial to hedge-rows skirting large fields, where it may be observed perched on the topmost twig uttering its oft-repeated monotonous note, which is more loud although not unlike that of the Yellow-hammer, *Emberiza citrinella.*

The ground is the situation chosen for the purpose of incubation, and in this respect it resembles the true Larks; the nest is composed of various grasses lined with hair and fibrous roots; the eggs are five or six in number, of a pinkish grey, streaked and spotted with reddish brown.

The sexes so nearly resemble each other in the colour of their plumage as to render a description of both unnecessary; the young also assume at an early age, with a trifling exception, the precise tints of the adult.

The whole of the upper plumage is brown inclining to olive, the centre of each feather being darker; the under surface yellowish white, with numerous stripes of dark brown running down the shaft of each feather; feet and bill brown.

The Plate represents an adult male of the natural size.

BLACK HEADED BUNTING.

Emberiza melanocephala, *(Scopoli)*.

Drawn from Life and on Stone by J. & E. Gould. *Printed by C. Hullmandel.*

BLACK-HEADED BUNTING.

Emberiza melanocephala, *Scopoli.*

Le Brunt crocote.

We are not able to enter into a minute detail respecting the habits and manners of this beautiful species of Bunting, as it has never been known to visit the British Islands, nor, as far as we are able to ascertain, either the northern or western portions of the European Continent; the middle and southern districts and the border-line which joins the Asiatic confines being its native locality.

M. Temminck informs us that it is very abundant in Dalmatia and all over the Levant, and common in Istria and in the environs of Trieste; he also states that it sings agreeably, and chooses hedgerows and low bushes for its place of incubation, building near the ground and laying four or five white eggs, thinly sprinkled with minute dots of a light ash-colour. Its food consists of seeds, grains, and occasionally insects.

In the male the whole of the head and the cheeks are deep black; the sides of the neck, throat, and the whole of the under surface of a fine king's-yellow; the back of the neck, the scapulars and back are of a rich rufous brown; the rump and tail-coverts inclining to yellow; the wings and tail light brown, each feather having a lighter edge; the beak ashy-blue; feet and tarsi light brown; length about six inches and a half.

In the female the whole of the upper surface is brown, the middle of each feather inclining to black; the throat yellowish white; the whole of the under surface inclining to a yellowish red.

In all our collections from Western India we receive this species in abundance, or if not this identical bird, one very closely allied to it, having all its characters, except that the feathers of the head instead of being entirely black are tipped with a grayish brown, and the rest of the plumage, which in the European species is so vivid and distinct, is less pure and decided.

The specimens from which our figures were taken form part of the collection of the Zoological Society of London, and were received from Berne. Switzerland we believe to be the western bounds of its locality.

In the annexed Plate we have figured a male and female in their spring plumage.

YELLOW BUNTING.
Emberiza citrinella, (Linn.)

Drawn from Nature & on Stone by J. & E. Gould.

Printed by C. Hullmandel.

YELLOW BUNTING.

Emberiza citrinella, *Linn.*

Le Bruant jaune.

While perched on the topmost branch of the roadside hedge displaying its richly coloured tints, this well-known bird would, were it less common, excite great interest in the passing traveller, as well as in those who lead exclusively a country life, and who therefore have it under their observation during all seasons of the year.

The male is most attractive in the early months of spring, his energies at this period having by the natural impulse warned him of the approaching breeding-time. Now, being mated, he may be seen mounted on the most slender twig pouring forth his simple song, which, although neither melodious nor varied, nevertheless has a natural simplicity which cannot fail to please, and it is doubtless cheerfully listened to by his less gaily attired mate, who prefers the more secluded bush or more dense parts of the hedge-row.

The Yellow Bunting is rather a late breeder, seldom commencing until the herbage is sufficiently grown to afford it a complete shelter from observation. The shelving side of a bank or tuft of grass is generally selected for the situation of the nest, which is most neatly constructed of dried grasses and moss, lined with finer grasses and hair: the eggs are four or five in number, of a pale bluish white, marked with spots and lines of chocolate red.

In winter the Yellow Bunting associates in considerable flocks, often in company of other granivorous birds, and spreads over fields and arable lands; in severe weather resorting to farm-yards and similar situations. It is, we believe, indigenous in every part of Europe, to which quarter of the world it appears to be strictly confined, as we have never seen any examples of it in collections from any other locality.

The young during the first autumn resemble the female, which, as we have above stated, is much less brilliant in all her markings than the male.

The male has the crown of the head, throat, chest, and under surface rich gamboge yellow, the flanks and under tail-coverts streaked with reddish chestnut; the upper surface rich brown inclining to olive, the centre of each feather being darker; primaries blackish brown with lighter edges; rump brownish orange; tail brownish black, the outer edges of the feathers yellow, and the inner web of the outer feather on each side largely blotched with white; legs and feet yellowish brown.

The upper surface of the female resembles that of the male, but the tints are less brilliant; the under surface also is not so bright, and is destitute of the rich chestnut streaks which adorn the male, these markings being brown.

The Plate represents a male and female of the natural size.

YELLOW-BREASTED BUNTING.

Emberiza aureola. *(Pall.)*

Drawn from Nature & on Stone by J & E Gould. *Printed by C. Hullmandel.*

YELLOW-BREASTED BUNTING.

Emberiza aureola, *Pall.*

La Bruant auréole.

THIS very beautiful Bunting has been more than once captured within the precincts of the European continent; it consequently becomes necessary for us to give a figure of it, and in so doing we introduce to our readers one of the most beautiful species of this group, so celebrated for their agreeable and well-contrasted colours. A specimen of the male, one of the very finest we have ever seen, was obligingly lent to us by T.B.L. Baker, Esq., of Hardwicke Court, Gloucester, a gentleman to whom we shall ever feel indebted for many acts of great kindness and liberality, and who has considerably facilitated the study of ornithology by the publication of a work entitled "An Ornithological Index", in which are enumerated the genera and species contained in the works of most of the present writers, and which he hopes will form a stepping-stone to a still more elaborate production by some more experienced ornithologist.

M. Temminck states that the native habitat of the Yellow-breasted Bunting is Kamtschatka, Siberia, and the Crimea; it has also been occasionally seen in the southern parts of Russia, and in other portions of the eastern boundaries of Europe.

The male is much more richly coloured than the female, and may be described as follows:

A band of black extends round the base of the beak and over the ear-coverts; the top of the head and the whole of the upper surface is of a rich chestnut, a band across the chest of the same colour; throat and under surface rich yellow marked with streaks of brown on the flanks; primaries and tail brown, the latter having the two outer feathers on each side marked with a large white spot near the tip; beak and tarsi brown.

The female is nearly devoid of the rich colouring which characterizes the male; the upper surface being dull brown tinged with green, the under surface olive yellow with the flanks marked as in the male.

The Plate represents a male and female of the natural size.

CIRL BUNTING.

Emberiza cirlus; *(Linn.)*

Drawn from Nature & on Stone by J & E. Gould.

Printed by C. Hullmandel.

CIRL BUNTING.

Emberiza Cirlus, *Linn.*

Le Bruant zizi.

For the discovery of this beautiful species of Bunting in our island, we are indebted to the industry and research of the late Colonel Montagu. It is now much more numerous than it formerly was ; but unlike its ally the well-known Yellowhammer (*Emberiza citrinella,* Linn.), which is distributed through the whole of our island, the Cirl Bunting is extremely local in its habitat, being seldom seen in the midland and northern counties. It is common in Devonshire, and all along our southern coast. In Sussex we have ourselves seen it in abundance, particularly in the neighbourhood of Chichester, where it annually breeds. It is much more shy and retiring than the Yellowhammer; its song is also different, more resembling that of the Chaffinch. It frequents nearly the whole of the southern provinces of Europe, and is especially abundant along the shores of the Mediterranean as well as in Italy and the southern parts of France. In general habits, manners, and nidification it closely resembles the Yellowhammer. Its nest is generally placed either beneath a low bush or at the foot of a large tree; it is composed of dried grass intermingled with vegetable fibres, and lined with hair. The eggs are in general more round than those of the Yellowhammer; in colour they are grey, marked with those peculiar zigzag lines of dark brown which are so characteristic of all the Buntings' eggs. Its food consists of various kinds of grain, to which insects are largely added, of which Montagu informs us grasshoppers are the greatest favourites.

The sexes offer a contrasted difference in the colour of the plumage, the male being adorned with a gorget of black and distinct facial markings.

The adult male has the crown of the head and back of the neck olive grey, the former exhibiting longitudinal dashes of black ; a yellow stripe from the base of the beak encircles the eye, and terminates on the side of the head; the throat is black in summer, but as winter approaches it becomes obscured with olive grey; below this black a yellow band extends across the throat; the whole of the upper surface is reddish brown, each feather having a greyish margin ; quills blackish brown; the breast, below the yellow gorget, is greenish olive; sides of the chest washed with ferruginous under fine yellow; two outer tail-feathers white for the greatest part of their inner web ; bill brown ; legs brownish flesh colour.

The adult female, which differs little from the young male of the year, wants the black throat and gorget of yellow; the head is olive green with dashes of brown ; the chest is yellowish grey streaked with brown ; the under surface dull pale yellow ; and the upper plumage is less vivid than in her more ornamented mate.

Our Plate represents a male and female of the natural size.

ORTOLAN BUNTING.
Emberiza hortulana: *(Linn:)*

Drawn from Life & on Stone by J & E Gould.

Printed by C. Hullmandel.

ORTOLAN BUNTING.

Emberiza hortulana, *Linn.*

L'Ortolan.

This bird has long been celebrated as one of the greatest delicacies of the table throughout the countries of France and Italy, for which purpose numbers are annually caught and artificially fattened. The South of Europe and the northern portions of Africa appear to be its natural habitat; it is nevertheless generally spread throughout continental Europe, even as far as Holland, Sweden and Russia. The British Isles are only occasionally visited; one of the examples, a male, now in the museum of the Natural History Society at Newcastle-upon-Tyne, having been taken on the Yorkshire coast. It is not improbable, however, that we should find this bird more frequent than it is believed to be, were it not overlooked from its similarity to the Yellow Bunting (*Emberiza citrinella*, Linn.).

Dr. Latham informs us that it is strictly migratory in its habits and is frequently taken in the spring and autumn at Gibraltar, whence we may suppose that the greater number pass over to Africa and make that continent their winter residence. It is during these migrations, when vast numbers are assembled together, that they are caught in traps, principally in Italy and the South of France, and are then kept by being placed in a dark room, and there fed with plenty of oats and millet-seed, upon which they quickly fatten. From the accounts of various authors, it would appear that they offer several variations of plumage, caused by peculiar diet and other circumstances: these varieties, being purely accidental, are not to be considered in the same light as the variations of plumage which occur in many other birds.

The nest of the Ortolan is constructed of fibres and leaves, and placed in the most convenient situation the locality may afford, most commonly in low bushes and hedges, but sometimes on the ground among corn. The eggs are five in number, of a reddish grey marked with streaks of brown.

The plumage of the male is much more lively than that of the female. The top of the head is greenish olive; an edging of white feathers forms the margin of the eyelid; ear-coverts brown; the throat, the sides of the face, below the eye, and the chest, are of a delicate yellow; the upper surface reddish brown, the feathers of the back and wings being dashed in their centre with black; the under surface pale tawny; beak and legs flesh-coloured.

In the female, the whole of the upper surface is greyish brown, with a number of small black lines on the head, the feathers of the back having their centres black also; the throat is pale yellow, and is bounded by a range of brown spots; the rest of the under surface is pale brownish red; the eyelid is edged with white as in the male.

The Plate represents an adult male and female of the natural size.

RUSTIC BUNTING.

Emberiza rustica, *(Gmel.)*

Drawn from Nature & on stone by J. & E. Gould. Printed by C. Hullmandel.

RUSTIC BUNTING.

Emberiza rustica, *Pall.*

Le Bruant rustique.

For fine examples of both sexes of this exceedingly scarce Bunting we are indebted to the Directors of the Museum at Frankfort; and although its native country is Siberia, Kamtschatka, and the adjacent islands, we are inclined to admit it among the Birds of Europe, on the assurance of some naturalists that it is frequently found within the limits of the north-eastern portions of the Continent. M. Temminck has also admitted it on the same grounds, though he himself has never received it in a recent state, and until he does, he prefers taking his account from the work of Pallas.

In the disposition of its colouring the Rustic Bunting resembles several other species of the genus *Emberiza*; but it departs in a trifling degree from that form; and in the stoutness of its bill and the shortness of its tail would appear to approach the Finches.

The female may be distinguished from the male by the absence of the black colour on the crown of the head and ear-coverts; in other respects their plumage is closely similar.

We have no information to communicate respecting its habits and manners, nor is its nidification or the colour of its eggs as yet ascertained.

The male has the top of the head, with the exception of a white line down the middle, and the space between the beak and the ear-coverts, black; a broad white streak passes over the eye, and down the sides of the neck and throat; the whole of the upper surface is rich brown, each feather having a darker mark in the centre; this brown colouring passes into rufous on the chest, which it surrounds like a collar; the wings are of the same colour as the back; secondaries tipped with white; primaries and tail brown, the two outer feathers of the latter white on their outer edges; the flanks red brown, each feather having the centre darkest; under surface white; legs and bill dull yellow brown.

The plumage of the female is somewhat paler and more obscure than that of the male, and the black which ornaments the head of the male is replaced by brown; the stripe over the eye and down the throat is yellowish white.

The Plate represents a male and female of the natural size.

LESBIAN BUNTING.

Emberiza Lesbia; *(Gmel.)*

Drawn from Nature & on Stone by J. & E. Gould. *Printed by C. Hullmandel.*

LESBIAN BUNTING.

Emberiza Lesbia.

Le Bruant de Mitilène.

The *Emberiza Lesbia* is one of the rarest species of the present genus, but at the same time one of the most universally distributed; it sparingly inhabits the eastern parts of southern Europe, and occurs but very rarely in Italy and Provence; it is also found in Greece, and we have seen it in collections from China. M. Temminck states that it is found also in Japan, where it is known under the name of "*Jamuzuzume.*"

In its habits and manners it doubtless closely resembles the other members of the family, and but little difference is perceptible between the sexes.

The head is greyish olive, with a stripe of dark brown down the centre of each feather; back of the neck and back reddish brown, with a broader and more conspicuous stripe down each feather, but becoming nearly imperceptible on the rump; wing-coverts chestnut, striped down the centre with blackish brown; secondaries blackish brown, bounded on each side with rufous and margined with pale brown; quills and tail brown margined with paler brown; ear-coverts deep reddish brown, beneath which is a broad stripe of buff; throat whitish; bounded on each side with numerous oblong spots of dark brown, which meet and cover the front of the breast; under surface buff, marked on each side immediately below the breast with several indistinct spots of chestnut, and on the flanks by stripes of dark brown on the centre of each feather; bill, legs, and feet pale brown.

The female only differs in having the spots on the sides of the throat and breast more numerous, and in having the whitish part of the throat less extensive.

We have figured a male and female of the natural size.

MEADOW BUNTING.
Emberiza cia, *(Linn.)*.

Drawn from Nature & on Stone by J & E. Gould.

Printed by C. Hullmandel.

MEADOW BUNTING.

Emberiza cia, *Linn.*

Le Bruant fou, ou de pré.

This species of Bunting, although common in the meadows bordering the Rhine, as well as in the southern parts of France, Italy, Spain, and adjoining the Mediterranean, does not appear to be distributed in the North as is the case with so many of its congeners, neither Holland nor England being among the places of its habitat.

The nearest-allied species among our native Buntings is the Reed Bunting (*Emberiza schœniculus*, Linn.), which it resembles, not only in its general habits and manners, but in the peculiar character of its markings, particularly about the head, and in the feebleness of the beak. The nearest extra-British species in alliance with it is the *Emberiza lesbia*: and it would appear that with both of these birds in certain stages of plumage it has been confounded; and not with these only, as will appear from the following translation of a note which we have taken the liberty of extracting from M. Temminck's *Manuel d'Ornithologie.* "Besides the double use which Buffon makes of this species in describing it under the name of Foolish Bunting, and Lorraine Bunting, he commits a second error in giving his description and *Ortulan de neige*, pl. 511. fig. 2, as the female of the *Ortolan de Lorraine.* The German authors are equally in error in enumerating under the synonym of *Le Bruant fou* the birds described and figured by Buffon under the names of *Gavoué* and *Mitiléne de Provence*; these form two distinct species. The French naturalists place the *Emberiza passerina* of Gmelin, *Syst.* i. p. 871. sp. 27., in the synonym of *Le Bruant fou,* while the description of Gmelin pourtrays very exactly an old female of *Le Bruant des roseaux.*"

The *Emberiza cia* offers in its sober tints a harmony of colours which renders it far from being the least pleasing of its genus. The food of this bird, as its feeble bill indicates, consists of the small seeds of farinaceous plants, such as millet, canary, &c., as well as insects of various species; in fact, as above stated, its manners and actions are in close unison with our well-known Reed Bunting. It constructs a nest in bushes and tufts of herbage, and not unfrequently on the ground: the eggs are five in number, of a whitish colour marked with a few lines of black.

The whole of the head and breast is ash coloured; three stripes of black occupy the face on each side, one passing above the eye, one through the eye to the occiput, and one encircles the lower part of the face from the angle of the beak; a greyish white stripe passes above the eye, bordered by the two lines of black; the whole of the upper surface is of a rufous brown, each feather having a dusky mark down the centre; the feathers of the shoulders are edged with light grey; the primaries brown; the three outer tail-feathers white, the remainder brown edged with reddish; the whole of the under surface pale rufous.

The female is destitute of the beautiful grey which ornaments the head and chest of the male, as well as the jet black lines, which are only faintly indicated on the cheeks; the head and chest are pale greyish brown; the throat dotted with dusky spots; the rest of the plumage resembles the male, except that it is more obscure.

The Plate represents a male and female of the natural size.

PINE BUNTING.

Emberiza pithyornus. *(Pall.)*

Drawn from Nature & on Stone by J & E Gould.

Printed by C. Hullmandel.

PINE BUNTING.

Emberiza pithyornus, *Pall.*

Le Bruant à couronne lactée.

In size this rare Bunting rather exceeds the Yellow-hammer (*Emberiza citrinella*, Linn.), which so frequently attracts the notice of the passing traveller through the British Islands. Its true habitat would appear to be the northern parts of Russia and Siberia, though, according to M. Temminck, it is frequently found as far south as the centre of Turkey, and the shores of the Caspian Sea, Hungary, Bohemia, and Austria are among the places of its resort. Dr. Latham states that it frequents the pine-forests of Siberia, and has the note of the Reed Bunting. Although it has not the brilliant yellow colouring which pervades the plumage of many of its tribe, the *Emberiza pithyornus* is very pleasing to the eye, from the harmonious arrangements of its rich but somewhat sober tints, in which respect, and in fact in its whole contour, it assimilates exceedingly to the Bunting-like Finches of the New World, such as the *Emberiza leucophrys*, Gm., (*Zonotrichia leucophrys*, Sw.); and in all probability, when the vast countries of Siberia, Kamtschatka, &c. have been more thoroughly investigated, that species, intermediate in form, will be found to complete this chain of affinities.

The sexes of the Pine Bunting may be distinguished from each other by the more obscure colouring of the female, and the total absence of the gorget and superciliary stripe of chestnut with which the male is adorned.

The plumage of the male is as follows:

A stripe of white passes along the top of the head to the occiput; on each side of this white stripe is another of black, and this is again succeeded by one of chestnut immediately over the eye; ear-coverts white; throat rich chestnut; below this is a half band of white succeeded by a broad band of dusky greyish chestnut across the chest; whole of the back, wings, and flanks rich brown, each feather being darkest in the centre; rump and upper tail-coverts pale chestnut; tail brown, each feather edged with reddish brown, and the two outer ones largely blotched with white; centre of the breast, belly, and under tail-coverts white; legs and bill yellowish brown.

The female is more obscure in all her markings; the ear-coverts are brown with a band of white beneath them; superciliary mark yellow white; throat white surrounded with small dark spots.

The Plate represents a male and female of the natural size.

CRETZSCHMAR'S BUNTING.

Emberiza cæsia, *(Cretzschm.)*

Drawn from Nature & on Stone by J & E. Gould.

Printed by C. Hullmandel.

CRETZSCHMAR'S BUNTING.

Emberiza cæsia, *Cretzschmar.*

Le Bruant cendrillard.

We have received beautiful examples of this rare bird from Dr. Cretzschmar of Frankfort, who has also obliged us with numerous other rarities from the fine collection under his charge. From the circumstance of so distinguished a naturalist having added this interesting bird to the Fauna of Europe, as an occasional visitant to the southern and eastern portions of that continent, we feel no hesitation in inserting it in the present work.

The true habitat of the *Emberiza cæsia* are the northern and eastern portions of Africa, in which countries it was observed in abundance by Dr. Rüppell. In the third part of his "Manuel d'Ornithologie," M. Temminck states that "it inhabits Syria and Egypt; is probably more common in the middle of Europe than it is supposed to be, where isolated individuals may have been taken for varieties of the *hortulana* and *cia*; it is found accidentally in Austria and Provence, an individual having been taken near Vienna in 1827.

We have never seen an example either from India or any of the islands of the Archipelago, which circumstance would lead us to conclude that it is almost exclusively confined to the portion of the globe above mentioned, and in which it will be necessary to seek for information relative to its peculiar habits and œconomy.

The plumage of the sexes is less contrasted than is generally observed in birds of this genus. In spring the male has the top of the head, back of the neck, ear-coverts, and chest grey; a narrow streak of the same colour passes from the chest to the base of the lower mandible; throat, cheeks, and a narrow band across the forehead light chestnut brown; upper part of the back brown, each feather having a darker centre; rump and upper tail-coverts brown without spots; the whole of the abdomen rich chestnut brown, more intense on the breast; wings dark brown; the secondaries and scapularies strongly edged with light brown inclining to chestnut; tail dark brown, the outer edges of the feathers chestnut, and the two outer ones on each side largely tipped with white on their inner webs.

The female has the chest marked with numerous small spots of black on a ground of brownish grey, which colour pervades the whole of the head; the remainder of the plumage resembles that of the male, only being much less intense in colour.

The Plate represents a male and female in their spring plumage, of the natural size.

MARSH BUNTING.
Emberiza palustris; (Savi.)

Drawn from Nature & on Stone by J.&E. Gould. Printed by C. Hullmandel.

MARSH BUNTING.

Emberiza palustris, *Savi.*

This rare species, which offers so close a resemblance in general colouring and habits to our well-known Black-headed Bunting, (*Emberiza melanocephala,* Scopoli,) exhibits nevertheless, in the robust structure of its beak, a departure from the typical characters of the genus, and either forms its extreme limits, or may be regarded as the representative of another genus; but its affinities are at present but little understood, the bird itself being very rare, and only to be met with in the southern and eastern provinces of Europe.

The best account of this bird is to be found in Professor Savi's "*Ornitologia Toscana,*" according to which eminent author it dwells in the marshes of Tuscany, but he has not yet been able to obtain a sight of its nest and eggs; if, however, we may judge from analogy, we may consider its habits and manners as very much resembling those of the Common Reed Bunting. Professor Savi further informs us that it inhabits the vicinity of stagnant waters covered with reeds and bulrushes, and that it feeds to a great extent upon the insects which lodge upon the culmens of the reeds.

The sexes offer the same relative differences that are observed in the Reed Bunting, the black head of the male being exchanged in the female for brown blotched with dashes of black.

In the male, the upper part of the head, cheeks, and throat are black; a white stripe begins near the angle of the beak, and extends round to the back of the neck; the whole of the upper surface is of a rich chestnut brown, the centre of each feather being largely blotched with black; the under surface is white, the flanks being marked with longitudinal lines of brown; bill black; tarsi brown.

The female, which closely resembles the male in her general plumage, is distinguished by the colouring of the head already alluded to; by the absence of the white stripe round the neck; and by the dull brownish white of the under surface, which is thickly dashed with longitudinal spots of deep brown.

The Plate represents a male and female of the natural size.

REED BUNTING.

Emberiza schœniculus; *(Linn.)*

Drawn from Nature & on Stone by J & E. Gould.

Printed by C. Hullmandel

REED BUNTING.

Emberiza Schœniculus, *Linn.*

Le Bruant de Roseau.

The situations to which the Reed Bunting gives preference are the edges of rivers, large ponds, and beds of osiers; though at certain times, particularly during severe weather, it quits its marshy abode and associates with the Yellow-hammer and other small granivorous birds, frequenting at such periods the open fields, and, when pressed by hunger, visiting even the farm-yard, in search of a more abundant supply. It appears to be indigenous in every portion of Europe, or if not in every portion, at least through the whole of the centre. Like some other species of its genus, its summer and winter plumage exhibits a remarkable contrast; the male being characterized during the former season by a jet black head and throat, rendered more conspicuous by the white stripe from the base of the bill, and the collar of the same colour round the back part of the neck: in winter the male loses the black plumage of the head and throat, and is then scarcely to be distinguished from the female. The assumption of the black colouring commences early in spring, and is fully accomplished at the approach of the breeding-season, which begins as soon as a sufficiency of fresh herbage and the young shoots of the willow have rendered the reed a covert dense enough to shelter the nest from observation. The nest is generally placed near the ground, on a low stump of willow or any entangled herbage: the eggs are five or six in number, and of purplish grey, streaked and spotted with dark red brown.

The Reed Bunting is not at all remarkable for its song, which consists of only a few simple notes delivered without either energy or execution.

In summer the male has the whole of the head, ear-coverts, and throat black, the two latter being separated by a white stripe, which extends from the base of the bill to the sides of the neck, where it meets a collar of the same colour extending from the back of the neck; the whole of the upper surface of a rich brown, the centre of each feather being of a darker hue; the two middle tail-feathers brownish black edged with brown, the outer feathers largely blotched with white at their extremity; under surface white clouded with brown; flanks spotted longitudinally with obscure dusky lines; bill black; feet and legs brown.

The female differs from the male in having the general plumage more obscure, and in the total absence of the black head and white collar which are so conspicuous in the male; her flanks are also more largely spotted with brown.

The Plate represents a male and female in their summer plumage, of the natural size.

1. HOUSE SPARROW.
Pyrgita domestica; (Cuv.)

2. TREE SPARROW.
Pyrgita montana. (Cuv.)

Drawn from Nature & on stone by J & E Gould.

Printed by C. Hullmandel.

Genus PYRGITA, *Cuv.*

Gen. Char. *Bill* strong, conical, longer than deep; upper mandible slightly curved; tip emarginate; culmen slightly raised; lower mandible compressed and smaller than the upper. *Nostrils* lateral, immediately behind the bulging base of the upper mandible, round, and nearly concealed by small plumes. *Wings*: the second quill-feather rather the longest. *Tarsi* nearly as long as the middle toe. *Toes* three before and one behind, those in front divided: claws sharp and curved, that of the hind toe rather larger than that of the middle. *Tail* square or very slightly forked.

COMMON SPARROW.

Pyrgita domestica, *Cuv.*

Le Gros-bec Moineau.

Of the four species of this group indigenous to Europe, no one is more extensively spread or more generally known than the Common Sparrow, a bird with which we are all so well acquainted that to enter into the details of its history seems almost superfluous. We are informed that in Italy and Spain its place is supplied by two species peculiar to those countries, viz. *Pyrg. Cisalpina* and *Pyrg. Hispaniolensis*, but with this exception it is undoubtedly spread over the whole of Central Europe; it also occurs in Northern Africa and in the hilly districts of India. In England it is stationary throughout the year, congregating in flocks in autumn and winter, but in summer dwelling and breeding either in small companies or in pairs. Accommodating itself to all situations, it breeds indifferently among the branches or in the holes of trees and under the eaves of houses, not unfrequently usurping the nest of the Common Martin (*Hirundo urbica*); but never far from the habitation of man, to whose presence it appears perfectly indifferent, hence we see it as abundant in the largest cities as in the smallest villages. The nest when placed in a tree is of a domed form, carelessly constructed of straw, grass, and any materials at hand, but always lined with feathers: the eggs are five or six in number, of a greyish white spotted with brown. The food of the Common Sparrow consists, during a great part of the year, principally of seeds and grain of different kinds, which in summer are in a great measure exchanged for insects and their larvæ, with which it invariably feeds its young. This bird is destroyed in vast numbers in many agricultural districts, on account of the supposed injury it inflicts upon the farmer by the destruction of his corn; but we much question whether this practice can be fairly justified, for we conceive that the injury it may inflict is more than counterbalanced by the benefit accruing from the havoc it commits among the insect tribes, which are in fact the real enemies of the farmer, the fruit-grower, and every cultivator of the land; and we ourselves incline to think that it would be better to protect the grain or even to sacrifice some portion of it, than utterly to exterminate a creature which has, no doubt, been wisely appointed to fill its place in the great scheme of creation.

The male Sparrow is really a pretty bird when seen undisguised by the smoke and dirt which disfigure its plumage in our larger towns and cities: the crown of the head is bluish grey, back of the neck and stripe from the eye rich chestnut; cheeks and sides of the neck greyish white; throat and chest black; upper surface rich brown dashed with black; a white bar across the shoulders; under surface greyish white; feet and bill black in summer and brown in winter.

The female has the upper surface dull brown; the under surface greyish brown; and the feet and bill brown at all seasons.

TREE SPARROW.

Pyrgita montana, *Cuv.*

Le Gros-bec Friquet.

Unlike the preceding species, which loves to dwell in the streets of our towns, this affects the open country, where every field and wood affords it food and a congenial habitat. In the British Islands it is extremely local in its range, being scarcely known in some counties, while in others, Essex, Cambridgeshire, &c., it is tolerably abundant. It is found in most parts of central and southern Europe, and we have also received it from the Himalaya mountains and from China. The food consists of seeds, grains, and insects. Like all the other members of this restricted genus it is devoid of song. The nest is constructed in the holes of stunted trees and pollards, and very closely resembles that of the Common Sparrow, as do the eggs also, except that they are smaller. The sexes offer no difference in the colouring of the plumage. The Tree Sparrow may be distinguished from the male of the common species by its being much smaller in size, and by its having the top of the head rich chestnut brown; a patch of black on the ear-coverts, and two narrow bars of yellowish white across the shoulders.

The Plate represents a male and female of the Common Sparrow, and an adult male of the Tree Sparrow.

1. SPANISH SPARROW.
Pyrgita Hispaniolensis; (Cuv.)

2. ALPINE SPARROW.
Pyrgita Cisalpina; (Cuv.)

Drawn from Life & on Stone by J. & E. Gould.

Printed by C. Hullmandel.

SPANISH SPARROW.

Pyrgita Hispaniolensis, *Cuv.*

Le Gros-bec Espagnol.

The two species illustrated by the present Plate, bear, as will be seen, so close a resemblance to our common domestic sparrow, as at first sight to be easily mistaken for that bird, and therefore require a more than common attention to the disposition of the colouring, &c., in order to establish their differences. We have to lament that we cannot say much respecting their habits and manners; as those who have had opportunities of seeing them in their native localities appear to have noticed them so little, that the accounts are of the most meagre description. They appear to fill up the same place in the situations they inhabit that the common species does here, but are more inclined to resort to the barren lands and rocky districts of the country, than to collect in the villages and towns.

Of the two species given in our Plate, the *Pyrgita Hispaniolensis* is the least known. Its true habitat appears to be the southern portions of Spain, Sicily, the Archipelago, and Egypt. We have omitted to figure the females of these two species, as they so closely resemble those of our own country as not to be distinguished by plumage alone, without an intimate knowledge of the examples under examination.

The top and back of the head is of a bright and strong chestnut; the back and shoulders black, each feather bordered with rufous; the throat, fore part of the neck and chest, black; the sides marked with long dashes of the same colour; belly white; line over the eye and the cheeks dirty white; beak black, and more lengthened than in our own domestic species, or that which follows.

ALPINE SPARROW.

Pyrgita cisalpina, *Cuv.*

Le Gros-bec cisalpin.

"The Alpine Sparrow," says M. Temminck, "is only seen in the southern countries on the other side of the great chain of the Alps and Apennines, never on the northern side of those mountains:" from these localities it appears to extend itself along the whole of Italy and the southern countries of Europe. It differs in its habits from our own species, inasmuch as it gives the preference to plains and open country instead of cities and villages.

In the male, the top of the head and back of the neck are of a pure bright chestnut in summer, becoming, after the autumn moult, of a redder tinge, every feather being then edged with rufous; the cheeks pure white; in other respects the colour is like that of our own bird.

The female is so like that of *P. domestica* that one description will apply to both, with the exception that in the present bird the head and back of the neck are of a lighter ash-colour, and that its tints are generally paler.

Of the nidification and eggs of the two species here figured we have been unable to obtain any information.

Our Plate represents a male of each species, and the head of the female of *P. cisalpina.*

DOUBTFUL SPARROW.
Pyrgita Petronia.

Drawn from Nature & on stone by J. & E. Gould.

Printed by C. Hullmandel.

DOUBTFUL SPARROW.

Pyrgita Petronia.

Le Gros-bec soulcie.

We have followed the example of many previous ornithologists in associating this bird with those forming the restricted genus *Pyrgita*, or true Sparrows, although we doubt the propriety of so doing, as we think that it possesses peculiar characters, which would entitle it to rank as the type of a separate genus: its strong conical bill, lengthened wing, and abbreviated tail are not in strict unison with the generic characters of *Pyrgita*; but we have refrained from separating it, being desirous of obtaining further information respecting its habits and manners, which doubtless differ in many particulars from those of the Sparrows the typical form of which is represented by the common species inhabiting England. Independently of the characters alluded to as differing from those of *Pyrgita*, we may add that in this genus the markings and colour of the plumage of the sexes are very different, while in the sexes of the present bird no outward variations are perceptible.

Dr. Shaw informs us in his General Zoology, vol. ix. part ii. p. 434, that "this species is found over the greatest part of Europe, in the southern portions of which it is migratory, but is nowhere so common as in Germany. It is not found in this country; it affects woods, and builds in the holes of trees, laying four or five eggs, and feeds on seeds and insects. These birds are very delicate, as numbers are often found dead in trees in the winter, during which time they assemble in flocks."

The top of the head is longitudinally banded with greyish white tinged with yellow, which colour pervades each of the feathers of the back and upper surface; wings brown; the secondaries and scapularies tipped with yellowish white; primaries and tail brown with the outer webs margined with yellowish white; the inner webs of all the feathers of the latter, except the two middle ones, having a large spot of white near the extremity; under surface dusky grey and white, mixed deepest on the flanks; upper mandible brown, lower one yellow at the base and brown at the tip; irides brown; feet brown.

We have figured a male of the natural size.

CHAFFINCH.

Fringilla Cœlebs; *(Linn.)*

Drawn from Nature & on Stone by J & E Gould.

Printed by C. Hullmandel.

Genus FRINGILLA.

GEN. CHAR. *Bill* concave, longer than deep, straight, and pointed; cutting edges entire, and forming a straight commissure. *Nostrils* basal, lateral, oval, partly hidden by the frontal plumes. *Tail* slightly forked. *Legs* having the tarsi of mean length, with the toes divided and adapted for hopping or perching. *Claws* sharp.

CHAFFINCH.

Fringilla Cœlebs, *Linn.*

Le Gros-bec pinson.

THIS ornamental Finch is so well known to all persons whose attention has been directed to the habits of our native birds, that we doubt whether we can offer any novelty relative to its history. It appears to be very generally distributed over every portion of Europe, in most parts of which it is stationary. "All the ornithologists," says Mr. Selby, " describe this species as permanently resident with us, and nowhere subject to that separation of the sexes, and the consequent equatorial movement of the females, which is known to take place in Sweden and other northern countries. The fact, however, is otherwise, as the experience of a series of years has evinced that these birds, in a general point of view, obey the same natural law in the North of England. In Northumberland and Scotland this separation takes place about the month of November, and from that period till the return of spring few females are to be seen, and those few always in distinct societies. The males remain, and are met with, during the winter, in immense flocks, feeding with other granivorous birds in the stubble lands, as long as the weather continues mild, and the ground free from snow; and resorting, upon the approach of storm, to farm-yards, and other places of refuge and supply." The remarks which we have quoted from Mr. Selby will apply to the habits and manners of this bird in the South of England. We have observed that during autumn and the early parts of spring our gardens and orchards are comparatively deserted by this handsome bird, and that it must then be sought for in the wide fields and hedge-rows, far removed from our immediate precincts. It pairs early in the spring, and again returns to enliven our gardens and orchards by its simple song and sprightly actions, when the work of nidification is soon commenced. The nest is of the neatest construction, being outwardly composed of the most delicate lichens, (generally obtained from the apple-tree,) interwoven with wool, and lined with feathers and fine hair; it is placed in various situations, such as the branch of an apple-tree, the whitethorn, or any other shrub or tree whose foliage affords it a sufficient shelter to protect the eggs, which are four or five in number, of a pinky white spotted with reddish purple.

The food of the Chaffinch is of a mixed nature, feeding in winter on grains and seeds, and in summer on most species of insects and their larvæ, which it devours with avidity.

The sexes, as is the case with most of the true Finches, offer a contrasted difference in their colouring; neither can the beautiful spring plumage remain unobserved, when compared with the sober livery of winter.

The male in spring has the bill of a fine blue grey; the crown of the head and nape rich grey; the centre of the back chestnut; rump greenish yellow; lesser wing-coverts white; quills black, edged with yellowish white; two middle tail-feathers grey, tinged with olive; three next, on each side, entirely black; the outer ones with a large white spot on their inner webs; the cheeks, neck, throat, and under surface chestnut brown; lower part of the belly and vent white; legs and feet brown. In the female the whole of the upper surface is olive brown, becoming richer on the upper tail-coverts; cheeks, throat, and under surface greyish brown; vent and under tail-coverts white; the wings and tail as in the male, but the white marks less distinct.

The young males in autumn resemble the females.

Our Plate represents the birds in their spring plumage, although we must acknowledge our inability to do justice to the rich and harmonious tints which pervade the feathers of the living bird, and which afford so much attraction and ornament to our lawns and shrubberies.

MOUNTAIN OR BRAMBLE FINCH.

Fringilla montifringilla: *(Linn.)*

Drawn from Nature & on Stone by J. & E. Gould. *Printed by C. Hullmandel.*

MOUNTAIN OR BRAMBLE FINCH.

Fringilla montifringilla, *Linn.*

Le Gros-bec d'Ardennes.

This species of Finch is dispersed in considerable abundance throughout every country in Europe, and, as its specific name implies, prefers high and mountainous districts. In many parts of the Continent it is stationary, while in others it is strictly migratory. In the British Islands the winter season alone is the period of its visits, where it makes its appearance at the end of the autumn, and retires again on the approach of spring. During summer it dwells and incubates in those extensive forests of fir and pine which abound in all high northern latitudes. Although few seasons pass without the presence of this elegant bird in the central portions of our island, nevertheless it must have been remarked that at certain periods it makes its appearance in some of our woods and stubble-lands in flocks, often associating with Chaffinches and other granivorous birds in innumerable quantities. As to situation, they appear to evince a decided preference to woods of beech, on the mast of which they for a time subsist, feeding also on various seeds and the shoots of tender vegetables, resembling in this and many other respects the Chaffinch (*Fringilla cœlebs*, Linn.), and like the latter is equally typical in form; and for beauty and elegance it is not surpassed by any other of its genus. Although it is very probable that a limited number of this species remain to breed in the northern parts of Scotland, yet we have never been able to verify the fact. It is said to incubate in forests of lofty pine and spruce, the nest being composed of moss and wool, lined with feathers and hair. The eggs are white, spotted with yellowish brown, four or five in number.

In the general style of colouring the two sexes are similar; the male, however, far surpasses his mate in the richness and contrast of his plumage. In summer the male is adorned with a different dress from that of winter, that portion of the plumage which is then brown being exchanged for black during the spring and breeding-season. The male bird in the accompanying Plate exhibits a state of plumage intermediate between these two seasons, both sexes having been taken immediately before their departure.

The male has the head, ear-coverts, nape, and upper part of the back black, each feather being edged and tipped with yellowish brown; scapularies barred across the centre of the wing with white; edges of the secondaries, throat, and chest bright ferruginous brown; rump and vent white; primaries black edged with yellowish red; bill black at the tip, yellow at the base; legs brown; irides hazel.

The female has the general markings and colours of the male, but in every respect much more obscure and dull.

The Plate represents an adult male and female of the natural size.

SNOW FINCH.

Fringilla nivalis; *(Linn.)*

Drawn from Life & on Stone by J. & E. Gould.

Printed by C. Hullmandel.

SNOW FINCH.

Fringilla nivalis, *Linn.*

Le Gros-bec niverolle.

This species of Finch approximates so closely in form and general style of colouring to one species of the genus *Plectrophanes*, that it has been with some difficulty we have decided upon following the arrangements of M. Temminck in still retaining it in the genus *Fringilla*. We find that this bird, as it departs from its more typical relations, exhibits the same differences, and assumes almost the same characters and general appearance, as the Snow Bunting, *Plectrophanes nivalis*: the construction of its bill, however, which more strictly resembles that of *Fringilla*, denotes its true situation, and a more beautiful link could not be conceived, uniting as it does in the most complete manner the species of two genera, viz. the Buntings and Finches. Still it cannot be denied that the Snow Finch has as great a claim to a new generic title as the Snow Bunting, possessing as it does characters so essentially different from the true Finches.

We are led to believe from its form and the imperfect accounts published respecting its history, that its habits are in a great measure terrestrial, although it chooses the most elevated situations, such as the Alps, Pyrenees, and other mountainous districts of Europe, the British Isles excepted. In these wild and barren regions, upon the very verge of perpetual snow and ice, it dwells in unmolested security, and there finds that food which nature has destined for its support. This, according to M. Temminck, would seem to be of a mixed nature, consisting of seeds of various kinds, often that of the fir cone, and various species of insects. It builds its nest in crevices of the rocks, laying four or five eggs of a light green, irregularly sprinkled with ash-coloured dots, intermingled with blotches of dark green.

The sexes offer but little difference in plumage; neither does the summer and winter dress exhibit much variation, the beak being more or less yellow in winter, but deep black in summer.

In the male the top of the head, the cheeks and back of the neck are of a blueish ash; the scapulars and the two secondary feathers nearest the body are deep brown; all these feathers being bordered with a lighter colour; the remainder of the secondaries, the wing-coverts and the coverts of the tail are pure white; tail white, with the exception of the two centre feathers, which are blackish, and the whole tipped with the same colour; quill-feathers deep black; the under parts are white or whitish according to age; feet brown. This description applies to the female also, except that we find in her the ash colour of the head tinged with rusty brown, and the quill-feathers brown instead of black.

We have figured a male in summer plumage, and a female in that peculiar to winter.

WINTER FINCH.

Fringilla hyemalis; *(Linn)*

Drawn from Nature & on Stone by J & E. Gould *Printed by C. Hullmandel.*

WINTER FINCH.

Fringilla ? hyemalis.

Le Bruant Jacobin.

THE natural habits of this little Finch lead it to extend its summer migrations further north perhaps than most other members of the *Passerine* Order, and it is consequently an inhabitant of the regions far within the arctic circle, is common in Greenland, and has within the last few years been added to the Fauna of Europe. In the third part of his 'Manuel' M. Temminck states that it occasionally visits Iceland, and may perhaps be considered a bird of periodical passage in this portion of Europe.

Like the Robin the Winter Finch evinces little fear of man, and readily admits his near approach even in fine weather, and in hard weather is " so gentle and tame," says M. Audubon, " that it becomes, as it were, a companion to every child," and is indeed as well known and as much cherished by every person in America as the Robin is in Europe. It usually lives in families of twenty or thirty, is very jealous of intrusion, and readily darts forth to repel the invader. It is particularly fond of grass-seeds, and grain and berries of all kinds. M. Audubon also states that in its habits and manners it much resembles the Sparrow, resorting for shelter during cold weather to stacks of corn and hay, but in fine weather evincing a preference for the evergreen foliage of the holly, cedar, low pines, &c.

Of its nest and eggs nothing is known.

According to M. Audubon its flesh is extremely delicate and juicy, on which account it is frequently exhibited for sale.

The male has the head, all the upper surface, eight middle tail-feathers, wing-coverts, primaries, and chest blackish grey; secondaries blackish brown margined with reddish brown; two outer tail-feathers on each side white; under surface white, with a tinge of rufous on the flanks and under tail-coverts; bill reddish white with a black tip; irides blackish brown; feet and claws flesh colour.

The female differs in being of a lighter grey tinged with brown.

We have figured a male and female of the natural size.

COMMON OR BROWN LINNET.
Linaria Cannabina; *(Swains.)*

Drawn from Nature & on stone by J & E. Gould.

Printed by C. Hullmandel.

Genus LINARIA, *Auct.*

GEN. CHAR. *Bill* straight, conical, entire; mandibles compressed in front, and forming a very sharp point. *Nostrils* basal, lateral, concealed by incumbent feathers. *Wings* long, acuminate; first, second, and third quill-feathers of nearly equal length. *Tail* more or less forked. *Tarsi* slender, short. *Feet* having the lateral toes of equal length; the hind toe with its claw as long as the middle one. *Claws* slender, acute, curved, that upon the hind toe larger, and in old birds much longer than the rest.

COMMON, OR BROWN LINNET.

Linaria cannabina, *Swains.*

Le Gros-bec Linotte.

THE seasonal changes of plumage to which the *Fringillidæ* are generally subjected is in no one of the tribe more strikingly exemplified than in the birds constituting the restricted genus *Linaria*, of which the Common Linnet is the largest, and offers the most contrasted changes, being in winter clothed in a sombre and nearly uniform dress of brown, which in spring is exchanged for a rich rosy red on the crown of the head and breast, and in autumn it resumes the sombre winter colour: this diversity of plumage has caused some confusion, and added numerous synonyms to the name of the Common Linnet, and its nearly allied species the Redpole.

The Linnet is strictly indigenous to the British Islands, over the whole of which, and Europe generally, it is plentifully dispersed. It associates in flocks, and feeds upon small seeds, particularly those procured from the wild cruciform plants, &c. Open districts, such as commons and furze fields, constitute its favourite localities. The thickest parts of the furze bushes are generally selected for the sites of incubation, and the building of the nest is commenced early in the spring: it is constructed of moss, small twigs, and the stalks of grass, interwoven with wool and lined with hair and feathers; the eggs are mostly four in number, of a bluish white speckled with purplish red colour. "In winter," says Mr. Selby, "these birds assemble in very large flocks, and descend to the sea-coast, where they continue to reside till spring again urges them to pair and seek their upland haunts."

The Linnet is not more highly prized for the lovely hues of its summer dress than for the sweetness of its simple song, on which account great numbers are annually captured and reared for the purpose of being kept in confinement.

The female does not possess the rich colouring that characterizes the male in summer. Mr. Selby having taken considerable pains to ascertain and point out the various changes which this bird undergoes, we take the liberty of availing ourselves of his very accurate description.

"Bill deep bluish grey; forehead and breast of a bright carmine red; throat and under part of the neck yellowish white streaked with brown; crown of the head, nape, and sides of the neck bluish grey, in many instances varied with a few darker streaks; back, scapulars, and wing-coverts chestnut brown, with the margins of the feathers palest; flanks pale brownish red; middle of the belly and the vent greyish white; quill-feathers black, with more or less white on the basal half of their webs, and forming a distinct bar across the wings when closed; tail considerably forked, with the two middle feathers wholly black and pointed; the rest black, margined both on their inner and outer webs with white; legs and toes brown.

"In younger individuals the red upon the breast and head is not so pure in tint, nor to the same extent as in the older birds; the grey upon the crown of the head and the neck is also more varied with spots and streaks."

The female is inferior in size to the male, and has the "head and upper parts of the body umber brown, the margins of the feathers passing into yellowish brown; wing-coverts chestnut brown; throat and sides of the neck yellowish white, streaked and varied with yellowish brown; breast and flanks pale reddish brown, streaked with umber brown; middle of the belly yellowish white.

"The winter plumage of the male (after the first year) is nearly as follows: crown of the head varied with large black spots, which occupy the centre of the feathers; back and scapulars chestnut brown, but deeply margined with pale yellowish brown; breast reddish brown, with the tips of the feathers reddish white; flanks with large oblong brown streaks."

We have figured a male and female of the natural size.

MOUNTAIN LINNET OR TWITE.
Linaria montana; *(Ray)*.

Drawn from Nature & on stone by J & E. Gould. *Printed by C. Hullmandel.*

MOUNTAIN LINNET, OR TWITE.

Linaria montana, *Ray*.

Le Gros-bec à gorge rouse ou de Montagne.

THE Twite, although possessing a longer tail than the Linnet, has a more delicate contour of body, and is, we think, a more diminutive bird: in this respect, however, our opinion is not in accordance with that of Mr. Selby, who states, "It is rather larger than the Common Linnet, being bulkier in the body and having a longer tail." It differs from the Redpole in its larger size and in the total want of that rosy red colour which characterizes the crown of the head and breast of that species during summer. The changes to which the Twite is subjected, although quite apparent to the ornithologist, are nevertheless of a less striking character than in any other species of the genus *Linaria*. The specimen from which our figure was taken is in the plumage of the breeding-season; in autumn and winter they are lighter in colour, and more tawny on the face and throat. In its general economy and food the Twite is very similar to the Linnet, in whose company it migrates southward when the more northern countries become frozen. During these migrations every portion of our island is visited, and great numbers are captured by the bird-catchers while in pursuit of the more favourite Linnet and Goldfinch.

The Twite is abundantly dispersed over the northern portions of Europe, even within the regions of the arctic circle; the high and mountainous districts of these countries constitute its favourite residence and breeding-place, and are, indeed, its true habitat. It also passes the summer, but in smaller numbers, on the uplands of Scotland, the Western, Orkney, and Shetland Islands. "The nest," says Mr. Selby, "is placed amidst the tops of the tallest heath, and is composed of dry grass and heather, lined with wool, fibres of roots, and the finer parts of the heath; and the four or five eggs it contains are of a pale bluish green colour, spotted with pale orange brown. It leaves the mountains in autumn, assembling in flocks, which associate and travel with the Common Linnet, and are taken with them by the London bird-catchers, who can readily distinguish when there are any *Twites* in a flock by their peculiar note, expressive of that word."

In the colouring of their plumage, the only difference between the sexes consists in the female wanting the pink mark on the rump; but in size she is somewhat more diminutive than her mate.

Bill pale yellow; crown of the head and upper surface, with the exception of the rump, which is reddish pink, dark brown, each feather being edged with yellowish buff; throat, face, and stripe over the eye buff; flanks and under surface greyish brown, each feather having a darker centre; primaries and tail blackish brown, each feather having the external edge white; tarsi dark brown.

The Plate represents an adult male in summer, of the natural size.

MEALY REDPOLE.
Linaria canescens; *(Mihi)*.

Drawn from Nature & on Stone by J. & E. Gould.

Printed by C. Hullmandel.

MEALY REDPOLE.

Linaria canescens, *Mihi.*

It is not without due reflection and the examination of a great number of specimens that we are induced to consider this bird as truly distinct from the Lesser Redpole; although, it must be confessed, that to a casual observer little would appear to distinguish it from that bird. Independently of a marked superiority in size, its conspicuous greyish white rump, the broad band across the wing, the lighter stripe over the eye, and the general paleness and mealy appearance of the plumage at once tend to bear us out in our opinion, the more so as these circumstances are not accidental, but occur regularly in all the individuals which we have had opportunities of examining. In our views on the subject we are borne out by the concurring opinions of many ornithologists of the present day who are deservedly eminent for the closeness and accuracy of their researches. The practical bird-catchers in the neighbourhood of London have no doubts on the subject, but have ever been in the habit of regarding the Mealy Redpole as truly distinct. They also assert that it differs from the Lesser Redpole in its habits, manners, and in the situations it frequents; and that during some winters it is so scarce as seldom to be taken, while at others it is so abundant that flocks of hundreds are frequently seen. About the year 1829 it was particularly abundant and was taken in great quantities, but since that period it has occurred in far less numbers, so much so that only one or two have been latterly taken by any one person during the season. Whether this species is truly a native of Europe, or whether those which occur in our island are arrivals from the northern portions of the American continent, is a matter of doubt; true it is, that the specimens brought home by Dr. Richardson, which furnished the descriptions given in the *Fauna Boreali-Americana*, are strictly identical with the bird before us. A further knowledge of this bird, and especially of the changes which it undergoes, will at a future period determine whether the specific term of *canescens* must eventually stand or fall.

The Plate represents an adult, taken in the month of October, of the natural size.

LESSER REDPOLE.

Linaria minor; *(Ray)*.

Drawn from Nature & on Stone by J & E. Gould. *Printed by C. Hullmandel.*

LESSER REDPOLE.

Linaria minor, *Ray.*

Le Gros-bec sizerin.

THE Lesser Redpole is a native of the northern portions of our island and all the higher latitudes of the adjacent continent; from these districts numbers migrate southwards on the approach of winter, spreading themselves over every part of England, and most of the southern districts of Europe. In habits and manners it is gregarious, and is often found in the company of Linnets and Aberdevines. Its food consists almost exclusively of the seeds of various plants and shrubs, giving a decided preference to those of the alder, hazel, and willow; hence it resorts habitually to low and swampy situations, where its favourite food abounds. In habits and manners it is lively and active, and displays the greatest agility and address in picking out the seeds and buds of the smaller branches; nor is it less to be admired for its great docility and tameness, being at all times captured without any difficulty, and soon becoming familiar. The song of the Redpole, though not loud, is nevertheless simple and agreeable. It is found to breed in tolerable abundance in Scotland and in the northern portions of Europe. Its nest, which is particularly neat and compact, is placed in a low bush of willow, alder, or hazel, and sometimes furze, and is composed of grass and moss intermixed with the down of the catkins of the willow; the eggs are four or five in number, very small, and of a pale bluish green spotted with orange.

The beautiful rosy tints which pervade the breast of the male during the whole of the summer, render this little favourite one of the most elegant of our native finches. We may here remark, that when in a state of captivity, it loses the livery of summer, and does not regain it on the approach of the same season as it would do in a state of freedom, a circumstance which should render us cautious in drawing any conclusions respecting the changes of the plumage of birds from those that are kept in confinement. The female does not at any season acquire the fine tints which characterize the male during spring and summer.

The young of both sexes during the first autumn resemble the female, and do not require any further description than to say that the entire colouring is somewhat more tawny, and the rump only slightly tinged with rosy red.

In summer the adult male has the tip of the bill black, with the base of both mandibles fine horny yellow; space between the bill and the eye, the chin, and throat blackish brown; crown of the head and rump blood red; neck and breast rosy red, inclining to carmine, but becoming less pure on the flanks, which are slightly streaked with brown; middle of the belly, vent, and under tail-coverts white; the whole of the upper plumage tawny brown, each feather having a darker centre; primaries dark hair brown edged with yellowish white; tail brown, each feather having a lighter edge.

The Plate represents an adult male and female of the natural size.

SERIN FINCH.
Serinus flavescens.

Drawn from Nature & on Stone by J & E. Gould.

Printed by C. Hullmandel.

Genus SERINUS, *Mihi.*

Gen. Char. *Beak* much abbreviated, convex, and blunt at the tip; the edges of the upper mandible somewhat inflected, as are those of the under at the base, as far as the angle, which is not very decided. *Nostrils* basal and partly hidden by small feathers. *Wings* reaching half way down the tail, and having the first four feathers nearly equal, the second being the longest. *Tail* deeply forked. *Toes* feeble; the inner the same length as the hind one. *Nails* small.

SERIN FINCH.

Serinus flavescens, *Mihi.*

Fringilla Serinus, *Linn.*

Le Gros-bec Serin ou Cini.

It must not be supposed that we are partial to the construction of new genera (which, we fear, is often done somewhat unnecessarily,) because, in the present instance, we have removed the bird before us from the systematic station it has hitherto occupied: the fact is, that on investigating its characters, we could not satisfy ourselves that the Serin Finch has been hitherto assigned to any genus with which it strictly agrees. Closely resembling the Siskin (*Carduelis spinus,*) in general form and colouring, it departs widely from that bird in the form of its beak, which, on the other hand, is neither that of *Coccothraustes* nor of *Fringilla.* Remarkable for its short, blunt, and equally convex form, as well as for being peculiarly small, it has some similarity to the beak of the Bullfinch, but wants the breadth and great lateral protrusion and roundness at the tip, which in that bird both the upper and under mandibles so preeminently display: besides which the style of plumage is also totally dissimilar. We trust that in these views we shall be borne out by the assent of other naturalists, to whom we submit our opinions with due deference.

The native habitat of the Serin Finch is limited to the southern portion of the European continent, where it is very abundant, especially in Italy, and the South of France and Germany, frequenting the borders of streams, where willows and alders afford it shelter. It is also common in copses and orchards, where it breeds, making its nest, which is of small dimensions, in low trees and bushes, of vegetable fibres and grasses lined with wool. The eggs are five in number, marked at the larger end with brown dots on a white ground. Its food, like that of the Finches in general, consists of seeds, such as hemp, plantain, &c.

The sexes differ in plumage, that of the male being distinguished by the greater predominance of rich yellow; it may be thus described:

Forehead, throat, circle round the eyes, breast, and rump fine yellow; back of the head and upper surface greenish olive dashed longitudinally with dusky brown; ear-coverts dusky olive; flanks olive grey with stripes of brown; abdomen white; quills and tail blackish brown; irides dark brown.

The female, with which the young male agrees very closely, wants the yellow forehead, and her chest is dull yellow, thickly spread over with longitudinal dashes of brown; the upper surface is less bright than in the male, and the rump has only a trace of the fine yellow.

We have figured a male and female of the natural size.

GOLDFINCH.

Carduelis elegans, *(Steph.)*

Drawn from Nature & on stone by J & E. Gould.

Printed by C. Hullmandel.

Genus CARDUELIS.

Gen. Char. *Bill* conical, longer than deep, compressed anteriorly, and drawn to a very acute point; culmen of each mandible narrow; tomia of the upper mandible angulated at the base, and slightly sinuated. *Nostrils* basal, lateral, and hidden by incumbent bristles. *Wings* of mean length; the first quill-feather rather shorter than the second and third; which are nearly equal, and the longest of all. *Tail* rather short and forked. *Legs* having the tarsi short; lateral toes of equal length. *Claws* curved and acute; hind toe tolerably strong, with the sole broad.

GOLDFINCH.

Carduelis elegans, *Steph.*

Le Gros-bec Chardonneret.

The present beautiful species, with one characterized by us from the Himalaya mountains under the name of *Carduelis caniceps*, and an undescribed species from China, should form, we conceive, a restricted genus, from which we would exclude the Siskin and several others which have hitherto been associated in the genus *Carduelis*.

The European continent appears to be the utmost range of the Goldfinch: it gives preference to high lands and mountainous districts during winter, particularly such as are wild and barren, and afford a plentiful supply of the thistle, plantain, &c., the seeds of which constitute its favourite food: at this period it is generally to be observed congregated in small flocks, flying through the air and suddenly settling among its favourite food. When the spring advances and the trees display a verdant appearance, the Goldfinch separates in pairs, each male taking a mate and quitting the wild and open country for woods, orchards, and gardens, and on the Continent to the rows of fruit-trees that border the road-side. As soon as the foliage becomes dense enough to conceal the nest, the task of incubation is commenced: the nest is placed in the fork of a branch, and is of the neatest construction, being composed of lichens, moss, and dried grasses, lined with hair, wool, and the seed-down of the willow and thistle; the eggs are four or five in number, of a bluish white spotted over with dashes of brown towards the larger end.

The sexes are so nearly alike in the colour of their plumage that the duller tints of the female are the only difference. The young, until the first change, are characterized by a plumage very different from that of the parents, the head being greyish brown, and having none of those beautiful and contrasted markings of scarlet and black which so strikingly ornament the adult: in this state of plumage they are termed Branchers by the London bird-catchers, by whom thousands are annually caught and caged for sale. The traffic in these birds and the adults, which are taken at every season of the year, forms no inconsiderable trade, although it must be acknowledged that the bird is more to be valued for its beauty than for its song, which is very inferior to that of the Linnet or Canary.

The adult has the forehead and cheeks rich orange scarlet; a black line passes from the base of the beak to the eye, the top of the head, and occiput, the latter having a white space between it and the scarlet of the cheeks; back and sides of the chest olive brown; wings black, each feather being tipped with white, and the centre crossed by a bright band of yellow; tail black tipped with white; under surface greyish white; beak horn-colour; legs and feet flesh-colour.

In the young the whole of the head, back, and sides of the chest are greyish brown; the wings resemble those of the adult, except that the band of yellow is neither so broad nor so bright, and the markings on the wings are brownish white instead of pure white.

The Plate represents an adult and a young bird of the natural size.

SISKIN OR ABERDEVINE.

Carduelis spinus, *(Steph.)*.

Drawn from Nature & on Stone by J & E. Gould.

Printed by C. Hullmandel.

SISKIN, OR ABERDEVINE.

Carduelis spinus, *Steph.*

Le Gros-bec tarin.

THE mild and docile disposition which this lovely little bird evinces while in captivity, in unison with its tame and harmless manners in a state of nature, secure for it a more than usual degree of friendship and interest. It is not in the cheerful month of May, when all nature is alive to the harmonies of our newly arrived summer visitors, and when the freshly emerged foliage of our woods and gardens presents a universal nosegay, that the little emigrant before us is to be observed; for at that time it has bidden us farewell, to visit more northern climes, whither it has retired for the purpose of breeding and rearing its young. Its native habitat appears to be the higher regions of the European continent, and it is only in its most northern portions that it has, with any degree of certainty, been known to incubate. M. Temminck states that it is found in Sweden, but not in Siberia, and that it passes periodically into France and Holland. At the close of autumn, in the month of November, when the groves are deserted by our southern visitors, who no longer find their wonted sustenance of fruits and insects, the Siskin migrates from its summer retreat again to visit its favourite localities till the following spring.

Most authors have enumerated the Siskin among the rarities of our native birds; on the contrary, there are few more common and few more universally dispersed, particularly where birch and alder abound. It seems to evince a great partiality for these trees, which generally grow by the sides of small streams and in low marshy lands: in such situations the Siskin may be observed in considerable flocks, often in the company of the Lesser Redpole, which it greatly resembles in its actions, feeding on the tender buds and seeds of the alder, and clinging to the outermost branches, much in the manner of the Tits, although compared with them it is much less expert and lively. We have never seen the Siskin feeding on the seeds of the thistle, dandelion, or other plants which form the principal sustenance of its nearly allied congener the Goldfinch, nor is its bill so perfectly adapted for procuring food of this peculiar nature, this organ being more abbreviated and less conical: the bird has also a much shorter tarsus.

Although we do not admit the propriety of separating the present, with one or two other nearly allied European species, and also several from other parts of the globe, from the Goldfinch, the type of the genus *Carduelis*, nevertheless we may mention, that the slight variation of form alluded to has a great influence over their natural habits and economy.

So much is the Siskin esteemed for its mild and docile disposition and pleasing song, that it is highly valued for the aviary, and indeed is yearly captured in considerable numbers, and sold in London, either for the purpose of pairing with Canaries or Goldfinches, or to be shut in a solitary prison to serenade the ears of some tenant of the garret.

The plumage of the sexes differs considerably. The male has his markings and colouring more contrasted and bright during summer: the black then becomes more pure and distinct, and the sides and under parts more vivid.

Much contradiction exists respecting the places the Siskin chooses for nidification. M. Temminck states that it constructs its nest on the highest branches of the pine, and in such a situation were nests seen by Sir W. Jardine and Mr. Selby near Killin: it is now ascertained to breed in some of the pine forests of the Highlands of Scotland. The eggs are four or five in number, of a pale blueish white, speckled with purplish red.

The male has the top of the head and throat black; over each eye runs a broad stripe of yellow; the back of the neck, back and shoulders of a yellowish olive, with longitudinal patches of brown; the lower part of the throat, chest and belly yellow; the thighs and vent grey, with elongated stripes of brown; a band of yellow across the wings, which are black; the outer edge of the quill-feathers slightly margined with yellow; the tail-feathers yellow at the base, and black at the extremities; bill light brown.

The female differs from the male in the absence of the black on the head and throat, and the fine yellow which pervades the breast, that part being grey, with longitudinal stripes of dark brown; in the whole of the upper surface being darker, and in the fine yellow at the base of the tail being almost wanting.

The Plate represents a male and female of the natural size.

CITRIL FINCH.

Carduelis citrinella.

Drawn from Nature & on Stone by J & E Gould. *Printed by C. Hullmandel.*

CITRIL FINCH.

Carduelis Citrinella.

Le Gros-bec venturon.

In its lengthened and conical bill, the Citril Finch offers a strict alliance to the beautiful Goldfinch so common in our island, whilst in the olive-yellow colouring of its plumage it is in close affinity with the Siskin or Aberdevine, and, as far as we have been able to ascertain its habits and manners, corresponds more with the latter than the former. Like the other members of its family, it is said to be a fine songster. It has never yet been seen wild in England or in the North of Europe; appears to be scarce in the central parts of France, and the southern portions of Germany; is more common in Switzerland and the Tyrol; and is very abundant in Greece, Turkey, Italy, and Spain: in all these countries it evinces a partiality to the high and mountainous districts covered with larch and fir, on the branches of which it builds its nest. It is said to lay four or five eggs, of a whitish colour, marked with numerous blotches of brown of various sizes. Its food consists of the seeds of the various plants that grow in alpine regions. Like most species of this genus the sexes of the Citril Finch offer but a slight difference in the colouring of their plumage.

The male has the face, crown of the head, throat and under surface greenish yellow inclining to olive; the occiput and back part of the neck grey; the rump, scapularies, and a bar across the wings, fine yellow with a tinge of green; the primaries, secondaries, and tail-feathers blackish brown, each feather being edged with greyish olive; legs brown; irides hazel.

The female is rather less in size, and her colours are not so vivid as in the male, particularly on the throat and under surface, which parts are grey instead of greenish yellow.

The Plate represents a male and female of the natural size.

HAWFINCH.

Coccothraustes vulgaris; *(Briss.)*

Drawn from Life & on Stone by J&E Gould.

Printed by J. Hullmandel.

Genus COCCOTHRAUSTES, *Briss.*

GEN. CHAR. *Beak* very stout, swollen, thick; the *upper mandible* straight, entire.

HAWFINCH.

Coccothraustes vulgaris, *Briss.*

Le Gros-bec.

THE Hawfinch appears to have an extensive range through the countries of Europe, especially its midland districts. In the British Isles it has until lately been regarded as a bird of considerable rarity, and principally as a winter visitor. Of late years it has certainly been more common, and we are inclined to suspect that this will be found to support an opinion we have long since formed, that certain birds which have for a number of years been scarce, suddenly become numerous and continue so for an indefinite period, when they again retire and are as scarce as before. It is not in the present bird alone that we have observed this singular phænomenon; we may instance for example the Godwits, of which the Black-tailed species, a few years ago, was so abundant in the London market as entirely to exclude the Bar-tailed, which has now taken the place of the former. Our much-esteemed friend Mr. Henry Doubleday, of Epping, has by his ardent research in British Ornithology made us better acquainted with the history of this bird than any other person. "The Hawfinch," says he, "is not migratory, but remains with us during the whole of the year:" and he assigns as a reason for its not being more frequently discovered, the fact of "its shy and retiring habits leading it to choose the most secluded places in the thickest and more remote parts of woods and forests; and, when disturbed, it invariably perches on the topmost branch of the highest tree in the neighbourhood." Epping Forest, where Mr. Doubleday discovered it breeding in considerable abundance, affords, from its solitude, a place at once congenial to its habits and retiring disposition.

We have known the Hawfinch to breed at Windsor, and a few other places; but certainly nowhere so abundant as on the estate of W. Wells, Esq., at Redleaf, near Penshurst, Kent, who lately informed us that he has, with the aid of a small telescope, counted eighteen at one time on his lawn. M. Temminck informs us that it evinces a partiality to mountainous districts, and that it is a bird of periodical passage in France, but irregularly so in Holland.

Its food consists of berries, seeds, and the kernels of stone-fruits, for the breaking of which its strong beak and the powerful muscles of the jaws are expressly adapted. In winter, its principal subsistence is the Haw, whence its common appellation.

According to Mr. Doubleday, this bird breeds in May and June; in some instances in bushy trees at the height of five or six feet, and in others near the top of firs, at an elevation of twenty or thirty feet; the nest is remarkably shallow and carelessly put together, being scarcely deeper than that of the Dove; in materials it resembles that of the Bullfinch, but it is by no means to be compared to it in neatness and compactness of construction; it is chiefly formed of sticks, interspersed with pieces of white lichens from the bark of trees, and is loosely lined with roots: the eggs are from four to six in number, of a pale greenish white, varying in intensity, spotted and streaked with greenish grey and brown.

The young birds before the moult, exhibit considerable difference in plumage from the adult: the throat, cheeks and head being of a dull yellowish colour with the under parts white, the flanks marked with small streaks of brown, and the general plumage of the upper parts being spotted with dirty yellow.

In the male, the beak and feet in winter are of a delicate flesh brown, the former becoming in summer of a clear leaden blue, the ends straw-colour, and in some instances white; the top of the head, the cheeks and rump of a chestnut brown; a narrow circle round the beak, and a broad patch on the throat are black; back of the neck ash-coloured; mantle and shoulders deep brown; the quills and secondaries, which latter appear as if cut off abruptly at their ends, are of a deep black with purple and violet reflections; most of the greater and the last row of the lesser wing-coverts are white, so as to produce a large central mark; the outer tail-feathers are blackish brown, the middle ones white on their outer and brown on their inner edges; the under parts of a light vinous red.

The female has the plumage of a paler hue, the white of the wing being more dull, the head more dusky, and the under parts less pure.

We have figured a male and female of their natural size.

GREEN GROSSBEAK.
Coccothraustes chloris; *(Flem.)*

Drawn from Nature & on Stone by J & E Gould.

Printed by C. Hullmandel.

GREEN GROSBEAK.

Coccothraustes chloris, *Flem.*

Le Gros-bec verdier.

The Green Grosbeak is abundantly dispersed over the whole of Europe, where it is strictly indigenous, and as far as our observation has gone is nowhere migratory. Its natural habits lead it to frequent gardens, orchards, shrubberies and cultivated lands, and it is one of the most familiar and docile of our native birds; its outspread wings and tail during flight attracting the eye with colours which are scarcely surpassed in beauty by any one of the *Fringillidæ*. When spring has clothed the vegetable world with foliage, the Green Grosbeak constructs its nest on a branch in the most leafy part of shrubs or hedgerows, often at a considerable distance from the ground, the nest being generally composed of leaves, moss, grass and small twigs, lined with wool, hair and a few feathers. The eggs are four or five in number, of a pale blueish white, speckled at the larger end with reddish brown. The young are distinguished from the adult during the greater part of the first autumn by the strong oblong dashes of brown which pervade the breast and under surface. This particular feature, together with the robust bill, short tail, and bulky body, characterizes it as a true Grosbeak (*Coccothraustes*), at the extreme limits of which genus we consider this bird should be placed, where it would appear to form a union with the true species of *Fringilla* as restricted by authors of the present day.

At the commencement of autumn the Green Grosbeak assembles in considerable numbers, with Chaffinches and Buntings, and being driven by the severities of the season from fields and gardens, retires to farm-yards, where a bountiful supply of grain yields it a subsistence.

The male differs from the female in having the plumage more brilliant, and by rather exceeding her in size.

The male has the whole of the upper surface of a bright olive-green, passing into yellow; the quills blackish grey with their outer webs bright gamboge yellow; the tail-feathers, with the exception of the two middle ones, which are grey margined with light yellow, and their exterior edges, which are greyish brown, are of the same fine gamboge yellow as the wings; under parts greenish, passing into sulphurous yellow; legs brown; bill white with a tinge of pink.

Our Plate represents the adult male, and young bird of the first autumn, of the natural size.

PARROT CROSSBILL.

Loxia pityopsittacus; *(Bechst)*.

Drawn from Life and on Stone by J. & E. Gould. *Printed by C. Hullmandel.*

Genus LOXIA.

GEN. CHAR. *Beak* moderate, strong, compressed, the two mandibles equally curved, hooked, and crossing each other at their tips. *Nostrils* basal, round, concealed, under hairs directed forwards. *Toes* three before and one behind, the former divided. *Wings* moderate, the first quill-feather longest. *Tail* forked.

PARROT CROSSBILL.

Loxia pityopsittacus, *Bechst.*

Le Bec-croisé perroquet.

THE Crossbills, although evidently allied in their general habits to the Pine Grosbeak (*Corythus enucleator*, Cuv.), exhibit many circumstances in their general œconomy which are as yet far from being satisfactorily understood. The rigorous climate of the regions they frequent, and the deep seclusion of the pine groves where they find food and shelter, alike prohibit the naturalist from minutely inspecting them throughout every portion of the year: hence, though it is well known that the plumage of every species undergoes singular and contrasted changes, still it is yet a matter of doubt whether these changes are the result of a double moult, or produced by a change of colour in the feathers themselves from one tint to another, the moult being but single. Capable of bearing extreme cold, it is only in the highest northern latitudes that they breed in spring or summer, building their nests and breeding in our temperate latitudes in the inclement season of winter, and returning, as spring comes on, to their retreats within the arctic circle.

Of this genus the Parrot Crossbill is one of the rarest. In England it has been taken so seldom, as scarcely to claim a place among our accidental visiters. In Poland, Russia and Germany it is a bird of passage, being spread throughout the pine forests in winter, and returning northwards with the return of spring. In France and Holland its visits are accidental.

The Parrot Crossbill may be considered as the type of the limited genus to which it belongs,—a genus at once distinguished by the singular formation of the beak, the curved mandibles of which cross each other so as to produce an appearance of having been unnaturally distorted. This mode of construction, however, is a wise provision of nature, for the purpose of enabling the bird to separate the hard scales of the fir-cones covering the seeds which constitute its principal subsistence. These seeds it obtains by bringing the points of the mandibles from their crossed position and placing them in apposition. The points thus brought together are insinuated between the scale and the body of the fir-cone, and the mandibles are then separated by a powerful muscular lateral effort. The seed is at the base of the inner side of the scale, and is removed by the hard tongue of the bird while the scale is held apart from the cone. In the present species the bill is strong, large at its base, and much crooked; in the other species its structure is more slight and the curve of the mandibles less decided.

According to M. Temminck, the colouring of the male in its adult state consists principally of greyish olive; the cheeks, throat and sides of the neck ash-coloured; on the head there is a number of brown dashes bordered with dull greenish; the rump is yellowish green, as are also the breast and under parts, but with a shade of grey; the sides are dashed with blotches of dark grey; quill- and tail-feathers dark brown edged with greenish; irides and tarsi brown; beak dark horn colour.

The young males of the year are greenish brown with dashes of brown on the head and back; the under parts whitish grey with longitudinal spots of brown; rump and tail-coverts tinged with green. After the first moult, to the age of a year, the plumage exhibits a singular change, being of a beautiful crimson red, more or less pure, as M. Temminck states, according as the individual approximates to the period of the second moult, which occurs in April or May, when the quills and tail-feathers are black edged with reddish. It is however, we suspect, still doubtful whether this state of plumage is indeed that of winter or of an immature condition: if so, it is not a little remarkable that in this respect the birds of this genus should form an exception to the general rule which gives the richest hues to maturity and the season of love.

The female differs little from the plain-coloured young males of the year. The upper parts are greenish grey with dashes of brown; the rump yellowish; the under parts ashy with a slight tinge of green passing into white towards the vent and under tail-coverts.

The figures in the Plate are of the natural size, and represent the variations in colour common to this species.

COMMON CROSSBILL.

Loxia curvirostra; *(Linn.)*

Drawn from Nature & on stone by J & E Gould.

Printed by C. Hullmandel

COMMON CROSSBILL.

Loxia curvirostra, *Linn.*

La Bec croisé commun, ou des Pines.

Although the Common Crossbill frequently visits our island in large numbers at opposite seasons of the year, it can scarcely be considered as a permanent resident; a few isolated instances, it is true, are on record of its having bred with us, but its natural habitat is undoubtedly the high northern regions of the old continent. Mr. Selby informs us that in the year 1821 immense flocks visited this kingdom and scattered themselves among the woods and plantations, particularly where fir-trees were abundant. "Their first appearance was early in June, and the greater part of the flocks seemed to consist of females and the young of the year (the males possessing the red plumage assumed from the first moult to the end of that year). Many of the females I killed showed plainly, from the denuded state of their breasts, that they had been engaged in incubation some time previous to their arrival; which circumstance agrees with the account given of the early period at which they breed in the higher latitudes. Since this period Crossbills have repeatedly visited us, but never in such numbers as in 1821."

We may here observe that in the minds of many naturalists some doubts still exist, and until lately in our own, as to whether the rich rosy red colouring assumed by this bird is characteristic of the breeding-season or the permanent livery of the adult male. During our recent visit to Vienna, we had an opportunity of observing both sexes in every stage, an examination of which afforded us abundant proofs that the red plumage is acquired during the first autumn, for we saw many lately fledged that had their plumage thickly spotted; others, that had partially lost their spotted appearance, and had partly assumed the red colouring; and others that had their feathers entirely tinted of this colour: while the adults were, as most ornithologists have stated, characterized by a plumage of olive green, which appears to be permanent. In the bird-market of Vienna multitudes of Crossbills are exposed for sale, with Swallows, Martins, and many others of the smaller birds, for the purposes of the table: of these the Crossbill appeared to be especially in request, doubtless from its superiority of size, and from the nature of its food rendering its flesh both sweet and well tasted, to the truth of which we ourselves can bear testimony.

The nest of the Crossbill is placed in the fork of the topmost branches of the fir and other trees, and is composed of moss and lichens, generally lined with feathers: the eggs are four or five in number, of a greyish white marked at the larger end with irregular patches of bright blood red, the remainder minutely speckled with the same colour. Its note is a kind of twitter, uttered while occupied in extracting the seed from the fir cone which constitutes its principal food, and for obtaining which its bill is expressly adapted. The fruit of the orchard is sometimes attacked by this bird, when they commit considerable devastation among the apples and pears by splitting them asunder for the sake of the seeds within. Among the branches it is extremely active and agile, clinging in every possible direction by means of its bill and claws, like the members of the genus *Psittacus*.

Of all the small birds, the Crossbill seems to be the least distrustful of man, and when flocks arrive in our island it is well known that numbers are taken by means of a birdlimed twig, attached to the end of a fishing-rod placed across their back.

The green plumage referred to above resembles so closely that of the adult Parrot Crossbill, that any lengthened description will be unnecessary; nor, after what has been said above, do we consider it requisite to give any further account of the young.

Our Plate represents an adult and a young bird of the year, of the natural size.

WHITE WINGED CROSS BILL.

Loxia leucoptera; *(Gmell:)*

Drawn from Life and on Stone by J. & E. Gould.

Printed by C. Hullmandel

WHITE-WINGED CROSSBILL.

Loxia leucoptera, *Gmel.*

This interesting species is considered to be entitled to a place in our catalogue of British Birds, a specimen having been shot within two miles of Belfast, in the month of January 1802, which circumstance stands recorded in the Transactions of the Linnean Society. We are not aware that any other instance of the occurrence of this species has been noticed in this country, and M. Temminck has not included it in his Manual of the Birds of Europe.

We are indebted for all we shall have it in our power to say of this species, to the various authors who have supplied us with histories of the ornithological treasures of North America, over nearly the whole of which vast continent it ranges during the summer; and it is therefore extraordinary that it should not have been found oftener in the analogous climates of the old continent.

"We can trace the White-winged Crossbill," says the Prince of Musignano, in his scientific continuation of Wilson's valuable work, "from Labrador westward, to Fort de la Fourche, in latitude 56°, the borders of Peace River, and Montague Island on the north-west coast, where it was found by Dixon. Round Hudson's Bay it is common and well known. It is common also on the borders of Lake Ontario, and descends in autumn and winter into Canada, and the northern and middle States."

Dr. Richardson found this bird inhabiting the dense white-spruce forests of the fur-countries, feeding principally on the seeds of the cones. It probably ranges as far as the 68th parallel, where the woods terminate, though it was not observed higher than the 62nd. In the countries where they pass the summer, they are seldom observed elsewhere than in pine swamps and forests, feeding almost exclusively on the seeds of these trees, and a few berries. They build their nest on the limb of a pine, towards the centre; it is composed of grasses and earth, and lined with feathers. The female lays five eggs, which are white, with yellowish spots. The young leave their nest in June, and are soon able to join their parents in their autumnal migrations. In September they collect in small flocks, and fly from tree to tree, making a chattering noise, and in winter they retire to the thickest woods of the interior. Like the other species of this genus, the subject of our Plate is liable to many changes of plumage which are not yet perfectly understood, every flock containing specimens of great variety of colours, from the general green appearance of the females, to the buff orange tinge which is by some considered to be characteristic of the adult male. Very young males before assuming the red at the age of one year, exactly resemble the females; being only more inclined to grey, and less tinged with olive, and having the rump greenish yellow. The male in his second year has the general plumage crimson red, the base of each feather darker, approaching to black on the head, round the eye and on the forehead; the rump a beautiful rose red. The adult male differs from the preceding, in having a light buff orange tinge where the other is crimson; pale beneath; wings and tail deep black, the two bars on the wings, the edges of the quills and tail-feathers being very conspicuous and pure white. In this state the bird is rare.

In the female the general tint is a greyish olive, the base of each feather slate colour, and the centre black, giving the bird a streaked appearance; the rump pale lemon colour; neck, throat and breast yellowish olive grey, the lower part of the belly also patched with black; wings and tail brownish black; middle and long coverts of the former broadly tipped with white, forming a double band across, so conspicuous as to afford the most obvious distinguishing character of the species; all the quills are slightly edged paler; irides hazel; bill dark horn colour; legs nearly black.

We have figured a male and female.

PINE GROSBEAK.

Corythus enucleator, *(Cuv.)*

Drawn from Life & on Stone by J. & E. Gould. *Printed by C. Hullmandel.*

Genus CORYTHUS, *Cuv.*

GEN. CHAR. *Beak* short, hard, thick, rounded in every part and slightly hooked at the point. *Nostrils* basal, lateral and rounded, covered with thickly set hair-like feathers. *Tarsi* short. *Toes* entirely divided. *Wings* more lengthened than in the genus *Pyrrhula*. *Tail* moderate and slightly forked.

PINE GROSBEAK.

Corythus eneucleator, *Cuv.*

Le Bouvreuil durbec.

THE Pine Grosbeak, hitherto classed among the Bullfinches, has been separated by Cuvier and advanced to the rank of a genus under the name of *Corythus*, which, as will be readily perceived, has characters sufficiently strong to warrant its legitimacy. The situation which this genus appears to hold is that of the connecting link between *Loxia* on the one hand and *Pyrrhula* on the other; agreeing with the former in its place of resort, habits, manners and style of colouring, and with the latter in the short and rounded beak.—The Pine Grosbeak, though not strictly a native of Great Britain, has been several times killed in our Island. Its true habitat appears to be within the Arctic Circle of both Continents, and we know it to be abundant in Norway, Sweden and Russia, inhabiting the secluded recesses of the almost untrodden pine forests of those countries, where it feeds upon the seeds of pine cones, as well as various kinds of other seeds and wild alpine berries. In the more southern provinces of Europe it appears to be merely an accidental visitor, and is rarely met with even in the North of Germany.

In another point also we trace a similarity between this bird and the Crossbill;—viz., in the changes which its plumage undergoes, passing, according to the seasons, from greenish yellow to a scarlet more or less pure. In the annexed Plate we have given a figure of the male and female in what we consider to be their adult plumage: on this point, however, we differ from M. Temminck, whose description we take the liberty of transcribing.

"The livery of the adult and aged male:

"Head, throat, and upper part of the neck of an orange red, becoming lighter on the fore-part of the neck; the breast and underparts of an orange-yellow; the feathers of the back, scapulars and rump, of a blackish brown in the middle with a large border of orange-yellow; wings and tail black, the former having two transverse white bands; all the secondary feathers bordered with white; quill- and tail-feathers edged slightly with orange; length seven inches nine lines.

"The male after its first moult, till a year old:

"Head, neck, throat, breast, part of the belly and rump, of a crimson red, the more strong and brilliant as the individual approaches its second moult. Feathers of the back and scapulars black in the middle, with a large border of crimson-red; sides, belly, and lower tail-coverts ash-coloured; two roseate bands cross the wings, and the secondary feathers are largely bordered with the same colour: the quill- and tail-feathers are all edged with light red.

"Adult and young female:

"The females of the year have only the top of the head and the rump reddish; when adult, they have those parts of a brown strongly tinged with orange, the back of the neck and cheeks edged with the same colour; the back and scapulars ashy brown; the under parts ash-coloured with a slight tinge of orange; the wings have two bands of greyish white; all the wing-feathers edged with greenish orange."

Young (females) are more obscure in their colouring. The nest is built on trees at a short distance from the ground; the eggs are white, without spots, and four or five in number.—We have figured a male and female of their natural size.

SIBERIAN GROSBEAK; (Lath).

Corythus longicauda

Drawn from Nature & on stone by J & E Gould.

Printed by C Hullmandel.

SIBERIAN GROSBEAK.

Corythus longicauda.

Le Bouvreuil à longue queue.

On comparing our specimens of this bird with others of the Pine Grosbeak, which is the type of the genus *Corythus*, we could not but observe that it offers a closer alliance to this peculiar form than to that of any other to which it has hitherto been assigned; we have therefore, although ever averse to multiplying the names of a species, judged it best to place it in the group to which it appears to us most nearly allied: it is true that the greater length of the tail in this species is not in strict accordance with the characters of *Corythus*, but this would seem to be the only point of difference.

The Siberian Grosbeak is found in the same localities as the Pine Grosbeak, namely, most of the high northern regions of the old Continent, and particularly Siberia, where, as M. Temminck states, it is extremely abundant. In winter it migrates to the more southern parts of Russia and Hungary.

In its general economy it resembles the Pine Grosbeak, and its food is said to consist of wild berries, the buds of trees, and other vegetable matters.

Of its nidification no certain information has been recorded.

As our Plate will show, the sexes are distinguished by the male being clothed in a richer-coloured dress than that of the female; but a still further knowledge of this rare bird is requisite to enable us fully to understand its various changes.

The male is characterized by having a red mark round the bill; the top of the head, cheeks, and throat clear rose red; chest, belly, and rump inclining to crimson; feathers of the back reddish brown in the centre bordered with red; lesser wing-coverts and edges of the secondaries white; primaries dark brown, with the edges lighter; the three lateral feathers of the tail white, the others black bordered with light rose colour; beak and feet brown.

It would appear from M. Temminck's statement that it undergoes a partial change of plumage at the autumnal moult, being then of a much lighter tint, and having all the feathers bordered with whitish. We have seen specimens in this state which strictly agree with M. Temminck's description.

The general plumage of the female is of a clear olive, with the exception of the wings and tail, which are like those of the male. As the specimen which we have figured from, and considered a female, is much more grey in its plumage and has several spots of blackish brown, we have reason to expect that it may be a bird not arrived at maturity.

The Plate represents a male and female of the natural size.

SCARLET GROSBEAK.

Erythrospiza erythrina; *(Bonaparte).*

Drawn from Nature & on Stone by J. & E. Gould.

Printed by C. Hullmandel.

SCARLET GROSBEAK.

Erythrospiza erythrina, *Bonaparte.*

Le Bouvreuil Pallas.

Having adopted the genus *Erythrospiza* as established by the Prince of Musignano, we feel convinced that the present bird will form one of this well-marked group, the members of which appear to be so widely distributed. The Scarlet Grosbeak must not be confounded with the *Fringilla purpurea* of Wilson, a bird to which it bears a resemblance both in habits and in style of colouring. A close examination of the two species will, however, at once satisfy the ornithologist as it respects their non-identity; and we would further remark, that the present bird appears to be strictly confined to the Old World, while the *Fringilla purpurea* is in like manner restricted to the American continent.

The Scarlet Grosbeak is one of those European birds which are obtained with great difficulty, and of which very few specimens exist in our museums; indeed, except our own, which came from Russia, we know of none in the public or private collections of Great Britain; yet it is a species far from being uncommon in high northern latitudes, and in some parts of Russia, where, according to M. Temminck, it habitually frequents gardens, and appears, from the little information we have been able to obtain respecting it, to differ little in manners from our well-known Bullfinch.

The male and female, as will be seen in the Plate, offer a decided difference in their colouring, the male being ornamented by a beautiful deep stain of scarlet over the whole of the plumage which is totally wanting in the female as well as in the young of both sexes; it is also probable that the male loses this distinguishing mark in winter and regains it in spring.

The male has the head, neck, and top of the back of a lively crimson, fading off below into a beautiful rose colour; the small feathers round the base of the beak and nostrils are also of a dull rose; the wings and tail brown, the feathers being edged with deep rose colour; beak and tarsi brown.

The female has all the upper parts of a brownish grey, with longitudinal dashes of a deeper colour, the throat and cheeks being blotched with brown; the under surface white, or nearly so.

The Plate represents an adult male and female of the natural size.

ROSY GROSBEAK.

Erythrospiza rosea.

Drawn from Nature & on Stone by J. & E. Gould.

Printed by C. Hullmandel.

ROSY GROSBEAK.

Erythrospiza rosea.

Le Bouvreuil Pallas.

We believe we may safely affirm that this beautiful species of Finch is strictly confined to the northern regions of the Old World, and that it is not found, as stated by some authors, on the continent of America. By Wilson it was considered synonymous with the *Fringilla purpurea*, which although bearing a strong resemblance in its general contour and colouring, differs both from it and *Erythrospiza erythrina*, in the form of the bill: the two latter birds have this organ shorter, and more swollen at the sides, approaching in these particulars to the typical *Pyrrhulæ*, or Buffinches.

Russia and Siberia constitute the true habitat of the present species, though it may occasionally be found in Hungary and the more central parts of Europe. It is considered one of the rarest European birds, and is consequently much sought after by collectors. The female is quite unknown to ourselves, and we are not aware of any description of that sex having been recorded.

Head, back, rump, upper tail-coverts, breast, and all the under surface of a rich rosy hue, with a stripe of dark brown down the centre of each of the feathers of the back; crown of the head and the throat ornamented with pinkish white silky feathers; wings brown, the lesser coverts terminated with pinkish white, and the greater coverts with pink, forming two bands across the wing; tail brown margined with pink; bill and feet light yellowish brown.

Our figure is of the natural size.

VINOUS GROSSBEAK.

Erythrospiza? githaginea.

Pyrrhula githaginea; *(Temm.)*

Drawn from Nature & on Stone by J & E. Gould. Printed by C. Hullmandel.

VINOUS GROSBEAK.

Erythrospiza? githaginea.

Pyrrhula githaginea, *Temm.*

Le Bouvreuil githagine.

We have never been able to obtain more than a single specimen of this rare Little Grosbeak, whose native habitat is doubtless the northern and central portions of Africa: M. Temminck states that it is found in Nubia and Syria, whence it accidentally passes into Provence and other parts of the south of Europe; it is also said to visit the islands of the Archipelago. In his description, and in the figure of this species, published in the 'Planches Coloriées,' M. Temminck has represented the bill and legs as being red; in the specimen from which our figure is taken those parts were light yellow, and it did not appear that this difference had been the effect of time or death, as in that case there would have been faint traces of the red colour still remaining, which there were not. In all probability this bird will require to be separated from the group in which we have placed it; but we have deferred assigning to it new generic characters until further acquainted with the species.

The female is said to differ from the male in the absence of the rich rosy tints which adorn the latter, and in being of a uniform light brown very slightly clouded with a rosy hue; and in the under surface and wings being clear Isabella brown.

The male has all the under surface light brown, clouded with clear rose, which is palest on the throat and round the base of the bill; crown of the head ash, becoming brown on the nape of the neck; back, wing-coverts, rump, and the external edge of the wing and tail-feathers slightly tinged with rose colour.

We have figured a male of the natural size.

BULLFINCH.

Pyrrhula vulgaris; *(Temm.)*

Drawn from Nature & on Stone by J. & E. Gould. *Printed by C. Hullmandel.*

Genus PYRRHULA.

GEN. CHAR. *Bill* very short, and thick at the base; both mandibles convex, particularly the upper one, the point of which overhangs the point of the lower; *culmen* rather compressed and advancing upon the forehead. *Nostrils* basal, lateral, round, concealed by short feathers. *Wings* rather short, the fourth quill-feather the longest.

BULLFINCH.

Pyrrhula vulgaris, *Temm.*

Le Bouvreuil commun.

THIS handsome bird is the only one of the genus *Pyrrhula*, as restricted by modern naturalists, which has been hitherto discovered in Europe, that is to say, provided we consider the Bullfinch found in Germany and some other parts of the Continent (which in relative admeasurements is nearly a fourth longer,) as a variety merely, and not truly a distinct species. In our examination of this bird, had we been able to detect any difference of markings, or to ascertain that any dissimilarity existed in their habits and manners, we should not have hesitated on the subject; at present we remain in doubt on this point, which those who have an opportunity of examining the bird more closely than ourselves, and in a state of nature, can alone determine.

The interesting little group of which the present species forms a typical example, appears to be confined exclusively to the regions of the Old World, more particularly its northern and mountain districts. The elevated range of the Himalaya has not only produced an additional example, published by us in our work on the birds of that range under the specific title of *erythrocephala*, but we have since received another species from the same locality, and which is at present undescribed: we allude to this fact here, as confirmatory of the justice of separating birds possessing well-defined forms, however limited their numbers may be, into distinct genera, assured that future researches will increase the catalogue of species.

The Bullfinch is a constant resident in our island, although we are informed by continental writers that it is strictly migratory on the Continent generally, over the whole of which, except in Holland, where it is somewhat rare, it is plentifully dispersed. The habits of the Bullfinch are somewhat shy and retiring, giving preference to secluded thickets and coppices.

Its food consists for the most part of berries, seeds, and the buds of trees; hence in the spring no bird is accused of greater mischief in orchards and gardens.

The nest of the Bullfinch is rather loosely constructed, flat in its general form, and composed of small sticks lined with fibrous roots, and wool: it is mostly placed in the forked branches of trees and shrubs. The eggs are four or five in number, of a bluish white spotted with reddish brown. Although not entirely devoid of song, it is by no means remarkable in its wild state for its musical powers: its call note is a plaintive monotonous whistle. In captivity it is much valued, not only for its beauty, but for its powers of imitation, being capable of learning and repeating tunes and even words.

The sexes offer, as the Plate will show, a considerable difference in their colouring.

The male has the top of the head, the circle round the eye, the throat, wings, and tail of a deep glossy black; the back of the neck and mantle ash-colour; the cheeks, neck, chest, and flanks fine red; rump and abdomen pure white; a band of greyish white crosses the wing; the beak and irides black; tarsi blackish brown.

In the female, the red of the chest, neck, and flanks is exchanged for dusky greyish brown; the white of the rump is less conspicuous, and the markings of the head are not so pure and decided.

The Plate represents an adult male and female of the natural size.

STARLING.

Sturnus vulgaris, *(Linn.)*

Drawn from Life and on Stone by J & E. Gould.

Printed by C. Hullmandel.

Genus STURNUS, *Linn.*

GEN. CHAR. *Bill* straight, depressed, rather obtuse, and slightly subulated. *Nostrils* basal, lateral, and partly closed by a prominent rim. *Wings* long, the first feather very short, the second and third the longest and equal. *Feet* with three toes before and one behind; the middle toe united to the outer one as far as the first joint.

STARLING.

Sturnus vulgaris, *Linn.*

L'Etourneau vulgaire.

THE species formerly arranged under the old Linnean genus *Sturnus* have been separated into several distinct genera, bearing their appropriate characters, and which now form an interesting family, the members of which are very generally dispersed over the globe. The value of such separations is obvious to the Ornithologist, as by a knowledge of the characters of each genus he is at once enabled to ascertain its true situation.

The number of species contained in the genus *Sturnus*, as now restricted, is very limited, and are strictly confined to the Old World; two of these species are natives of Europe.

The Starling is a social and familiar bird, and were it less common would be highly esteemed, its habits and manners, and the variety of its plumage at certain ages and seasons being very interesting. Its range is extensive, being dispersed in considerable abundance over Europe; it is also found at the Cape of Good Hope, and from its being so numerous on the northern coast of Africa, we doubt not that they traverse the whole of that continent. We have received it from the Himalaya Mountains, and have ascertained that it is found as far east as China. It is a bold and spirited little bird, but soon becomes reconciled to confinement, where it not only sings sweetly, but may be taught to articulate words and even sentences. In a state of nature it is very harmless, and renders great service to the farmer, by clearing his pastures and fields from grubs, worms, and various other insects, on which it almost exclusively subsists. When in search of food, it runs along the ground with great celerity, prying and peeping with a cunning eye under every loose sod and tuft of grass.

The Starling congregates in large flocks during autumn and winter, and may be often observed in the company of rooks, daws, and fieldfares. On the approach of evening many of these flocks unite, and before going to roost this immense body may be seen traversing with undulating sweeps and evolutions the immediate neighbourhood of their resting-place. They prefer for this purpose secluded and warm situations, such as thickly set reed-beds, coppices, or plantations of fir. They pair early in the spring, and then spread themselves over the face of the country in search of a convenient breeding-place, some selecting the holes of trees, others old towers and ruins, and others the deserted nests of rooks, &c. They lay four or five eggs of a delicate pale blue. The young during the first autumn are characterized by the stage of plumage represented in the upper bird of our Plate; they begin to change in October, which is effected by a moult, and in the course of a week or two after are adorned with feathers, the whole of which, with the exception of the primaries and tail, are terminated with a large white or reddish white spot; the rest of the feathers being of a rich green with bronze reflections, as in the lower bird of the Plate. From this their spotted plumage they gradually change to that of the centre bird, having a fine yellow bill, and spotless lanceolate feathers upon the breast and underparts,—a state of plumage which is certainly not attained till the third year; and between these two latter stages, birds may be found in the same flock which exhibit plumage in every intermediate state. It may be observed, that as the feathers become elongated the white spot at the tip becomes less and less, till in the old bird it is lost. The males and females at the same age offer but little difference in plumage; the male, however, is generally the most brilliant in his markings.

We consider that these birds breed at a year old, although their plumage, as above stated, afterwards undergoes a considerable change.

The plumage of the old male in spring is peculiarly beautiful, not so much from its variety of colours as from the glossy metallic hues with which it seems burnished, exhibiting ever-changing reflections of purple and golden green; the upper wing-coverts marked with small triangular whitish spots; the lower coverts and the tail slightly edged with white; beak yellow; feet reddish flesh-colour:—the centre figure in our Plate exhibits the bird in this stage.

SARDINIAN STARLING.

Sturnus unicolor; *(Marm.)*

Drawn from Nature & on Stone by J. & E. Gould. Printed by C. Hullmandel.

SARDINIAN STARLING.

Sturnus unicolor, *Marm.*

L'Etourneau unicolore.

THE *Sturnus unicolor* does not possess that wide range of habitat which characterizes the preceding species, the *Sturnus vulgaris*. It is dispersed over the warmer parts of Spain, Sardinia, and the rocky shores of the Mediterranean generally. In these situations it may be observed in small numbers throughout the year, building in the recesses of the rocks, in the absence of which, old towers and ruins offer it an asylum equally suited as a place of repose, and a situation where it may raise its progeny: in fact, its general habits and manners bring it in close connexion with the *Sturnus vulgaris*; and if it is not seen congregated in almost countless flocks, it must be attributed to the limited number of the species, rather than to any difference in manners. We have not, with any degree of satisfaction, been able to trace the extent of the range which this species takes in Northern Africa: it would appear, however, to be somewhat limited, as we have never received or seen it in any collection from that continent, with the exception of Egypt and Abyssinia. On close examination, its plumage presents to the eye many rich and resplendent lights; and if not so gay as our pert and prying Starling, its general contour of body, clothed all over with long silky plumes, fully compensates for the deficiency.

The outward sexual differences are but trifling: the male may be always distinguished by the elongated feathers of the throat, which in fine adults are carried to an extreme, and which are displayed in the most beautiful manner when the throat is distended by their simple whistling strain.

The young birds bear so close a resemblance to the young of the common species, that a description will be unnecessary; if any difference exists, it is that the prevailing colour is darker.

The moult of the first autumn is characterized by the feathers being slightly tipped with white, which is totally lost in the following spring.

The adult male has the whole of the plumage of the body, wings and tail of a shining black, which is enlivened by reflections of purple and violet; beak blackish brown at the base, the point yellow; feet light brown.

The Plate represents a male and female in the adult livery and of the natural size.

ROSE COLOURED PASTOR.
Pastor roseus. *(Temm.k)*

Drawn from Life & on Stone by J. & E. Gould.   Printed by C. Hullmandel.

Genus PASTOR, *Temm.*

Gen. Char. *Bill* conical, elongated, cutting, very compressed, slightly curved, the point notched. *Nostrils* basal, ovoid, partly closed by a membrane and clothed with small feathers. *Feet* robust. *Toes* three before and one behind, the external toe united at its base to the middle one. *Tarsi* longer than the middle toe. *Wings* having the first *quill-feather* very short, or almost obsolete, the second and third equal and longest.

ROSE-COLOURED PASTOR.

Pastor roseus, *Temm.*

Le Martin roselin.

The birds composing the genus Pastor are exclusively inhabitants of the older-known portions of the globe, and especially its more eastern and warmer regions. The species are pretty numerous : the only one, however, which is known to visit the more temperate countries of Europe is the present beautiful and elegant example, the true habitat of which appears to be the western parts of Asia and the North of Africa, particularly Egypt and along the course of the Nile. From these districts it migrates regularly into the southern provinces of Italy and Spain ; seldom occurring further northward, and visiting our Island only occasionally at uncertain intervals ; but from the circumstance of its having been shot some few times in the British Islands, it has a claim, with many others equally scarce and equally peculiar to the warmer portions of the Continent, to a place in our Fauna. Several well authenticated accounts of the capture of this bird have appeared from the pens of Mr. Selby, Pennant, and Bewick ; but the only example within our personal knowledge was one shot in the month of May by our esteemed friend Mr. John Newman, of Iver Court near Windsor, in whose possession it now remains, exhibiting that beauty and richness of plumage which we have endeavoured to convey in our illustration.

In its manners it closely resembles our Common Starling (*Sturnus vulgaris*), congregating in the same manner in flocks, and frequenting pasture-lands for the sake of the grasshoppers and other insects which there abound ; often attending flocks and herds, and even perching upon the backs of cattle for the purpose of disengaging the larvæ which are bred beneath the skin. We are also informed that it abounds in Egypt, particularly those parts which are subject to the overflow of the Nile, attracted doubtless by the myriads of insects, locusts, &c., which the heat and moisture call into life; and for the services it thus renders to the natives it is held by them in great esteem. In addition to insects, it also feeds upon fruits and berries.

We are informed that the *Pastor roseus* chooses holes in trees, rocks or old buildings for the place of nidification, laying five or six eggs, the colour of which we have not been able to ascertain.

The only difference of the sexes in plumage consists in the more obscure tints of the female, and her rather smaller size ; the crest also is less silky and flowing ; the young, however, differ much,—and we would here point out another circumstance which indicates the close relationship between the present bird and the Starling. We have mentioned above, the great similarity in their habits, manners and food ; we now find an analogous and similar change of plumage in the young : and we would here suggest to those who are more especially interested in the Ornithology of Great Britain, whether the Solitary Thrush of Bewick, which has hitherto been taken for the young of the Starling, may not be that of the Rose-coloured Pastor ? We mention this as a query, because there are characters detailed in Bewick which the young of the Starling does not possess in any state, and which more nearly agree with the young of the present bird.

The colour of the Rose-coloured Pastor is very rich and delicate ; the beak and legs more or less flesh-coloured ; the head, throat and crest, together with the neck, black with violet reflections ; back and under parts of a delicate rose-colour ; wings and tail black with greenish reflections ; irides brown.

The young in the first autumn have the whole of the upper parts of the body of a uniform yellowish brown ; the wings and tail rather darker; the throat and under surface whitish; the former being marked longitudinally with brown blotches, and the head offering no indication whatever of a crest.

We have figured a male in full plumage, and a young bird of the year before its autumn moult ;—both of the natural size.

NUTCRACKER.

Nucifraga caryocatactes; *(Briss.)*

Drawn from Nature & on Stone by J. & E. Gould.

Printed by C. Hullmandel.

Genus NUCIFRAGA.

Gen. Char. *Bill* conical, longer than the head, straight; the upper mandible having the *culmen* rounded, overhanging the lower, both terminating in an obtuse and depressed point. *Nostrils* basal, round, open, concealed by hairs directed forwards. *Toes* three before and one behind, the two outer being united at their base. *Tarsus* longer than the middle toe. *Wings* long and pointed, the first quill-feather being the shortest, and the fourth and fifth the longest.

NUTCRACKER.

Nucifraga caryocatactes, *Briss.*

Le Casse noix.

We are sorry that it is not in our power to give a detailed account from personal observation of the manners of this singular and interesting bird, which with one other from the Himalaya mountains form the only known species of the present genus, which seems to connect the order with several other groups, of which we may enumerate that of *Picus* among the *Zygodactylous* birds; and Mr. Vigors considers it to assimilate in some degree to that extensive family the *Sturnidæ*, especially to the genera *Cassicus* and *Barita*: it must be acknowledged, however, that some other interesting form seems to be required in order to make the link of approximation complete.

The native habitat of the Nutcracker is the mountain woods of Switzerland and Germany, and indeed the greater portion of Europe, in the northern parts of which it is strictly migratory.

Its claim to a place among the birds of the British Islands rests on a few rare instances of its having been captured in this country. Its habits and manners accord with what we might expect from its peculiar form, bearing a marked resemblance to those of the Woodpeckers: like them, it ascends the trunks of trees, strikes the bark with its bill in order to dislodge the larvæ of insects which lurk beneath, and upon which it feeds, together with worms, fruits, nuts, the seeds of pine, &c.

It incubates in the holes of decayed trees, frequently enlarging the cavity to the necessary size, and lays five or six eggs, of a yellowish white.

The sexes, as in the *Corvidæ* in general, offer no external difference of plumage; the female is, however, somewhat smaller than the male, and perhaps a little more obscure in her markings.

The whole of the plumage is of a deep reddish brown, inclining to umber; the body varied, except on the head and rump, with large spots of white, occupying the centre of every feather; wings and tail brownish black, with green reflections, the latter being tipped with white, the two middle feathers excepted; bills and legs brownish black.

We have figured the bird of the natural size.

JAY.

Garrulus Glandarius; *(Briss.)*

Drawn from Nature & on Stone by J & E Gould.

Printed by C. Hullmandel.

Genus GARRULUS.

GEN. CHAR. *Bill* shorter than the head, conical, slightly compressed, straight at the base, rather deflected towards the tip, which is faintly emarginated; the lower mandible of nearly equal thickness, and having its culmen equally convex with that of the upper; commissure straight; head crested. *Nostrils* basal, lateral, hidden from view by short setaceous plumes. *Wings* rounded, with the first quill-feather short; the fourth, fifth and sixth of nearly equal length, and the longest in the wing. *Tail* square or slightly rounded. *Legs* weaker than in the genus *Corvus*. *Tarsi* longer than the middle toe; the outer toe joined at its base to the middle one, and longer than the inner; hind toe strong, with a dilated sole. *Claws* stout, moderately curved and sharp; that on the hind toe stronger and longer than any of the rest.

JAY.

Garrulus glandarius, *Briss.*

Le Geai.

THIS common but extremely ornamental bird is dispersed over the greater portion of the wooded districts of Europe, and together with one from the Himalaya mountains, and another which we have seen, truly distinct from either, form a small but well-defined group, which appears to range intermediate between the latter group and the Pies (*Picæ*), to which the generic title *Garrulus* should be strictly limited, to the exclusion of the Blue Jay of America, and its nearly allied congeners, together with the *Garrulus lanceolatus* of the Himalaya mountains. Thus circumscribed, the true Jays will be found to be exclusively peculiar to the Old World.

The Common Jay of Europe is a noisy, shy, and crafty bird, eluding observation by resorting continually to the more dense parts of woods and thick hedgerows, and is almost entirely arboreal in its habits, seldom going on the ground, and when it does, it is among thickets and bushes, which conceal it from view. Its chief subsistence consists of fruits, berries, and leguminous seeds, while the season lasts, together with the larvæ of insects, worms, grubs, &c., and occasionally the young and eggs of birds. Its propensities render it extremely mischievous in gardens stocked with fruit trees and leguminous vegetables.

The Jay is a permanent resident in our island, as well as in the temperate portions of Europe. It breeds in the most secluded coppices and woods, constructing its nest in the fork of a tree; the nest being formed externally of small twigs, generally of the birch, and lined with fibres, roots, &c. The eggs are four or five in number, of a pale blue, blotched with brown, but the markings are so numerous and minute as to produce a uniform dull grey.

At certain seasons the Jay assembles in small flocks, probably containing the brood of the year, which associate during the winter, until spring leads them to separate into pairs, and commence the great work of incubation.

There exists no visible difference in the plumage of the male and female, and the young at an early age closely assimilates to the adult in colouring. In captivity, this bird becomes a favourite, from its pert and familiar manners, and its aptness in learning words and even sentences.

Bill black, from the base of which a large moustache of the same colour extends over the cheeks; the top of the head is covered with a short full crest, the feathers of which are brownish grey, with a central dash of black, exhibiting as they pass to the occiput faint transverse bars of blue; the whole of the upper surface, as well as the under, is, with the exception of the upper and under tail-coverts (which are white), of a rich vinous or reddish ash colour; wings ornamented with a beautiful blue speculum barred with black; the shoulders chestnut barred with dusky brown; the primaries are silvery white on their outer edges; the secondaries are black, except the first three or four feathers, which are white at their base; tail black, the two middle feathers exhibiting faint indications of blue bars at their base; irides blueish grey; tarsi brown.

The Plate represents the bird of the natural size.

SIBERIAN JAY.

Garrulus infaustus *(Temm.)*

E. Lear del et lith.

Printed by C. Hullmandel.

SIBERIAN JAY.

Garrulus infaustus, *Temm.*

Le Geai imitateur.

THE northern portions of Europe, namely Norway, Sweden, and Siberia, constitute the habitat of this interesting bird, which offers to the naturalist many points for further investigation. With two closely allied species from the north of the American continent, it would seem to form a genus, approximating we admit to that of the true Jays, but still removed from it by certain modifications of character. The general form is less robust, the bill more feeble and shorter, and the feathers more plume-like and disorganized. We are not, however, prepared to institute a new genus, but provisionally assign the present bird a place in that of *Garrulus*. The Siberian Jay, like its American relative (*Garrulus Canadensis*), has a full share of that prying curiosity and imitative qualities which distinguish the race. Its manners are bold and inquisitive, and its actions quick and lively. Confined entirely to the northern latitudes, it is totally unknown in the temperate and southern districts of Europe; and its soft and downy plumage is no doubt well calculated to protect it from the effects of the extreme cold of a Siberian winter; which, as the bird is not migratory, it must in all respects be fitted to endure. In these dreary regions, where the human population is thin and scattered, the Siberian Jay relieves the woods and thickets of part of their loneliness, and attracts the notice of the traveller by its familiarity and restlessness.

Its food consists of wild berries and fruits, to which insects, their larvæ and worms are also added.

Of its nidification little is known; but in this respect we may naturally conclude that it resembles its allied congener the *Garrulus Canadensis*, which is an early breeder, even before the snow is off the ground; constructing a nest of sticks and grass, in a fir-tree in the recesses of the woods, and laying five blue eggs.

The head is covered with a crest of short blackish feathers; those which cover the nostrils, and those also around the base of the beak are yellowish white; the upper surface is olive brown; the shoulders and outer tail-feathers fine rufous; the quills and two middle tail-feathers brown; the throat and under surface of a lighter tint than the back, changing insensibly to a pale rufous, which becomes more decided on the thighs and under tail-coverts; beak and tarsi black. Length eleven inches.

We are not aware that it undergoes any periodical changes in its plumage, which is alike in both sexes.

The Plate represents an adult bird of the natural size.

MAGPIE.

Pica caudata. *(Ray)*.

Drawn from Life & on Stone by J. & E. Gould.

Printed by C. Hullmandel.

Genus PICA.

GEN. CHAR. *Beak* strong, compressed laterally, slightly arched, and hooked at the tip. *Nostrils* basal, open, protected by a covering of bristly feathers directed forwards. *Feet* with three toes before, and one behind, entirely divided. *Tarsus* longer than the middle toe. *Wings* rounded. First *quill-feather* very short; the fourth the longest. *Tail* long and graduated, the two middle feathers proceeding beyond the rest.

MAGPIE.

Pica caudata, *Ray*.

La Pie.

OUR celebrated countryman Ray appears to have clearly appreciated the generic characters of this bird, which he considered sufficiently distinct to warrant his separating it from the genus *Corvus*, to which Linnæus and the naturalists of his school have since referred it. We, however, agree with Ray in considering the difference it exhibits in manners, habits, and general appearance, sufficient to entitle it to be ranked under a separate genus.

The Magpie is one of the most ornamental birds which grace our country; the elegance of its shape, and the glossy black of its plumage, ever varying with reflections of green, contrasted with the purity of the white, render it altogether the most conspicuous bird of our parks and meadows:—bold and spirited, full of life and animation, ever noisy, prying, and inquisitive; the first to give warning of the approach of the fox or hawk, and the first to lead the teasing crowd which collect to harass the marauding intruder. Eminently distinguished by a keen dark eye, an air of cunning, intelligence, and familiar boldness, he has ever been an amusing favourite in captivity; but his propensity for thieving has tarnished his good name. An unwelcome visitor where game is preserved, no bird can be of greater annoyance, or more injurious; one of his favourite objects of search being the eggs of other birds; nor are the unfledged young safe from his attacks. His rapacity however is not confined to the park or the preserve alone, but leads him frequently to venture within the immediate precincts of man, for the purpose of committing depredations on the young broods of domestic poultry. Omnivorous to a great extent, his usual food consists of the larvæ of insects, grubs, snails, and worms; but he does not refuse carrion, grain, or fruits.

This bird is common, not only throughout Europe and the temperate parts of Asia, but also in the United States and the northern regions of America; generally dwelling in pairs throughout the greater part of the year, but congregating in considerable numbers as the breeding season approaches, when they are clamorous and animated, displaying a variety of motions and actions indicating their excitement, and well calculated to show off their plumage and form.

The only difference between the sexes appears to be the rather smaller size of the female.

There is a peculiar circumstance respecting the nidification of the Magpie, which has led to a suspicion among some naturalists that there are in reality two distinct kinds. The fact to which we allude is the different and indeed opposite situation which, without any apparent cause, these birds select for their nests;—in some cases a hedge-row, in others the topmost branches of a lofty tree: but as in their general manner and plumage we can trace no dissimilarity, this circumstance alone, unsupported by others, does not warrant us in making any such distinction.

The degree of art displayed by this bird in the construction of its nest has been noticed by the observers of nature in all ages: it is, indeed, framed and contrived with every attention to security and convenience; not that it is in reality concealed, for its size and situation render it eminently conspicuous.

The nest is externally constructed of sticks and twigs interwoven with great labour, becoming more compact as the building proceeds; within these twigs is disposed an internal coating of mud, and that again is neatly lined with fine grasses. The body of the nest is surmounted by a dome of wickerwork, having an aperture just large enough to admit the parent bird, who generally sits with her head to the hole, ready to quit the nest on the slightest alarm.

The female lays six or seven eggs, mottled all over with ash-brown on a ground of greenish white.

The young soon assume the plumage of the adult, and follow the parent birds till the end of autumn.

Our Plate represents an adult male. The head, throat, neck, upper part of the chest, and back, of a deep black; wing-feathers on the inner webs white, on the outer, shining green; tail-feathers graduated, of a greenish-black, with bronze reflections; scapulars, breast and belly, pure white; beak, irides, legs and feet, black.

AZURE WINGED MAGPIE.

Pica cyanea, *(Wagler)*.

Drawn from Life & on Stone by J & E. Gould *Printed by C. Hullmandel*

AZURE-WINGED MAGPIE.

Pica cyanea, *Wagler.*

It is with great pleasure that we here present, for the first time, a figure of this beautiful and elegant Magpie; a bird which has escaped the notice of most of the authors who have expressly treated on the Ornithology of Europe; and even M. Temminck, who has devoted so much attention to this department of the science, makes no mention of it in a work characterized by accuracy and research. It is only in that useful and little-known book, the "*Systema Avium*" of Dr. Wagler, that any correct notice is to be found respecting it, and even his account is very slight: furnishing us with few details as respects its habits or manners, he merely informs us that it is a native of Spain, arriving in flocks in April, frequenting bushes and willow groves, and is distinguished, like our common species, by its impudence and clamour.

We are personally indebted for the loan of the fine specimen from which our figure was taken, to the liberality and kindness of Captain S. E. Cook, who observed the species to be pretty abundant in the neighbourhood of Madrid, from whence he procured it in a recent state, with several other birds equally rare and valuable.

The beak and legs are black; crown of the head, occiput and ear-coverts, black with shining violet reflections; the whole of the back and rump ashy rose-colour; throat white; the under surface the same as the back, with the exception of its being a few shades lighter; wings and tail delicate azure blue, the primaries, with the exception of the two first which are wholly black, white on their outer web for about half their length from the tip; tail graduated, each feather tipped with white, the two middle ones more obscurely so: total length from twelve to fourteen inches. The sexes do not differ in external appearance.

Our Plate represents an individual in its finest plumage, and of the natural size.

ALPINE CHOUGH.

Pyrrhocorax Pyrrhocorax.

Drawn from Nature & on Stone by J & E Gould. *Printed by C. Hullmandel.*

Genus PYRRHOCORAX.

GEN. CHAR. *Beak* shorter than the head, conical, and somewhat bent towards the tip, with a slight notch at the point. *Nostrils* basal, lateral, and conical, with fine hairs directed forwards. *Tarsi* and *toes* strong and robust. *Nails* strong and hooked. *Wings* long, the fourth and fifth quill-feathers the longest.

ALPINE CHOUGH.

Pyrrhocorax Pyrrhocorax.

Le Choquard des Alpes.

IN all large families like that of the *Corvidæ*, we seldom fail to meet with various anomalous and isolated forms, which appear to stand out from the general group, amalgamating with none of the principal or more numerously filled sections into which the family is divided, but appearing like links of a chain connecting the family with others widely aberrant from it. Though we cannot in every instance trace a due succession of these links, the continuity of the chain being often interrupted, these forms seem like radiations from a given centre, branching out in lines tending in some instances towards even opposite points. The Nutcracker, for example, which belongs to the family of *Corvidæ*, indicates in its form, habits, and manners, an approximation to the *Picidæ* too strong to be overlooked by the discerning naturalist: the Red-legged Chough is by many regarded as tending towards the *Promeropidæ*, while the present bird claims an affinity with some of the *Merulidæ*. In the instances we have here adduced, we may observe that each example is the type and sole known representative of their respective genera with the exception of the Nutcracker, the genus to which it is assigned containing two species.

The natural situations which the Alpine Chough inhabits are the high rude and precipitous elevations of the Alpine districts of central Europe. During the summer it seldom descends far below the line of perpetual snow, but in severe winters it is sometimes driven from its inaccessible heights to the lower mountain ranges, more perhaps in order to obtain food than to avoid the severity of the cold.

Berries, grains, insects, worms, &c., constitute the food of the Alpine Chough; it is, indeed, almost omnivorous in its appetite.

Its nest is usually made in a cleft or fissure of the rock, and sometimes in the chinks of the walls of old buildings among the Alpine heights. The eggs are from three to five in number, of a dull white blotched with yellowish brown.

When adult, the plumage of this bird is of a uniform black; the beak orange; the tarsi and toes vermilion, the under sides of the latter being black; irides dark brown.

Both sexes are alike. In the young of the year the black is less pure; the beak is blackish, the base of the under mandible being yellow; and the tarsi are black. After the first moult the beak becomes yellowish, and the tarsi pass by shades of brown to red, their colour in the female being more obscure.

We have figured an adult of the natural size.

CHOUGH.

Fregilis graculus: *(Cuv.)*

E. Lear del et lithog

Printed by C. Hullmandel

Genus FREGILUS.

GEN. CHAR. *Bill* longer than the head, strong, arched and pointed. *Nostrils* basal, oval, hidden by small closely set feathers. *Head* flat. *Wings* long, first quill-feather short, fourth and fifth the longest. *Tail* square, or slightly rounded. *Feet* strong. *Toes* four, three before, one behind, the outer toe united at its base to the middle one. *Claws* strong, very much curved, that of the hind-toe the largest.

CHOUGH.

Fregilus graculus, *Cuvier*.

Le Pyrrhocorax coracias.

THE Chough is readily distinguished from the true Crows by the peculiar form of the beak: its habits and œconomy, as might be expected, are also somewhat different. In this country the Chough is found on the rocky coasts of Cornwall, Devonshire and Glamorganshire, at the Isle of Anglesea, and the Isle of Man. A few pairs may be seen about the high cliffs between Freshwater-gate and the Needle rocks of the Isle of Wight. In the North, they frequent the high and rocky coast about St. Abb's Head, and most of the islands of Scotland, where they breed at high elevations. The Swiss Alps and rocky portions of the most lofty mountains of the European continent, as well as the Himalaya, are among its favourite localities.

In such elevated situations, the strong toes and large curved claws of this bird are of essential service, in securing for it a firm hold against the rugged and perpendicular surface of the highest cliffs, among the inequalities of which it forms a nest of sticks lined with wool and hair, in which it deposits three or four eggs, not very unlike those of the Jackdaw, but longer, of a greenish white ground spotted with darker green and ash-brown. These birds are also said to build about the upper parts of high churches and towers near the coast. Their food consists of insects principally, with grain and berries.

The Chough is lively, restless, noisy and cunning, easily attracted by showy or glittering substances; is tamed without difficulty if taken young, and exhibits under confinement a variety of amusing tricks and actions.

The whole plumage is black, glossed with purple, green, and dark blue; the irides hazel; beak and legs vermilion red; claws black. Young birds of the year have their plumage dull black, with a bill less brilliant.

We have figured an adult male rather less than the natural size: the females are rather smaller.

RAVEN.

Corvus corax; *(Linn.)*

E. Lear del. et lithog.

Printed by C. Hullmandel.

Genus CORVUS.

Gen. Char. *Bill* strong, conical, cultrated, straight at the base, but bending slightly towards the tip; *nostrils* at the base of the bill, oval and open, covered by reflected bristly feathers. *Wings* pointed; the first feather being much shorter than the second and third, and the fourth the longest. *Legs* and *feet* strong, plated, with three toes before and one behind. *Claws* strong and curved. *Toes* divided. *Tarsus* longer than the middle toe.

RAVEN.

Corvus corax, *Linn.*

Le Corbeau noir.

The Raven is so extensively diffused, and is in consequence so universally known, that the name at once reminds us of its general character. The largest and strongest of its genus, and bold as well as cunning, it is always an object of suspicion to shepherds and husbandmen, from its daring attacks upon the young or weak among their flocks and herds, and in times of superstition was regarded as a bird of ill omen, its hoarse croaking being supposed to announce some impending calamity.

With a quick, searching eye, and a keen sense of smell, the Raven is ever on the watch to satisfy his appetite, and no sooner does the defenceless state of an animal, and the absence of the herdsman, afford a chance of success, but the Raven is there upon the ground. At first he makes his approach obliquely and with great caution. He is shy of man and of all large animals in motion, because, as it has been aptly observed, though glad to find others' carrion, or to make carrion of them if he can do it with impunity, he takes good care that none shall make carrion of him. If no interruption occurs, he makes his first attack upon the eye, afterwards feeds at his leisure, retires to a small distance to digest his meal, and then returns again.

The Raven is met with in almost every part of the globe. Rocks on the sea shore, mountain ridges and extensive woods are its most usual haunts: and are all equally favourable to its habits, occasionally it visits open plains and large fields, especially when they are used as pasture. Like the other birds of this genus the raven is not particular in selecting food, but eats indiscriminately small mammalia, eggs, reptiles, dead fish, insects, grain and carrion; they have also been seen feeding their young out of the nests of a rookery.

The male and female are frequently observed together, and they are said to pair for life. There is no difference in the plumage of the sexes, and they are subject to only one moult. They build on high trees, or if near the shore, in the crevices of the most inaccessible parts of rocks, and use the same nest, formed of sticks, wool and hair, for years in succession. The eggs, four or five in number, of a blueish green blotched with brown, are produced very early in spring. The female during incubation, which lasts about twenty days, is regularly attended and fed by the male bird, who not only provides her with abundance of food, but relieves her in turn, and takes her place on the nest. The young birds are driven away as soon as they are able to provide for themselves. If taken young, the Raven is easily domesticated, and becomes very tame and familiar, imitating different sounds correctly, and has often been taught to pronounce a variety of words distinctly. They are also noted for carrying away and hiding pieces of polished metal.

The whole of the plumage is black, the upper part glossed with blue; feathers on the throat narrow and pointed; tail rounded at the end; beak, legs and toes black; claws black, strong and curved.

Our figure represents an adult bird, one fourth less than the natural size.

CARRION CROW.

Corvus corone, *(Linn)*

E. Lear del et lith.

Printed by C. Hullmandel.

CARRION CROW.

Corvus Corone, *Linn.*

La Corneille noir.

We are induced to believe that the range of habitat of this well-known species is not so extensive as is generally supposed, but that most of the birds received from distant countries, although very similar, are specifically distinct not only from the Carrion Crow of Europe but also from each other, and that although these differences are not apparent to the casual observer, they will be found on a critical examination to be sufficiently important.

The Carrion Crow is very generally distributed over the British Islands, where it is a permanent resident; it also appears to be equally dispersed over the western portion of the European continent, but is rarely found so far east as Hungary and many parts of Austria. In its habits, manners, and general economy the Carrion Crow is nearly allied to the Raven; like that bird it wanders about in pairs, evincing the greatest wariness of disposition and shyness on the approach of man, which may, however, be partly attributed to the persecution it meets with from almost every one. The Crow is a more powerful and robust bird than the Rook, from which it may readily be distinguished by the greenish metallic hue of its plumage, and by its thickened and more arched bill, which is never deprived of the bristly feathers that cover the face and nostrils. It is also clearly destined by nature to fulfill a very different office; for, while Rooks congregate in immense flocks and disperse themselves over cultivated districts in search of insects, grubs, and grain, the Crow, as before observed, wanders about in solitary pairs, or at most in parties of six or eight, in search of all kinds of carrion, upon which it feeds voraciously; and hence it may be frequently observed on the banks of the larger rivers, which constantly afford it a supply of putrid animal matter; to this kind of food are occasionally added eggs, the young of all kinds of game, and it is even so daring when pressed by hunger as to attack very young lambs, fawns, &c. When once mated, it would appear that Crows never again separate, and if unmolested in their chosen breeding-place, the same pair generally return every year not only to the same locality but to the same tree. The nest is usually placed in a fork near the bole, is of a smaller size than that of the Rook, and is constructed of sticks and mud, lined with wool and hair. The eggs are five or six in number, of a greenish ground, blotched all over with thickly set patches of ash-coloured brown.

It is perhaps one of the most destructive birds the preserver of game has to contend with, and in consequence the poor Crow being sadly persecuted uses the utmost vigilance and cunning to evade the pursuit of his great enemy the gamekeeper.

The sexes offer no difference in the colour of the plumage, and they assume the full colouring from the nest.

The whole of the plumage is black, the upper surface being glossed with blue and greenish reflexions; bill, legs, and feet black, the scales on the two latter being in laminæ, or plates.

The figure is of the natural size.

HOODED CROW.

Corvus cornix, *(Linn.)*

Drawn from Nature & on Stone by J. & E. Gould. *Printed by C. Hullmandel.*

HOODED CROW.

Corvus cornix, *Linn.*

La Corneille mantelée.

This fine species of Crow is not indigenous to England, but is now ascertained to be a permanent resident in many districts of Scotland, where, according to Mr. Selby, it breeds in trees, rocks, or sea cliffs, as may accord with their situation, the nest being formed of sticks, and lined with soft materials. That gentleman further informs us that in those districts where it is found, there is no diminution of its numbers during the winter months; and we may reasonably conjecture that those individuals who pay their annual visit to the midland and southern counties of England during the autumn months are accessions from Norway and Sweden: we are strengthened in our opinion upon this point from the circumstance of their appearing at the same time as the Woodcock and many others of our Northern visitors. Although the Hooded Crow is plentifully dispersed over many districts in England, it must be allowed that its choice of places is extremely local: it frequents the shores of the sea, the banks of large rivers, extensive downs, and such arable lands as are devoid of hedgerows and woods. On the Continent it may be observed in all the mountainous districts. It is common in the Alps and Apennines, but nowhere more so than in Norway and Sweden. In its habits and manners it bears a strict resemblance to the Carrion Crow: like that bird it wanders about in pairs, or at the most three or four together. Their omnivorous appetite enables them to subsist upon all kinds of carrion, which they devour with avidity. Those that take up their positions upon the coast or about armlets of the sea find a plentiful supply in the remains of dead fish and crustacea, to which are added worms and various species of mollusca. In the inland districts they eat worms, beetles, and whatever offal may fall in their way.

The Hooded Crow is abundantly dispersed along the banks of the Thames, and all such rivers as are under the influence of the tides.

During the period of incubation they are said to be very destructive to the eggs and young of the Red Grouse, and will even attack lambs and sheep. The eggs are four or five in number, of a greenish ground colour, mottled with dark brown.

The sexes are alike in plumage, and the young attain at an early age the colouring of their parents.

The head, throat, wings, and tail are black, with purple and green reflections; the remainder of the body is smoky grey, the shafts of each feather being darker; legs and bill black; irides dark brown.

The Plate represents an adult male of the natural size.

JACKDAW.

Corvus monedula; *(Linn.)*

Drawn from Nature & on Stone by J. & E. Gould.

Printed by C. Hullmandel.

JACKDAW.

Corvus monedula, *Linn.*

Le Choucas.

The Jackdaw, still more bold and familiar than the Rook, which approaches so near the residence of man during the period of incubation, advances under the very roofs of our dwellings, as if to solicit for itself and its sooty progeny some especial care and protection; it also lives in towers, old castles, and deserted ruins, the loneliness of which it enlivens with its noisy animated actions and gregarious habits.

Its range of habitat, although not equal to that of the Raven, is nevertheless widely extended, the bird being dispersed over every part of Europe, and the contiguous portions of Asia and Africa.

During the seasons of autumn and winter, the Jackdaw associates with the Rook, in whose society it appears to dwell in amity, feeding with it by day and retiring with it at night to the rookery or the accustomed roosting-place. On the approach of spring it separates from the Rook, and again bends its way towards its favourite place of incubation. Independently of the situations alluded to, it nestles in rocks and the holes of trees, and in some instances in rabbit-holes in the ground. The nest is composed of sticks and lined with wool: the eggs are four or six in number, of a pale greenish blue spotted all over with blackish brown.

Omnivorous in its appetite, the Jackdaw feeds on fruits, pulse, and grain, to which are added, grubs, snails, worms, and even carrion. In its disposition it is thievish and mischievous: easily domesticated and familiar, it may be taught to articulate words with distinctness.

The sexes are alike in the colouring of their plumage, and do not undergo any change either in winter or summer.

The young during the first year are more uniform in their colouring than the adult: the silvery grey of the head and neck is not attained until the bird is three or four years old.

The adult has the top of the head black, with violet reflections; back part of the head and neck silvery grey, the feathers of these parts being long and silky; the whole of the upper surface greyish black, the primaries and secondaries having blue and violet reflections; feet and bill black; irides greyish white.

The Plate represents an adult male and female of the natural size.

ROOK.
Corvus frugilegus; (Linn.)

J. Lear del et lith.

Printed by C. Hullmandel

ROOK.

Corvus frugilegus, *Linn.*

Le Freux.

This familiar bird appears to be distributed over the greater part of Europe, giving preference to those cultivated portions which afford it a supply of granivorous food, upon which it partially subsists, and for which it is generally condemned by the husbandman as an injurious and destructive neighbour; though, were the habits of the Rook carefully investigated, we doubt not it would be satisfactorily proved that he amply repays the farmer for the few grains he steals, by the destruction of immense numbers of grubs and insects which he devours in the course of a single year, thus rather claiming our gratitude for his services than deserving our enmity: it must be acknowledged, too, that its presence helps to enliven our fields and pastures.

The Rook is very fastidious in its choice of a place for performing the duties of incubation, frequently leaving the trees of the forest for those situated near our dwellings, and, in some instances, even taking up its abode in the midst of towns and cities.

The adult Rook may at all times be readily distinguished from its near ally the Crow by the naked face and gular pouch, which parts have been divested of their feathers by the constant thrusting of its mandibles into the earth in search of food; its wings are also more lengthened and pointed, and the hue of the upper portion of its plumage is more inclined to purple.

The Rook is gregarious, and in no country is to be observed in greater numbers than in the British Islands, which afford it an asylum congenial to its peculiar habits and mode of life. It commences the work of nidification in the month of March, constructing a large nest of sticks, lined with a coating of clay and fine grasses. The eggs are five in number, of a blueish green blotched with darker stains of brown. The young for the first ten or twelve months do not lose the feathers which cover the nostrils; and during this period they so nearly resemble the Crow that a more than usually minute examination is required to discover the difference, though a careful attention to the peculiar form of the bill will obviate any difficulty.

The sexes are so strictly similar in the colouring of their plumage that actual dissection is requisite to distinguish them.

Bill and feet black; the whole of the plumage black glossed with changeable hues of green and violet purple; feathers on the back of the neck long and filamentous.

The Plate represents an adult, rather less than the natural size.

GREAT BLACK WOODPECKER.

Picus martius, *(Linn)*.

Drawn from Life & on Stone by J. & E. Gould. *Printed by C. Hullmandel.*

Genus PICUS.

GEN. CHAR. *Beak* long, straight, pyramidal; and cutting towards the point. *Nostrils* basal, oval, inclosed by membrane; covered with hairs, directed forwards. *Tongue* long, taper, capable of protrusion, armed with a horny tip. *Toes* four; in pairs; antagonizing; the front pair united at their base. *Tail,* twelve feathers; graduated short; shafts stiff and elastic. *Wings,* third and fourth *quill-feather* longest.

GREAT BLACK WOODPECKER.

Picus Martius, *Linn.*

Le Pic noir.

THE Woodpeckers form a family more numerous perhaps than any other in the whole range of Ornithology; and, if we except Australia and the South Sea Islands, are equally extended over the old and new portions of the globe. Abundant, however, as the species may be, they are so united by a pervading similarity of habits, food, manners, and even colouring, as to constitute a group pre-eminently natural and well defined;—hence the description of one species is to a great extent applicable to all: still, however, as is the case in all natural families, differences sufficiently characteristic exist to warrant a subdivision into groups more or less typical.

Among the true or typical Woodpeckers, may be placed the *Picus Martius;* at once exhibiting the generic characters in their highest degree of developement, it exceeds in size all its congeners of the Old World, and indeed is inferior only to the Ivory-billed Woodpecker of the United States of America.

However plentiful it might have been when our Island was less cultivated than at present, and covered with extensive forests, certain it is that this bird is now so seldom to be met with, if at all, as scarcely to come under the designation of a British species. According to M. Temminck it is rare even in France and Germany, and must be sought for in the more northern regions of Europe, as Norway, Sweden, Poland, Russia, and also Siberia, to which in the present day its habitat is almost entirely confined.

At the head of a family of true Climbers, the habits of the Great Black Woodpecker are in conformity with its wants and its means of supplying them. We need hardly say that it is on the bark of trees more exclusively that the Woodpecker finds its food, and to this end are its powers and organs adapted. If we examine the toes of the present species, which are to be taken as illustrative of form in the whole of the family, (with the exception of a single limited group,) we find them long and powerful, furnished with strong claws, admirably adapted for grasping or clinging to the rough inequalities of the bark: besides this, they are placed in pairs, so as in some measure to antagonize; but not, as generally stated, two before and two behind, for one pair is lateral, and diverges from the other at an acute angle, so as to be applied to the convexity of the tree, and thus render the grasp close and firm. The tail is composed of stiff feathers, the shafts of which taper gradually from the base to the extremities, which curving inward when pressed against a tree, not only form a fulcrum for the support of the body, but by their elasticity tend to propel it forwards. This provision, the more needed from the posterior situation of the legs, is admirably calculated for ascending; and having explored the bark by a spiral course, the Woodpecker flies off to the next tree, to repeat the same process.

The flight of the present species is undulating, seldom protracted to any extent, but limited to a transit from tree to tree in the seclusion of its native woods.

Its food consists of the larvæ of wasps, bees, and other insects: in addition, however, it devours fruits, berries, and nuts with avidity.

The female selects the hollows of old trees, in which she deposits two or three eggs of an ivory whiteness.

The two sexes differ but little in plumage,—the crimson crown distinguishing the male, that colour being in the female confined to the occiput; the rest of the plumage is a deep jet-black; the irides yellowish-white; the naked circle round the eye and the feet black; the bill horn-colour, black at the tip.

The young males are characterized by the irides being of a light ash colour; the crown of the head is marked with alternate spots of red and black, which give place gradually to the bright uniform crimson of maturity.

The length of an adult bird is about fifteen inches; and our Plate represents a male and female of their natural size.

GREEN WOODPECKER.

Picus viridis; *(Linn.)*

Drawn from Life & on Stone by J. & E. Gould.

Printed by C. Hullmandel.

GREEN WOODPECKER.

Picus viridis, *Linn.*

Le Pic vert.

The present bird represents a group of the great family of *Picidæ* or Woodpeckers, which appears to hold an intermediate station between the species of the American genus *Colaptes*, distinguished by their slender arched bills and terrestrial habits, and those which exhibit a closer approximation to the typical form, whose habits, manners, and food, confine them entirely to trees.

The present group appears to contain about eight or ten well-marked species, all peculiar to the old continent, but of which number only two, viz. the *Picus viridis* and *Picus canus*, Linn., are common to Europe, where they appear to fill the same relative situation that the species of the genus *Colaptes* do in America.

This familiar and well-known bird is not only frequent in every part of Great Britain, but is equally spread throughout the whole of Europe, with the exception of the marshy and low lands of Holland, frequenting woods and forests, where its presence may be generally discovered by its clamorous note, or its restless disposition in proceeding from tree to tree in search of insects. This kind of food it takes by inserting its long and retractile tongue into the crevices of the bark in which they lodge, but is not less frequently seen on the ground in search of ants, snails, worms, &c., nor will it refuse fruits, walnuts and berries. It deposits its eggs,—which are of a smooth shining white, and from four to six in number,—in the holes of trees partially occasioned by decay and enlarged by its own exertion. The Green Woodpecker remains with us the whole of the year, and having attained its adult stage of plumage undergoes no subsequent variation. The top of the head, the occiput, and moustache or stripe on the cheek, are of a brilliant red ; the face black ; the upper surface fine green ; the rump tinged with yellow ; the under parts pale greyish green ; quill-feathers brown, crossed with bars of yellowish white ; tail brown, barred transversely with a lighter colour ; bills and legs greyish green ; irides white.

The female differs from the male externally only in being rather less in size and in the absence of the red moustache, which colour is supplied by black.

The young have only traces of red on the head ; the moustache is indicated by black and white feathers ; the general colour is paler and more obscure, the back being marked with ash-coloured blotches, and the under parts with brown zigzag bars ; irides dark grey.

We have figured an adult male and a young bird in the plumage of the first autumn.

GREY HEADED GREEN WOODPECKER.

Picus canus, *(Gmelin)*.

Drawn from Life & on Stone by J & E Gould

Printed by C. Hullmandel.

GREY-HEADED GREEN WOODPECKER.

Picus canus, *Gmel.*

Le Pic cendré.

We have reason to think with M. Temminck, that the present species has been often confounded with the *Picus viridis*; but it may be distinguished by its rather smaller size, and the grey colour of its head, the red mark on the top of which is more circumscribed, while in the female it is entirely wanting. In general habits, however, the two species are altogether similar, and may be taken as examples of a group including some species from the Himalaya mountains, and other parts of the old world, which present a departure in several characteristics from the more typical Woodpeckers, of which we have given the *Picus martius* as a representative. The subjects of this group appear more terrestrial in their habits, searching for their food on the ground, and less exclusively confined to the trees as climbers. We shall not, however, here enter into the details of these dissentient peculiarities, which occupy a more prominent place in the description of the *P. viridis*, but confine ourselves to the bird before us.

The native localities of the Grey-headed Green Woodpecker would appear to be Norway, Sweden, Russia, and, more or less, the whole of the northern portions of Europe; and, as we are informed by M. Temminck, is also an inhabitant of the northern parts of America. Although it may be deemed presumption to doubt the assertion of so great a naturalist, we cannot help expressing our belief that neither this bird nor any of the species of the group to which it strictly belongs are to be met with on any part of that continent; its place there being occupied by a genus similar in habits and manners, to which the title of *Colaptes* is assigned, and which possesses essential external differences. In France and Switzerland it is very scarce; and we believe is never found in Holland or in the British Islands.

Its food, like that of the Green Woodpecker, consists of insects in general, more especially ants and their larvæ, occasionally feeding on fruits and nuts. Its nidification is also the same; the female depositing four or five eggs of a pure white in the hollow of a tree.

The beak is greenish-yellow, becoming dark at its edges; legs black; irides very light red; forehead crimson; a black mark extends between the eye and the beak; the occiput and space between the crimson forehead and eyes grey; on the cheeks, which are cinereous with a slight tinge of green, a narrow black line extends downwards; the back bright green; upper tail-coverts yellowish; wings olive-green; quill-feathers darker, with distinct yellowish-white spots along their outer edges; tail dark olive-brown, the two middle feathers having traces of obscure transverse bars; the under parts, like the cheeks, cinereous tinged with green.

The general colour of the female is the same as in the male, with the exception of the head, the crown of which is entirely grey; the black mark between the eye and beak less apparent, and that on the cheeks smaller.

The young males at a very early age, even while in the nest, are to be known by the crimson forehead and the black mark on their cheeks; but the young females at this period have no trace of these lines, and they are not acquired till some time afterwards.

Length between eleven and twelve inches.

Our Plate represents a male and female of the natural size, in their adult plumage.

WHITE-RUMPED WOODPECKER.

Picus leuconotus: *(Temm:)*.

Drawn from Nature & on Stone by J & E. Gould. *Printed by C. Hullmandel.*

WHITE-RUMPED WOODPECKER.

Picus leuconotus, *Bechst.*

Le Pic leuconote.

Of the group of true Woodpeckers distinguished by the alternate black and white of their plumage, the present is by far the largest and most handsome; we may also add, that it is the rarest, never occurring in England, nor in the southern portions of the European continent.

Its true habitat appears to be Siberia and the adjacent parts of Russia, whence it occasionally emigrates as far as the North of Germany; but this is only in severe winters.

This fine species is so little known even to naturalists, that we are not aware of any other specimens in England besides the examples we possess, a circumstance to be attributed to the little intercourse kept up with the remote and desolate region where it abides. It is there, doubtless, by no means uncommon; and we think we may venture to affirm, without fear of contradiction, that there exist in that country many birds and quadrupeds of which there is no record in the pages of science.

The *Picus leuconotus* does not appear to offer any difference from its congeners in its general habits and manners. It is said to be partial to woods of the latest growth, but, according to M. Temminck, is not found in the depths of the pine-forests. Its food consists of various insects and their larvæ. It incubates, as is usual with these birds, in the holes of trees without forming any nest; the eggs are white, and four or five in number.

It may always be distinguished from the Greater Spotted Woodpecker of England by the blotches along the flanks, by the pure white of the rump and the more extended crimson of the abdomen.

The male has the crown of the head crimson; the forehead yellowish white; the cheeks, back of the neck, rump, and chest white; a black moustache stretching from the base of the bill to the occiput; and the sides of the chest and flanks marked with longitudinal black dashes; the upper part of the back, shoulders, and middle tail-feathers black: the wings and outer tail-feathers barred with black and white; the abdomen and tail-coverts crimson; the irides red.

The female has the crown of the head black instead of crimson; in other respects she resembles the male.

We have figured a male and female of the natural size.

GREAT SPOTTED WOODPECKER.

Picus major; (Linn.)

Drawn from Nature & on Stone by J & E Gould. Printed by C. Hullmandel.

GREAT SPOTTED WOODPECKER.

Picus major, *Linn.*

Le Pic épeiche.

This familiar species of the group of Spotted Woodpeckers enjoys a range of habitat more extensive, perhaps, than any other of its European relatives, there being no wooded districts, especially in the central portions of Europe, where it is not extremely common. In England it abounds in forests, woods, large parks, and gardens. The group to which it belongs, although occasionally descending to the ground in search of food, are far more arboreal in their habits and manners than the Green Woodpeckers represented by the *Picus viridis, caniceps,* and several others from the Himalayan mountains. They exhibit great dexterity in traversing the trunks of trees and the larger decayed limbs in quest of larvæ and coleopterous insects which lurk beneath the bark, and to obtain which they labour with great assiduity, disengaging large masses of bark, or so disturbing it by repeated blows as to dislodge the objects of their search. Besides searching trees of the highest growth, they are observed to alight upon rails, old posts, and decayed pollards, where, among the moss and vegetable matter, they find a plentiful harvest of spiders, ants, and other insects ; nor are they free from the charge of plundering the fruit-trees of the garden, and in fact commit great havoc among cherries, plums, and wall-fruits in general.

Their flight is rapid and short, passing from tree to tree, or from one wood to another, by a series of undulations. In their habits they are shy and recluse, and so great is their activity among the branches of trees, that they seldom suffer themselves to be wholly seen, dodging so as to keep the branch or stem between themselves and the observer.

The sexual differences in plumage in most of the Woodpeckers consist in a difference of colour or marking about the head, the males and females resembling each other in every other respect. The male of the present species, it will be observed, is only to be distinguished by a narrow occipital band of scarlet. It is somewhat singular, however, that the young of both sexes, for the first three or four months of their existence, have the whole of the brow scarlet (as may be seen on referring to the Plate), and in this state so closely resemble the *Picus medius* as to have been mistaken for that bird, a circumstance which has led to the supposition that the *P. medius* was indigenous to this country, whereas it is strictly confined to the Continent.

We need hardly say that the *Picus major* resembles its congeners in its mode of nidification and in the colour of its eggs, which are of a glossy whiteness. They are deposited in the hole of a tree, often excavated and enlarged to a considerable depth; generally producing four or five young, which, with the exception of the crown of the head, as before noticed, resemble their parents in their colours and markings.

The top of the head, a line from the base of the bill descending down the sides of the neck, the back of the neck, mantle, rump, and four middle tail-feathers are black ; wings blackish brown with irregular bars of white ; forehead brownish white ; cheeks, spots on the lower part of the sides of the neck, the scapularies, and under surface white, the latter having a tinge of brown, especially on the abdomen ; the occiput in the male and the under tail-coverts in both sexes scarlet ; the outer tail-feathers white, with one or more imperfect lines of black ; bill dark horn-colour ; tarsi deep lead-colour ; irides purplish red.

The Plate represents a male and female, and their young, of the natural size.

MIDDLE SPOTTED WOODPECKER.

Picus medius, (Linn.)

Drawn from Nature & on Stone by J & E Gould.

Printed by C. Hullmandel.

MIDDLE SPOTTED WOODPECKER.

Picus medius.

Le Pic mar.

THE present bird is more intimately allied to the *Picus leuconotus* than to any other species yet discovered: it is, however, although a third smaller than the Common Woodpecker of England, known under the name of *Picus major*, with the young of which the present species has often been confounded; this error has arisen from the circumstance of the young individuals of the *P. major* having the whole of the crown of the head scarlet. In fact, the plumage which characterizes both sexes of that species for the first four or five months of their existence would answer minutely to that of the present species when mature, with the exception of the longitudinal dashes on the breast, which are wanting in *P. major*, and which are always present in *P. leuconotus* and *P. medius*. It is somewhat singular that the female of the present bird, unlike most others of its genus, so nearly resembles the male, as in most instances to be scarcely distinguishable except by internal examination.

The *Picus medius* inhabits the borders of woods, parks, and gardens, and, although never found in the British Islands, is very abundant in many parts of the Continent, especially the southern provinces. M. Temminck states that it is very seldom found in Holland, and we have never seen it from the North of Europe; neither have the collections from Africa or Asia, as far as we have examined, afforded a single example of this bird. Its food consists of insects, which it takes solely from the sides or trunks of trees, the crevices of which it diligently examines in search of them: besides insects, it also feeds during the season on various fruits and berries. It lays its eggs, which are of a glossy white, in the holes of trees.

The colouring of the plumage is as follows: A frontal band of ash colour occupies the space between the bill and the crown of the head, which is scarlet, the occiput being furnished with somewhat elongated feathers of the same colour; the neck and chest are white, with the exception of an obscure band, which passes down the sides of the neck; the back and wings are black, the scapularies being white, and the quills marked with bands of the same colour; the flanks are rich rose colour, with longitudinal blotches of black occupying the centre of each feather; under surface and vent crimson; the middle feathers of the tail black; the outer ones white, banded with black; irides reddish brown, encircled by a light ring; tarsi lead colour.

The Plate represents a male and female of the natural size.

LESSER SPOTTED WOODPECKER.
Picus minor; *(Linn.)*

Drawn from Nature & on Stone by J & E. Gould.

Printed by C. Hullmandel.

LESSER SPOTTED WOODPECKER.

Picus minor, *Linn.*

Le Pic épeichette.

The present elegant species of Woodpecker, which has received from the older ornithologists the specific title of *minor*, is, indeed, the least of all the European Woodpeckers; but if we include India and other portions of the globe, we find species considerably smaller, rendering its appellation erroneous, unless we consider it as strictly in reference to its European congeners. As far as we have been able to ascertain, this portion of the globe forms the restricted habitat of this species, over the whole of which, however, it is pretty generally distributed, confining itself to the precincts of woods, parks, and orchards. In England it is far more abundant than is generally supposed; we have seldom sought for it in vain wherever large trees, particularly the Elm, grow in sufficient numbers to invite its abode: its security from sight is to be attributed more to its habit of frequenting the topmost branches than to its rarity. Near London it is very common, and may be seen by an attentive observer in Kensington Gardens, and in any of the parks in the neighbourhood. Like many other birds whose habits are of an arboreal character, the Lesser Spotted Woodpecker appears to perform a certain daily round, traversing a given extent of district, and returning to the same spot whence it began its route. Besides the Elm, to which it is especially partial, it not unfrequently visits orchard-trees of large growth, running over their moss-grown branches in quest of the larvæ of insects, which abound in such situations. In its actions it is very lively and alert. Unlike the Large Woodpecker, which prefers the trunks of trees, it naturally frequents the smaller and more elevated branches, which it traverses with the utmost ease and celerity: should it perceive itself noticed, it becomes shy, and retires from observation by concealing itself behind the branch on which it rests; if, however, earnestly engaged in the extraction of its food, its attention appears to be so absorbed that it will allow itself to be closely approached without suspending its operations. When spring commences, it becomes clamorous and noisy, its call being an oft-repeated single note, so closely resembling that of the Wryneck as to be scarcely distinguishable from it. At other times of the year it is mute, and its presence is only betrayed by the reiterated strokes which it makes against the bark of trees.

Like the rest of its genus, it deposits its eggs in the holes of trees; the eggs being four or five in number, and pure white.

The sexes offer no other difference than that the female has the crown of the head white, whereas in the male it is of a fine scarlet.

The young attain the plumage of adults immediately after they leave the holes in which they were reared.

The adult male has the crown of the head scarlet; the cheeks, stripe over the eye, sides of the neck, and under parts dull white; an irregular black band passes from the beak down the sides of the neck; back of the neck, upper part of the back, rump, and middle tail-feathers black; wings and centre of the back barred with black and white; outer tail-feathers white, obscurely barred with black; faint longitudinal dashes of the same colour are also observable on the breast.

The Plate represents an adult male and female of the natural size.

THREE TOED WOODPECKER.

Picus tridactylus, *(Linn.)*

Drawn from life & on stone by J & E. Gould. *Printed by C. Hullmandel.*

THREE-TOED WOODPECKER.

Picus tridactylus, *Linn.*

Apternus tridactylus, *Swains.*

Le Pic tridactyle.

Mr. Swainson has applied the generic term *Apternus* to this Three-toed Woodpecker; and we refer the reader to the second volume of the North American Zoology, page 301, for a full explanation of the views of the scientific author in his systematic arrangement of this most extensive and characteristic family. The principal distinguishing feature of this genus is the absence of the hind-toe; a deficiency, however, which does not occasion any very material difference in the habits of the bird before us, which bear a close resemblance to those of the typical group.

The present species is by no means uncommon in the northern parts of the European Continent, the vast forests of the mountainous parts of Norway, Sweden, Russia and Siberia, forming its principal habitat; it is also found among the Alps of Switzerland, is but an accidental visitor in France and Germany, and has never been taken, we believe, in the British Islands.

It subsists, like most of the Woodpeckers, on insects and their larvæ, as well as fruits and various wild berries. It chooses holes in trees for its breeding place, which if too small it readily enlarges, the female laying four or five eggs of a pure white.

The male and female present the usual differences of colour which characterize the family.

In the male, the forehead is variegated with black and white; the top of the head is golden yellow; the occiput and cheeks glossy black; from the base of the bill a black stripe extends to the chest, between which and the eye runs a bar of white; a narrow white line also extends to the occiput from behind the eye; throat and chest white; back, sides and under parts barred with black and white, the bars of the under surface being more regular though the black is less deep; wings brownish black, with white spots on the quill-feathers; the four middle tail-feathers black, the rest alternately barred with black and white; the upper part of the tarsi covered with feathers; the superior mandible brown; the inferior dirty white, as far as the point; irides obscure blue. Length nine inches.

In the female, the top of the head is of a glossy or silvery white, interspersed with fine black bars. The rest of the plumage the same as in the male.

In very old males the yellow of the head is more bright, and the white of the under parts predominates, but never loses the black transverse bars.

Our Plate represents a male and female of their natural size; the generic name *Apternus*, Swains., being inadvertently omitted.

WRY NECK.

Yunx torquilla *(Linn.)*

Drawn from Life & on Stone by J. & E. Gould.

Printed by C. Hullmandel.

Genus YUNX.

Gen. Char. *Beak* short, straight, conical, produced towards the point: edges of the *mandibles* smooth. *Nostrils* situate at the base, partly closed by membrane. *Toes,* two before, united at their origin; *hind toes* two, disjoined. Second *quill-feather* the longest.

WRYNECK.

Yunx Torquilla, *Linn.*

Le Torcol.

Closely allied in form to the numerous tribe of Woodpeckers, it is not a little singular that the species comprehended in the genus *Yunx* should be limited to two only, of which, until lately, the present bird was alone recognised. Our acquaintance with a new species from South Africa, deposited in the Museum of the Zoological Society, is due to the acumen of N. A. Vigors, Esq., who has introduced it to science under the name of *Yunx pectoralis,* from the distinguishing rufous colour of the breast. The Wryneck derives its name from its peculiar habit of elongating the neck, which at the same time it writhes from side to side with serpent-like undulations, now depressing the feathers so as to resemble the head of a snake, and again half-closing the eyes, swelling out the throat, and erecting its crest, when it presents an appearance at once singular and ludicrous.

Among our most interesting and attractive birds, this little harbinger of spring delights us, not by the splendour of its hues, but by the chasteness of its colouring, and by the delicate and unique dispositions of its markings, which from their intricacy and irregularity almost defy the imitative efforts of the pencil.

Among our migratory birds the Wryneck is one of the earliest visitors; arriving at the beginning of April, generally a few days before the Cuckoo (whose mate, from this circumstance, it has been called), when his shrill monotonous note, *pee pee pee,* rapidly reiterated may be heard in our woods and gardens. The localities of this bird appear to be very limited; the midland counties being those to which it usually resorts in England. M. Temminck informs us that it is seldom found beyond Sweden, and is rare in Holland, occupying in preference the central portions of Europe. We are able to add to this information, by stating that it is abundant in the Himalaya Mountains, whence we have frequently received it as a common specimen of the ornithology of that range, with other birds bearing equally a British character.

In manners, the Wryneck is shy and solitary; and were it not for its loud and well-known call, we should not often be aware of its presence; its unobtrusive habits leading it to close retirement, and its sober colour, which assimilates with the brown bark of the trees, tending also to its concealment.

In confinement, however, or when wounded, this little bird manifests much boldness; hissing like a snake, erecting its crest, and defending itself with great spirit.

It breeds with us soon after its arrival, the female selecting the hole of a tree, in which she deposits her eggs, to the number of eight or nine, of an ivory white. The young soon assume the plumage of the parent birds, which exhibits scarcely any sexual differences.

The food of the Wryneck, like that of the weaker-billed Woodpeckers, consists of caterpillars and other insects, especially ants and their larvæ, to which it is very partial. In the manner of taking its food this bird makes but little use of the bill itself; its long cylindrical tongue, capable of being protruded to a considerable distance, and lubricated with an adhesive mucus, for the secretion of which an extensive salivary gland is provided, being the chief instrument. This it inserts between the crevices of the bark, or among the loose sandy earth of the ant-hill, protruding and withdrawing it so rapidly, with the insect adhering, as almost to deceive the eye.

Leaving England in the early part of the autumn, the Wryneck passes over to the southern districts of Europe, and probably extends its journey to Asia, where it finds a genial climate, and food still abundant.

The prevailing colour of this elegant little bird consists of different shades of brown, inclining to gray on the head, the rump, and the tail, but of a bright chestnut on the larger wing-coverts and the primaries; the whole beautifully variegated with delicately shaped markings of a deep brown, which give it a mottled appearance. Breast wood-brown, penciled with slender transverse tracings; abdomen dirty white, speckled with minute dark triangular spots; bill yellowish-brown; irides chestnut; feet and legs flesh-coloured.

The annexed Plate represents the male and female of their natural size; the latter in the act of leaving the hole in the tree, in which we may suppose her to have formed a nest.

COMMON NUTHATCH.
Sitta Europea; *(Linn.)*

Drawn from Nature & on Stone by J & E. Gould. *Printed by C. Hullmandel.*

Genus SITTA.

Gen. Char. *Bill* straight, cylindrical, slightly compressed; subulated, acuminated. *Tongue* short, horny, and armed at the point. *Nostrils* basal and rounded, partly hidden by reflected bristles. *Feet* with three toes before and one behind, the outer toe being joined at its base to the middle one; hind toe of the same length as, or longer than, the middle toe, with a long and hooked claw. *Tail* of twelve feathers. *Wings* rather short; the first quill very short, the third and fourth the longest.

COMMON NUTHATCH.

Sitta Europea, *Linn.*

La Sitelle torchepot.

As far as our recollection serves us, the continent of Europe is the only division of the globe to which this species belongs; nevertheless, the members of the genus *Sitta*, although limited in number, are widely dispersed, but appear to be more particularly attached to the northern and higher latitudes, or to such portions of the tropical countries as from their elevation enjoy a cold or temperate climate. The present species with *Sitta rupestris* (which in the Plate and descriptive letter-press has by an oversight been named *rufescens*) are the only ones which inhabit Europe, while the mountain ranges of India afford us several others, as do also the northern regions of America; nor should we omit the islands of the Indian Archipelago and the continent of New Holland, which if they do not produce a Nuthatch precisely similar to our own in form, at all events possess a group so closely allied to the true Nuthatches as to assure us that their general economy is nearly identical.

The habits by which the species of the genus *Sitta* are characterized are not a little singular, and in many respects agree with those of the Woodpecker; they differ, however, in this remarkable circumstance, that the Nuthatch is not only capable of running up the trunk of a tree with great agility and quickness, but of descending also, head downwards, with equal facility, a manœuvre which the Woodpecker is incapable of performing. As the feathers of the tail are short and very soft, this instrument is of no use as an agent in climbing; and in this respect the Nuthatch differs, not only from the Woodpeckers, but also from the Creepers, to whom the tail is of main importance. The position with the head downwards appears to be to the Nuthatch that which is most easy and natural. It not only assumes this attitude when alighting on the trunk or limb of a tree, but hammers at the bark or splits a nut in a chink in the same position.

The sexes offer no distinguishable difference in the colouring of their plumage, which is also assumed by the young of the year. Insects, nuts, and various berries constitute their food. Their incubation is performed in the holes of decaying trees.

The present beautiful bird is spread throughout the greater part of Europe, and is common in many of the wooded districts. In our own island it is abundant in some localities, while in others it is seldom to be met with. Woods and plantations are its favourite haunts, especially where aged oak and other forest-trees overshadow the underwood. Active and alert, it is ever in motion, now flitting from tree to tree, now traversing the bark in quest of food, or hammering at some decayed part in order to dislodge the insects which have mined their way beneath. The strokes of its bill are smart and strong, and may be heard for a considerable distance; it is thus that it shivers the hard covering of the hazel-nut, which it first fixes in some chink or fissure, and works at it with the head downwards; apparently to increase the mechanical effect of the blow. In the spring the call-note of the Nuthatch is a clear shrill whistle; at other times the bird is silent. The nest consists of a few dried leaves, which constitute a bed in the hole of a tree for the reception of the eggs, which are from five to seven in number, and of a greyish white spotted with reddish brown. The female is assiduous in her task, and defends her nest with her bill and wings, hissing at the same time in token of anger and distress. In winter the Nuthatch often resorts to orchards and gardens in search of food, but does not migrate. The colouring is as follows:

The whole of the upper surface of a fine blueish grey; the quills and base of the tail-feathers, except the two middle ones, being black; the outer one on each side having a black spot near the tip; a black band passes from the bill through the eye; and down the sides of the neck, where it ends abruptly near the shoulders; throat whitish; the under surface rufous brown, becoming of a chestnut on the flanks; bill and tarsi black; irides hazel.

We have figured a pair of these birds of the natural size.

DALMATIAN NUTHATCH.
Sitta rufescens; *(Temm?)*

Drawn from Nature & on Stone by J & E. Gould

Printed by C. Hullmandel.

DALMATIAN NUTHATCH.

Sitta rufescens, *Temm.*

It is with much pleasure we are enabled to introduce a second European species of the limited and well-defined genus *Sitta*, which, we believe, is now figured for the first time. In size it exceeds the common species, and indeed all its congeners.

The Dalmatian Nuthatch is an inhabitant not only of the country from which it takes its name, but also the whole of the south-eastern portion of Europe generally; indeed, to this section of the globe it appears to be strictly limited.

It may be observed that among all the collections of birds from India we have had opportunities of examining, the species in question has never occurred, although the range of the Himalaya presents us with two others totally different either from the Dalmatian or the Common European Nuthatch.

In its general style of colouring as well as in its form, habits, and manners, it exhibits a striking resemblance to the *Sitta Europæa*.

Of its nidification and the number and colour of its eggs we have been unable to obtain any information, yet doubtless they differ but little from those of its immediate and well-known ally. The magnitude of this bird, together with its robust and lengthened bill, and the black mark on each side the neck, sufficiently distinguish it from all known species.

The sexes do not differ in their colouring.

The whole of the upper surface is of a beautiful ash grey; a dark line begins at the base of the upper mandible, passes over the eyes, down the neck, and bends across the shoulders; the quill-feathers are blackish brown; the throat and breast white, passing into pure chestnut on the flanks and lower part of the abdomen; bill black at the tip, and horn-colour at the base; tarsi lead-colour.

The Plate represents an adult of the natural size.

ASIATIC NUTHATCH.
Sitta Asiatica; *(Temm.)*

Drawn from Nature & on Stone by J & E. Gould.

Printed by C. Hullmandel.

ASIATIC NUTHATCH.

Sitta Asiatica, *Temm.*

M. Temminck has kindly forwarded to us a fine example of this elegant species of Nuthatch for the purpose of illustration, accompanied by a note stating that it was from Russia, and would form a portion of the supplement to the third part of his "Manuel." We can only regret that we have not been able to acquire any further information respecting it.

It is rather smaller in size than the common species (*Sitta Europæa*), and is much lighter in the general tone of its colouring.

The crown of the head and all the upper surface are light grey; wings greyish brown, the primaries being darker; outer tail-feathers white at the base and dark brown towards the tip; the remainder, except the two middle ones, which are grey, dark brown at their base; a stripe over the eye; the chest and upper part of the abdomen white; lower part of the abdomen and under tail-coverts dull rufous; a black stripe commences at the base of the bill and runs through the eye to the shoulders; bill black; feet brown.

Our figure is of the natural size.

COMMON CREEPER.

Certhia familiaris, *(Linn.)*

Drawn from Nature & on stone by J. & E. Gould.

Printed by C. Hullmandel.

Genus CERTHIA, *Ill.*

Gen. Char. *Bill* of mean length, curved, triangular, compressed, slender, and sharp-pointed. *Tongue* short. *Nostrils* basal, pierced horizontally, naked, and partly covered by an arched membrane. *Feet* with three toes before and one behind, which last is strong, and longer than the middle toe; the outer toe united at its base to the middle one. *Tail* wedge-shaped, composed of twelve stiff, sharp-pointed, and deflected feathers. *Wings* having the first quill short, and the second and third shorter than the fourth, which is the longest of all.

COMMON CREEPER.

Certhia familiaris, *Linn.*

Le Grimpereau.

The genus *Certhia* as now restricted will contain but two species, the bird here figured (which is the only one hitherto discovered in Europe), and one from the Himalaya mountains, characterized some years since by Mr. Vigors under the name of *Certhia Himalayana*. This new species bears a strong resemblance to its European congener, from which it may be distinguished by the markings of brown across the tail-feathers; it is also a trifle larger.

The Common Creeper appears to be very generally dispersed over the whole of the Continent, but according to M. Temminck it becomes more rare as we approach the northern parts of Russia and Siberia; which may be reasonably accounted for, by the diminished number of insects in all high latitudes.

It is a stationary species in the British Islands, where it is very generally dispersed, but is of course more plentiful in the neighbourhood of wooded districts, plantations, &c. It also frequents gardens and orchards, where its presence may generally be detected by its weak shrill cry, which is not unlike that of the Golden-crested Wren (*Regulus auricapillus*).

It is an excellent climber, ascending the boles of trees with great rapidity, in search of insects, upon which it solely subsists. Its stiff and elastic tail, together with its long hind toe and curved claw, presents a structure peculiarly adapted for ascending trees.

Its nest is constructed in the hole of a decayed tree, and is formed of grass and mosses, with a lining of feathers: the eggs, which are from seven to nine in number, are white speckled with reddish brown.

Head and upper surface yellowish brown intermingled with black, brown, and greyish white; rump pale chestnut red; first four quills dusky; the remainder have a broad reddish white band in the middle, and the tips white; tail greyish brown; a whitish streak passes over the eyes; throat, breast, and under surface white, passing into ochreous yellow on the vent; upper mandible dusky, lower yellowish white; legs and toes yellowish brown.

The sexes are alike in plumage.

We have figured an adult bird of the natural size.

HOOPOE.

Upupa epops. *(Linn.)*

Drawn from Life & on Stone by J & E. Gould.

Printed by C. Hullmandel.

Genus UPUPA, *Linn.*

GEN. CHAR. *Beak* very long, slightly arched, slender, triangular and compressed. *Nostrils* basal, lateral, ovoid, open, surmounted by the feathers of the forehead. *Toes* three before, and one behind; the external and middle ones united as far as the first joint. *Nails* short, a little bent, except in the hind one which is straight. *Tail* square, consisting of ten feathers. *Wings* moderate; fourth and fifth *quill-feathers* the longest.

HOOPOE.

Upupa epops, *Linn.*

La Huppe.

THERE are few birds more elegant in their appearance or more singular in their manners than the Hoopoe; and although it is not a resident in the British Isles, nor strictly a periodical visitor, we are, from its frequent occurrence, enabled to give much information respecting its natural habits and modes of life. The genus to which it belongs is extremely limited in the number of its species, three only being at present recognised. Our European example, the *Upupa epops*, may be regarded as a migratory bird, and its natural range is very extensive. It is found over nearly the whole of Africa; India and China may also be enumerated among the countries it inhabits, as specimens received from the latter and the Himalaya mountains sufficiently testify. In continental Europe, it is spread from the southern to the northern extremities, but is more abundant in the former, where it appears to be a bird of regular and periodical passage; being, however, regulated in these migrations by the abundance of the food upon which it subsists, viz., the larvæ of scarabæi, together with other insects which live near moist and humid grounds, not even rejecting tadpoles, small frogs, and worms. In the British Islands, as we have already observed, its occurrence is very irregular, being scarce in some seasons, and much more frequent in others; and when it does visit us, its animated motions and foreign appearance, unfortunately for the bird, bring round it a host of persecutors. There are, however, a few instances on record of its having bred among us. The southern coast of England, as we might most naturally expect, is that on which it makes its first appearance, generally in the month of May; hence they disperse themselves over the Island, and are often met with in the most unexpected localities; but the situations most preferred are thick hedgerows, copses, and isolated trees or bushes, in the neighbourhood of low marshy lands: they seem to have but little care respecting their concealment, generally perching on the most conspicuous branch, erecting and depressing the beautiful fan-like crest as if to attract observation: but though it perches upon trees, it is not, as its peculiar legs and feet indicate, a bird ordained by nature to be an exclusive inhabitant of the woods and groves, its feeble toes being ill adapted for clasping with strength and firmness. Its flight is slow and undulating, similar to that of the Woodpeckers.

To enumerate its frequent capture in England would neither add to science nor to a knowledge of its habits; still we beg to mention an instance, which came within our knowledge, of one shot by L. Sullivan, Esq. on the 28th of September 1832, in his own pleasure-grounds at Broom House, Fulham, Middlesex; and we are led to suppose, from the lateness of the season, that it had incubated in the neighbourhood. It chooses for the site of its nest a variety of situations, as opportunity may serve; holes in trees, crevices in rocks, fissures in walls or masonry, holes in the ground or dungheaps, being among the places it has been observed at different times to occupy: the eggs are five in number, clouded with dark grey on a light grey ground.

The young soon assume the adult plumage, which is precisely similar in either sex.

The ground colour of the head, neck, and shoulders is of a beautiful fawn; a double row of long feathers surmounts the head, beginning at the base of the beak and ending at the occiput, capable of being thrown up perpendicularly, so as to form a fan-like crest; each of these feathers is tipped with black; the wing-coverts and scapulars are banded alternately with black and white; the quills are black with a white oblique band; rump white; tail black banded across the middle with white; the flanks and under tail-coverts light greyish fawn dashed with obscure lines of brown.

We have figured two adult birds of the natural size.

WALL CREEPER.

Tichodroma phœnicoptera, *(Temm.)*

Drawn on Stone by E. Lear.

Printed by C. Hullmandel.

Genus TICHODROMA, *Ill.*

GEN. CHAR. *Beak* very long, slender, slightly arched, cylindrical, angular at the base, and depressed at the point. *Nostrils* basal, pierced horizontally, naked, partly closed by a membrane. *Toes* three before, the external united at its base to the middle one, and one behind with an elongated nail. *Tail* slightly rounded with feeble shafts. *Wings* large and rounded.

WALL CREEPER.

Tichodroma phœnicoptera, *Temm.*

La Tichodrome echellette.

THE form and plumage of this beautiful bird would induce most persons to suppose it a native of a tropical clime ; it is, however, strictly an inhabitant of Europe, although its local distribution appears to be confined exclusively to the middle and southern portions of the Continent. Unlike most of the smaller birds, it frequents the naked and precipitous parts of the most elevated mountains, such as the Alps of Switzerland, the Apennines, and Pyrenees. Among these towering rocks, where the ruins of castles and fortresses are not unfrequent, this pretty bird is seen flitting from crevice to crevice, enlivening the solitude of the scene by its presence. In the choice of its food it is curious and peculiar, being particularly partial to spiders and their eggs, which with various species of insects and their larvæ constitute its diet : for these it is incessantly on the search, not however creeping up and down the sides of the rock or the face of the wall, as is the case with the true *Certhia*, but hopping or flitting from one crevice or projection to another ; hence we see the tail-feathers feeble and not furnished with stiff springy shafts, since they are not required to aid the bird in the same manner as they do in the Woodpeckers or Creeper. The grasp of its long and slender toes is peculiarly tenacious ; the least roughness, or any hold however slight, is therefore sufficient to afford a resting-place. Connecting the habits and the situation which this bird occupies with the means bestowed upon it, we cannot but see how suitably it is endowed ; the slender bill, its tenacious feet, its broad and rounded wing, giving a flitting character to its mode of flight,—all combining to qualify it for its mountain habitat. The moult is double, occurring in spring and autumn, and the two sexes are alike in plumage except during the breeding season, when the throat of the male is black, and the crown of the head of a somewhat darker grey. Before the autumn moult comes fairly on, the feathers of these parts are exchanged, and the markings disappear ; the two sexes are then undistinguishable.

The head, neck, back, and upper surface generally are of a delicate grey ; the under parts of a darker tint of the same colour ; the whole of the wing-coverts and the outer edge of the greater quills for half their lengths, of a lively crimson ; the remainder of the quill-feathers black, each having two spots of white on the inner web, so as to form a double bar when the wing is expanded ; tail black tipped with white ; beak, irides, and tarsi black.

We have figured a male and female in full plumage, and of the natural size.

COMMON CUCKOO.
Cuculus canorus; *(Linn.)*

Drawn from Nature & on stone by J. & E. Gould. *Printed by C. Hullmandel.*

Genus CUCULUS, *Linn.*

GEN. CHAR. *Bill* rather compressed, slightly curved, of mean length; gape wide; lower mandible following the curve of the upper. *Nostrils* basal, round, margined by a naked and prominent membrane. *Wings* of mean length, acuminate; the first quill-feather short, the third the longest. *Tail* more or less wedge-shaped. *Tarsi* very short, feathered a little below the knee. *Feet* with two toes before, and two behind, the outer hind toe partly reversible; the anterior toe joined at the base, the posterior ones entirely divided.

COMMON CUCKOO.

Cuculus canorus, *Linn.*

Le Coucou gris.

IN stating that the Cuckoo is a migratory bird, we add nothing to what is already well known; it is in fact the most celebrated harbinger of returning vivification, and its familiar call is always hailed with pleasure as the token of returning spring and the fresh awakening of Nature from her winter's sleep.

As is the case with most of our summer visitants, the food of the Cuckoo consists principally of insects, especially of caterpillars, larvæ, &c., a proof that its winter sojourn is in climates where this kind of diet is ever to be obtained; hence Africa, a place of winter residence for so many of our migratory birds, affords to this species, among the rest, a welcome retreat. Its range extends over nearly all parts of Europe, and a great portion of Africa and Asia; specimens received from the Himalaya mountains and other parts of India, being strictly similar to those taken in our own island. The Cuckoo does not construct a nest for the reception of its eggs, but deposits them in those of other birds of a much smaller size and of insectivorous appetites: the species most commonly chosen as the foster parents of its offspring are the Titlark, Hedge Sparrow, &c. In the nest of these birds it deposits a single egg; but whether it lays only one or more, is a point at present not ascertained, but it is most probable that it lays several, and deposits them in as many different nests. Shortly after the young Cuckoo is excluded from the shell, with the offspring of its foster parent, it attains to so much strength as to be able to eject them from the nest, itself remaining the sole occupant; and in fact, from its large size and ravenous appetite, it is as much as these substituted parents can do to supply it with food. Mr. J. E. Gray, of the British Museum, from observations made by himself, asserts that the Cuckoo does not uniformly desert her offspring to the extent that has been supposed, but, on the contrary, that she continues in the precincts where the eggs are deposited, and in all probability takes the young under her protection when they are sufficiently fledged to leave the nest. They retire in August, at least the adults, which in their migration always precede the young. The birds of the year quit this country in September.

The sexes may at all times be distinguished by the male being the largest and most robust, and by having the whole of the neck and chest of a fine grey, while the female has the sides of the chest obscurely rayed with markings of brown.

On dissecting this bird in the early months of spring, we cannot fail to observe a great dilatation of the throat, the membrane covering which internally is of a fine rich orange: the cause of this we have not been able to determine satisfactorily; it may be connected with the organs of voice. The circumstance of the stomach of the Cuckoo containing a lining of numerous hairs, was for a long period a matter of great curiosity to naturalists; but these are now considered to be a deposition of the hairs from the larger caterpillars upon which it feeds, and which it swallows whole.

The young birds differ much from the adults, having at first the upper surface of deep brown margined and spotted with reddish brown, the feathers on the forehead margined with white, and a patch of the same colour at the back of the head, the throat and under surface yellowish white transversely barred with black, the irides brown, and the legs pale yellow. Young females are more reddish brown, and have only a faint indication of the white patch at the back of the head.

The adults have the head, neck, breast, and upper surface bluish grey, which is deepest on the wing-coverts; the under surface, thighs, and under tail-coverts white transversely barred with black; the inner webs of the quill-feathers marked with oval white spots; the tail black with small oblong white spots along the shafts, and the tips white; the bill blackish brown at the tip and yellowish at the base; the gape and eye-lids rich orange; the irides gamboge yellow, and the legs and feet lemon yellow.

We have figured an adult, and the young bird in its first autumn, of the natural size.

GREAT SPOTTED CUCKOO.
Cuculus glandarius, *(Linn.)*

Drawn from Nature & on stone by J & E. Gould.

Printed by C. Hullmandel.

GREAT SPOTTED CUCKOO.

Cuculus glandarius, *Linn.*

Le Coucou Geai ou, Tacheté.

The crested head, lengthened and powerful tarsi, together with the more elegant form exhibited in the bird before us, indicate very clearly that a further subdivision of the family is requisite: the reason why we have figured it under the generic name of *Cuculus*, and not under that of *Coccyzus*, applied to it by some authors, is that the bird to which the latter title was first applied possesses characters different from either the present bird or the true Cuckoos.

We do not in this place feel disposed to enter largely into a consideration of the divisions of this family, and therefore defer adding a new generic name to the Great Spotted Cuckoo until we have had an opportunity, which we hope will occur at no distant period, of revising the whole group, when not only this, but several other species will be brought under investigation.

So little is known of the habits and manners of this bird that it is still uncertain whether, like the species common to England, its eggs and offspring are confided to the care of other birds, or whether it constructs its own nest and performs the process of incubation in the ordinary way; which paucity of information is occasioned by its being so sparingly dispersed over the continent of Europe that no opportunities have occurred of observing the most interesting portion of its economy,—its nidification.

Its true habitat is the wooded districts skirting the sultry plains of North Africa; but the few that pass the Mediterranean find a congenial climate in Spain and Italy, further north than which they are rarely seen.

That valuable work the "Planches Coloriées" of M. Temminck contains an accurate description of this bird in all its various changes of plumage, a portion of which we venture to extract. This Cuckoo, which is larger than the common species, is characterized by a crest comprised of filamentous feathers, by a very long graduated tail, by the linear and tubular form of the nostrils, and by the comparatively strong bill and feet. In the old male the crest, all the head, and the cheeks are ash colour, more or less deep according to age; the stems of the feathers of these parts are brown, and the base of the webs whitish; a band of blackish ash commences at the regions of the ears, passes under the occiput, and extends to the nape of the neck; the back, the rump, the scapularies, and the coverts of the wings are of a greyish brown tint, slightly clouded with a greenish lustre, the tips of all these feathers having a white spot, which varies in size and purity according to age; the young and birds of the middle age have these spots more extended and better defined than the adults and old birds; the primaries are of a dark brown, edged with grey, and terminated with white; the feathers of the tail are ash brown ending in pure white; throat and chest reddish white; the abdomen and under tail-coverts pure white; the feet are dark brown inclining to yellow on the under surface; the bill is brownish black at the point; the base of the under mandible reddish yellow.

The plumage of the middle age differs from that of the adult in having the head and crest of a much darker colour and the whole of the upper surface more inclining to reddish brown with slight reflexions of green; the primaries are rufous, tinged with greenish brown towards the points, which are pure white; the throat and chest are clear reddish brown; the under surface as in the adult male.

The young of the year is still darker in its plumage; the crest is short; the feathers of the back and secondaries are of a reddish brown; the two middle tail-feathers are slightly tipped with white; the front of the neck and the chest are deep rufous; all the other inferior parts are reddish white; feet and beak lead colour; irides grey.

The Plate represents a male of the natural size, nearly adult.

AMERICAN CUCKOO.
Coccyzus Americanus; *(Bonap.)*

Drawn from Nature & on Stone by J & E. Gould. Printed by C. Hullmandel.

Genus COCCYZUS, *Vieill.*

GEN. CHAR. *Bill* of moderate length, rather strong, arched, the culmen convex, the base compressed. *Nostrils* basal, elongated. *Wings* short. *Tail* long, cuneiform. *Tarsi* and middle toe long and equal.

AMERICAN CUCKOO.

Coccyzus Americanus, *Bonap.*

Le Coucou Cendreillard.

FOUR examples of this American species having been taken in Great Britain, namely two in Ireland, one in Wales, and one in Cornwall, we have no hesitation in admitting it to a place in this work.

The first notice we are acquainted with of the occurrence of this bird appeared in the Field Naturalist's Magazine of Mr. Rennie. Mr. Ball of Dublin Castle, in a letter to the editor of this Magazine, made known the capture of the first specimen, which was shot near Youghal, in the county of Cork, in the autumn of 1825; and the second was shot at a later period at Old Connaught near Bray. The Cornwall specimen was the subject of a private communication, and the fourth was obtained on the estate of Lord Cawdor in Wales, during the autumn of 1832. This last example has now by the liberality of His Lordship been deposited in the British Museum, and one if not both of the Irish specimens were exhibited at the Zoological Society by Mr. Thompson of Belfast in June 1835.

"This bird," says Mr. Audubon, "I have met with in all the low grounds and damp places in Massachusetts, along the line of Upper Canada, pretty high on the Mississippi and Arkansas, and in every State between these boundary lines. Its appearance in the State of New York seldom takes place before the beginning of May, and at Green Bay not until the middle of that month." The most frequent note of this bird sounds so much like the word cow, frequently repeated, that it has obtained the general appellation of Cow-bird; and from being particularly vociferous before rain, it is in some States called the Rain-crow. Unlike our English Cuckoo this American species builds a nest and rears its young with great assiduity, but it sometimes robs smaller birds of their eggs, and its own egg, which cannot be mistaken from its singular colour, is occasionally found in another bird's nest. Mr. Audubon says "that its own nest is simple, flat, composed of a few dry sticks and grass, formed much like that of the Common Dove: the eggs are four or five in number, of rather an elongated oval form, and bright green colour. The young are principally fed with insects during the first weeks, and they rear only one brood in a season, unless the eggs are removed or destroyed."

The appearance of these different examples of an American species in this country has caused some speculation. M. Temminck, unwilling to consider it as a migration from America to Europe, thinks it probable that the bird may yet be found in the north of Europe.

The upper mandible is dark brown, the under one yellow, the irides hazel, prevailing colour of the head, neck, back, wings, wing-coverts and two middle tail-feathers light greenish brown; the other tail-feathers are black, with the ends white, the outer tail-feather on each side is white on the outer web; the tail graduated; all the under surface of the body greyish white; the legs and toes blue.

We have figured this bird of the natural size.